Monographs in Computer Science

Editors

David Gries
Fred B. Schneider

Monographs in Computer Science

Mark Burgin

Super-Recursive Algorithms

 Springer

Mark Burgin
Department of Mathematics
UCLA
Los Angeles, CA 90095
U.S.A.
mburgin@math.ucla.edu

Series Editors:
David Gries
Cornell University
Department of Computer Science
Ithaca, NY 14853
U.S.A.

Fred B. Schneider
Cornell University
Department of Computer Science
Ithaca, NY 14853
U.S.A.

Library of Congress Cataloging-in-Publication Data
Burgin, M.S. (Mark Semenovich)
    Super-recursive algorithms / Mark Burgin.
        p. cm. — (Monographs in computer science)
    Includes bibliographical references and index.
    ISBN 0-387-95569-0 (alk. paper)
        1. Recursive functions. 2. Algorithms. I. Title. II. Series.

    QA9.615.B87   2005
    511.3'52—dc22   2004041748

ISBN 0-387-95569-0   Printed on acid-free paper.

Printed in the United States of America.   (TXQ/MV)

9 8 7 6 5 4 3 2 1   SPIN 10891097

springeronline.com

To my parents and grandparents

Contents

Preface

This book introduces the new realm of superrecursive algorithms and the development of mathematical models for them. Although many still believe that only recursive algorithms exist and that only some of them are realizable, there are many situations in which people actually work with superrecursive algorithms. Examples of models for superrecursive algorithms are abstract automata like inductive Turing machines as well as computational schemes like limiting recursive functions.

The newly emerging field of the theory of superrecursive algorithms belongs to both mathematics and computer science. It gives a glimpse into the future of computers, networks (such as the Internet), and other devices for information interchange, processing, and production. In addition, superrecursive algorithms provide more adequate models for modern computers, the Internet, and embedded systems. Consequently, we hope (and expect) that this theory of superrecursive algorithms will, in the end, provide new insight and different perspectives on the utilization of computers, software, and the Internet.

The first goal of this book is to explain how superrecursive algorithms open new kinds of possibilities for information technology. This is an urgent task. As Papadopoulos (2002) writes, "If we don't rethink the way we design computers, if we don't find new ways of reasoning about distributed systems, we may find ourselves eating sand when the next wave hits." We believe that a theory of superrecursive algorithms makes it possible to introduce a new paradigm for computation, one that yields better insight into future functioning of computers and networks. This form of computation will eclipse the more familiar kinds and will be commercially available before exotic technologies such as DNA and quantum computing arrive.

Another goal of this book is to explain how mathematics has explicated and evaluated computational possibilities and its role in extending the boundaries of computation. As we do this, we will present the theory of algorithms and computation in a new, more organized structure.

It is necessary to remark that there is an ongoing synthesis of computation and communication into a unified process of information processing. Practical and theoretical advances are aimed at this synthesis and also use it as a tool for further development. Thus, we use the word computation in the sense of information processing as a whole. Better theoretical understanding of computers, networks, and other information processing systems will allow us to develop such systems to a higher level. As Terry Winograd (1997) writes, "The biggest advances will come not from doing more and bigger and faster of what we are already doing, but from finding new metaphors, new starting points." In this book, we attempt to show that such new metaphors already exist and that we need only to learn how to use them to extend the world of computers in ways previously unimaginable.

Algorithms and their theory are the basis of information technology. Algorithms have been used by people since the beginning of time. Algorithms rule computers. Algorithms are so important for computers that even the mistakes of computers result mostly from mistakes of algorithms in the form of software. Consequently, the term "algorithm" has become a general scientific and technological concept used in a variety of areas. The huge diversity of algorithms and their mathematical models builds a specific "algorithmic universe". However, the science that studies algorithms emerged only in the twentieth century.

Since the emergence of the theory of algorithms, mathematicians and computer scientists learned a lot. They have built mathematical models for algorithms and, by means of these models, discovered a principal law of the algorithmic universe, the Church–Turing thesis, and it governs the algorithmic universe just as Newton's laws govern our physical universe. However, as we know, Newton's laws are not universal. They are true for processes that involve only ordinary bodies. Einstein, Bohr, Dirac, and other great physicists of the twentieth century discovered more fundamental laws in the microworld that go beyond the scope of Newton's laws. In a similar way, new laws for the algorithmic universe have been discovered that go beyond the Church–Turing thesis. The Church–Turing thesis encompasses only a small part of the algorithmic universe, including recursive algorithms. This book demonstrates that superrecursive algorithms are more powerful, efficient, and tractable than recursive algorithms, and it introduces the reader to this new, expanded algorithmic universe.

Consider the famous Gödel theorem on the incompleteness of formal arithmetic. In the context of recursive algorithms, this theorem has absolute and ultimate meaning, vitally restricting the abilities of mathematicians and mathematics. In the context of superrecursive algorithms, the Gödel theorem becomes relative, stating only differences in abilities based on superrecursive and recursive algorithms. That is, the theory articulates that, for sufficiently rich mathematical theories, such as arithmetic, superrecursive algorithms allow one to prove much more than conventional methods of formal deduction, which are based on recursive algorithms (Burgin, 1987). When Gödel proved his theorem, it was a surprise to most mathematicians. However, from the superrecursive perspective, the Gödel theorem is a natural result that simply reflects the higher computational and deductive power of superrecursive algorithms.

Although the main concern of this book is superrecursive algorithms and hypercomputation, a variety of other problems are also analyzed. They include general problems such as: What is an algorithm? What is a description of an algorithm?

How do we measure computational power, computational efficiency, computational equivalency for computers, networks, and embedded systems and their mathematical models? What are the structures, types, and functioning of information processing systems? Can we provide a systematization of mathematical models of computational processes and their algorithmic representations? How do they affect computer, network, and information-processing architectures?

The organization of this book

This book begins with models of conventional, recursive algorithms, with an overview of the theory of recursive algorithms given in Chapter 2. We then present even less powerful, but more tractable and feasible, subrecursive algorithms, giving an overview of their theory in Chapter 3. We consider some classes of subrecursive algorithms that are determined by restrictions in construction; for instance, finite automata. Subrecursive algorithms defined by restrictions on the resources used, e.g., Turing machines with polynomial time of computation, are mentioned only tangentially.

Our approach has a three-fold aim. The first aim is to prepare a base for superrecursive algorithms; an exposition of conventional algorithms helps to understand better the properties and advantages of new and more powerful algorithmic patterns, superrecursive algorithms.

The second aim of our approach is to give a general perspective on the theory of algorithms. Computer scientists and mathematicians have elaborated a huge diversity of models. Here we try to systematize these models from a practical perspective of computers and other information processing systems. As far as we know, this is the first attempt of its kind.

The third aim of our approach is to achieve completeness, making the book self-contained. This allows a reader to understand the theory of algorithms and computation as a whole without going to other sources. Of course, other sources may be used for further studies of separate topics. For instance, Rogers (1967) has more material about recursive algorithms, and Hopcroft, Motwani, and Ullman (2001) contains more material about finite automata and context-free grammars.

But this book allows a reader to better comprehend other theories in computer science by systematizing them, extending their scope, and developing a more advanced perspective based on superrecursive algorithms.

After considering conventional models of algorithms, we introduce models of superrecursive algorithms. In Chapter 4 we consider the computational power of superrecursive algorithms. In Chapter 5 we consider the efficiency of superrecursive algorithms, which is represented in the theory by a kind of complexity measure. The book culminates in a positive reevaluation of the future development of communication and information processing systems.

The exposition is aimed at different groups of readers. Those who want to know more about the history of computer science and get a general perspective of the current situation in the theory of algorithms and its relations to information technology can skip proofs and even many results that are given in the strict mathematical form. At the same time, those who have a sufficient mathematical training and are interested mostly in computer and information science or mathematics can skip preliminary deliberations and go directly to the exposition of superrecursive algorithms and automata. Thus, a variety of readers will be able to find interesting and useful issues in this book.

It is necessary to remark that the research in this area is so active that it is impossible to include all ideas, issues, and references, for which we ask the reader's forbearance.

Theories that study information technology belong to three disciplines: information sciences, computer science, and communication science. All such theories have a mathematical foundation, so it is no surprise that mathematics has its theories of computers and computations. The main theory is the theory of algorithms, abstract automata, and computation, or simply, the theory of algorithms. It explains in a logical way how computers function and how computations are organized. It provides means for evaluation and development of computers, nets, and other computational systems and processes. For example, a search for new kinds of computing resulted in molecular (in particular, DNA) and quantum computing, which are the most widely discussed. At this point, however, both of these paradigms appear to be restricted to specialized domains (molecular for large combinatorial searches, quantum for cryptography) and there are no working prototypes of either. The theory of algorithms finds a correct place for them in a wide range of different computational schemes.

Acknowledgments

Many wonderful people have made contributions to my efforts with this work. I am especially grateful to Springer-Verlag and its editors for their encouragement and help in bringing about this publication. In developing ideas in the theory of algorithms, automata, and computation, I have benefited from conversations with many friends and colleagues who have communicated with me on problems of superrecursive algorithms, and I am grateful for their interest and help.

Credit for my desire to write this book must go to my academic colleagues. Their questions and queries made significant contributions to my understanding of algorithms and computation. I would particularly like to thank many fine participants of the Logic Colloquium of the Department of Mathematics, the Computer Science Department Seminar, the Applied Mathematics Colloquium, Seminar of Theoretical Computer Science of the Department of Computer Science at UCLA, and the CISM Seminar at JPL for extensive and helpful discussions of the theory of superrecursive algorithms, which gave me so much encouragement. Discussions with A.N. Kolmogorov from Moscow State University and Walter Karplus from UCLA gave much to the development of ideas and concepts of the theory of superrecursive algorithms. I would also like to thank the Departments of Mathematics and Computer Science in the School of Engineering at UCLA for providing space, equipment, and helpful discussions. Finally, teaching the theory of computation in the Department of Computer Science at UCLA has given me a wonderful opportunity to test many new ideas from the theory of algorithms.

Super-Recursive Algorithms

1

Introduction

> *Nisi credideritis, non intelligitis.*
> *(Unless you believe, you will not understand.)*
> Saint Augustine, 354–430

People live in a very complex world. They devote their intelligence to understanding and accommodating this complex environment. To cope with complexity, people inevitably create more and more complex mechanisms and technologies. According to one of the basic principles of cybernetics suggested by Ashby (1964), *to achieve complete (relevant) adaptation/control, the complexity/variety of a system must be of the same order as the complexity/variety of its environment.*

This implies that we need more powerful computers as technical devices for information processing, as well as more powerful theories as abstract devices for intellectual activity. In the process of achieving more power, computers are becoming more and more complex. However, in spite of the exponential growth of computing, storage, and communication power, scientists demand more.

The following examples by Ian Foster (2002) vividly illustrate the situation. A personal computer in 2001 is as fast as a supercomputer of 1990. But 10 years ago, biologists were happy to be able to compute a single molecular structure. Now, they want to calculate the structures of complex assemblies of macromolecules and screen thousands of drug candidates. Personal computers now ship with up to 100 gigabytes (GB) of storage, as much as an entire 1990 supercomputer center. But by 2006, several physics projects, CERN's Large Hadron Collider (LHC) among them, will produce multiple petabytes (1015 byte) of data per year. Some wide area networks (WANs) now operate at 155 megabits per second (Mbps), three orders of magnitude faster than the state-of-the-art 56 kilobits per second (Kbps) that connected US supercomputer centers in 1985. To work with colleagues across the world on petabyte data sets, scientists demand communication at the level of tens of gigabits per second (Gbps).

The world of information processing is very complex and sophisticated. It involves interaction of many issues: social and individual, biological and psychological, technical, organizational, economical, political. The complexity of the world of modern technology is reflected in a study of Gartner Group's TechRepublic unit (Silverman, 2000). According to this study, approximately 40% of all internal IT projects are canceled or unsuccessful. Overall, an average of 10% of a company's IT department produces no valuable work each year. An average canceled project is terminated after 14 weeks, when 52% of the work has already been done. In addition, the study states that companies spend an average of almost $1 million of their $4.3 million annual budgets on failed projects.

Companies may be able to minimize the chances of cancellation if they have relevant evaluation theory and consult people who know how to apply this theory. All developed theories have a mathematical basis, so it is no surprise that mathematics helps science and technology in many ways.

To advance the field of information processing, we have to use existing theories and develop new ones to reflect the complexity of existing systems and guide our search for more powerful ones. As stressed in the Preface, the theory of algorithms is basic for this search. This theory explains in a logical way how computers, networks, and different software systems function and how they are organized. This book describes recent achievements of the theory of algorithms, helping to better understand the field of information processing and to find new directions for its development.

Algorithms were used long before computers came on the scene and in a formalized form came from mathematics. That is why theory of algorithms was developed in mathematics. With the advent of computers, the theory of algorithms changed its emphasis. It became a theory of the computer realm. This computer realm is the main perspective for this book. We analyze how different mathematical models of algorithms represent properties of computers and networks, how models reflect programming technique, and how they provide new paradigms for computation. We hope also to expand our understanding of how different systems function in an organized and consistent way.

1.1   Information processing systems (IPS)

There was no "One, two, three, and away!",
but they began running when they liked,
and left off when they liked,
so that was not easy to know when the race was over.

Lewis Carroll, 1832–1898

The computer realm consists of not only of computers but many other systems. The unifying feature of the systems models of which we consider in this book is that they all process information, so they are *information processing systems*. For simplicity, we call them IPS. Initially, the theory of algorithm dealt with the von Neumann computer, a one-processor computer with memory. Over the years, with the advent of parallelism and then the internet, the theory grew to encompass them. Now, even that is not enough. A modern theory of algorithm has to deal with supercomputation based on clusters of computers and grid computation. Moreover, Bell and Gray (1997) predict that stand-alone computers will evolve to become parts of everything. However, they will still be IPS. IPS will continue to expand and evolve as long as civilization exists.

Because of the complexity of IPS, we need to explicate its main types, to describe its principal components and how they relate to each other. Types of information processing imply a typology of IPS and explain what is possible to be done with the information being processed or, in other words, what information operations exist. To have an organized structure, we consider information operations on several levels. The most basic is the microlevel. It contains the most fundamental operations that allow one to build other operations on higher levels. Basic information operations are determined by actions that involve information. There are three main actions:

1. Changing the place of information in the physical world.
2. Changing information itself or its representation.
3. Doing nothing with information (and protecting it from any change).

These actions correspond to three basic information micro-operations:

1. *Information transition.*
2. *Information transformation.*
3. *Information preservation.*

There are three main forms of each operation:

1. Information transition.
   a) Transition outside from inside of a system is called *emission*.
   b) Transition inside a system from outside is called *reception*.
   c) Transition between two similar points/systems is called *pure transition* or *equitransition*.
2. Information transformation.
   a) *Substance transformation* is a change of information itself in a direct way.
   b) *Form transformation* is a change of information representation.
   c) *External transformation* is a change of the context of information.

Definition 1.1.1. The *context of information* is the knowledge system to which this information is related.

According to the general theory of information (cf. (Burgin, 1997)), changing the context of information, we change information itself in an indirect way.

3. Information preservation.
   a) Information storage.
   b) Information storage with protection from change/damage.
   c) Information storage with protection from change/damage and restoration of damaged portions.

Actually storage is never pure because it always includes, at least, data transition to special storage places, which are traditionally called memory. In some cases, storage includes information transformation. An example is the dynamic storage in neural networks (see Section 2.4). On the macro level, we have much more information operations. The most popular and important of them are:

◇ *Computation*, which includes many transformations, transitions, and preservations.

◇ *Communication*, which also includes many transformations, transitions, and preservations.

Examples of other information operations are information acquisition, information selection, information search, information production, and information dissemination. Now computers perform all of them and computer power is usually estimated with respect to these operations.

There are important relationships between these operations. For instance, it is possible to compute through communication. This is the essence of the connectionist paradigm for which the main operation is transmission of signals, while their transformation is an auxiliary operation. The most explicit realization of the connectionist paradigm is the neural network. Any conventional computation, for example, computation on a PC, also demands information transmission, but in this case transmission is an auxiliary operation. However, the Internet has both connectionist and transmission capabilities, and computations like grid computation will combine both paradigms.

The storage of data has been always a basic function of conventional computers, and hierarchical systems of memory have been developed over time. However, an IPS can memorize data even without a separate component for this function. It is possible to store information by computation or communication. Such dynamic storage is described for artificial neural networks in Section 2.4.

IPS have two structures — static and dynamic. The *static structure* reflects the mechanisms and devices that realize information processing, while the *dynamic structure* shows how this processing goes on and how these mechanisms and devices function and interact. We now discuss the static structure, which has two forms: synthetic and systemic, or analytic.

1.1.1 Static synthetic structure

Any IPS, we denote it by **W**, consists of three components:

◇ *Hardware*, which consists of physical devices of the IPS.
◇ *Software*, which contain programs that regulate the IPS functioning.
◇ *Infware*, which represents information processed by the IPS.

In turn, the hardware of an IPS has its three main parts: the *input device(s)*, *information processor(s)*, and *output device(s)*, which are presented in Figure 1.1 and give the first level approximation to the structure of IPS.

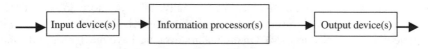

Figure 1.1. Triadic structure of IPS

The theory of algorithms has traditionally concentrated on the central component of IPS, paying more attention to the problem of how abstract automata and algorithms work rather than what is the result of this work. However, the triadic structure of an IPS implies that all three components are important. Neglecting any one of them may cause us to have inadequate understanding of IPS, which can hinder the development of IPS.

To elaborate on this, consider the following. Initially, computers were able only to print results. Contemporary computers can display their results on a printer, a monitor, and even different audio devices. Computers and embedded devices send their signals to control a diversity of mechanisms and machines. Contemporary machines now have not only a keyboard and a mouse but also trackballs, joysticks, light pens, touch-sensitive screens, digitizers, scanners, and more. The theory must take this into account.

It is interesting to remark that while information processing in quantum computers has been well elaborated, researchers have found that input and especially output appear to be much more complicated issues. Reading the obtained result of a computation is a crucial problem for building future quantum computers. This problem remains unsolved (cf. Hogg, 1999).

Awareness of criticality of input and output components has resulted in the development of the practical area of human-computer interaction (HCI). People began to comprehend the interactive role of computers (in particular) and IPS (in general): a substantial amount of computers are built for working with and for people. Interaction becomes crucial not only in utilization of computers and their software, but also for computer and software design (Vizard, 2001).

The same understanding in the theoretical area resulted in inductive, limit, and interactive directions in the theory of algorithms. The first two directions advanced computational potential by developing output techniques (Burgin, 1983, 1999, 2001; Gasarch and Smith, 1997; Hintikka and Mutanen, 1998), while the latter approach achieved similar results by making the principal emphasis on input and output components as the basis for interaction between IPS (Hoare, 1984; Goldin and Wegner, 1988; Milner, 1989; Wegner, 1998; van Leeuwen and Wiedermann, 2001). This extends computing power of algorithms and provides mathematical models, which are more adequate for representing modern computers than classical models such as Turing machines or cellular automata.

We have been discussing the input and output components of an IPS. We now turn to the processor component. The processor component is itself traditionally partitioned into three components: *control device(s)*, *operational device(s)*, and *memory*. On the one hand, there are IPS in which all three devices are separate and connected by information channels. Abstract devices of this type are Turing machines, pushdown automata, and random access machines. In many cases, this involves a sophisticated architecture. On the other hand, it is possible that two or even all three components coincide. Abstract devices of this type are neural networks, finite automata, and formal grammars. The latter possess only an operating mechanism in the form of transformation rules.

The structure of an IPS allows us to classify three main types of IPS according to their computer architecture: the *centralized computer architecture* (CCA), *controlled distributed computer architecture* (CDCA), and *autonomous distributed computer architecture* (ADCA).

Remark 1.1.1. There is a distinction between computer architecture and computing architecture. The computer architecture of an IPS (computer) reflects organization of the IPS devices and their functioning. The computing architecture of information processing represents the organization of processes in the IPS. One computer architecture allows one to realize, as a rule, several computing architectures.

Definition 1.1.2. The CCA is a structure of an IPS with one control and one operational device.

The classical Turing machine with one tape and one head embodies the centralized computer architecture and realizes the sequential paradigm of computation.

Remark 1.1.2. Usage of a single operational device implies that operations are performed sequentially. However, this does not preclude parallel processing. Indeed, it is possible that a single portion of data consists of many parts. For example, it may be a vector, matrix, or multidimensional array. Vector and array machines (Pratt, Rabin, and Stockmeyer, 1974; Leeuwen and Wiedermann, 1985) are mathematical models for such parallel information processing by an IPS with CAA. Computation of such machines is explicitly sequential and implicitly parallel.

A complement for centralized computation is distributed computation, which is substantiated by *distributed computer architecture*. It has two types: CDCA and ADCA.

Definition 1.1.3. The CDCA is an IPS structure with one control and many operational devices.

The Turing machine with many heads is an example of a *controlled distributed computing architecture*, and realizes the parallel paradigm of computation.

Remark 1.1.3. The control device in an IPS with CDCA organizes the work of all operational devices. It can work in two modes. In one, called SIMD (*single instruction multiple data*), the control device gives one and the same instruction to all operational devices. In the other mode, called MIMD (*multiple instruction multiple data*), the control device processes separate control information for each operational device or their clusters.

Definition 1.1.4. The ADCA is an IPS structure with many control and many operational devices.

The neural network demonstrates the *autonomous distributed computing architecture* and realizes the concurrent paradigm of computation.

Remark 1.1.4. There are different ways for an IPS with ADCA to organize relations between control and operational devices. The simplest way is when one control device and one operational device are combined into a single block. In a more elaborated IPS, one control device is connected to a cluster of operational devices. Another case exists for when there are no restrictions on connections. Moreover, the system of connections between control and operational devices may be hierarchical with a sophisticated structure.

1.1.2   Static systemic structure

The system approach (Bertalanffy, 1976) explains that a system consists of elements and connections between the elements. Taking into account that any system is connected to other systems, we come to the structural classification, suggested by Bell and Gray (1997) for cyberspace. Accordingly, any IPS **W** consists of three parts:

◇ *Autonomous* IPS that belong to **W** (such as computers, local networks, etc.).
◇ *Networking technology* of **W**, which connects autonomous IPS from **W** and allows them to communicate with each another.
◇ *Interface* (transducer) *technology* of **W**, which connects **W** to people and other systems from the environment of **W**.

In turn, each of these parts has hardware, software, and infware. All of them give specific directions of the IPS development. For example, we may discuss the progress of computer (IPS) hardware or innovations for interface software.

The dynamic structure of an IPS also has two forms: hierarchical and temporal.

1.1.3   Hierarchical dynamic structure

The static structure of IPS influences their dynamic structure. Consequently, we have the following hierarchical structure:

◇ Processes in autonomous IPS.
◇ Processes of interaction of autonomous IPS.
◇ Interface processes with external systems.

We make a distinction between computational or information processing architecture and computer or IPS architecture. The former represents organization of a computational process. The latter defines computer/IPS components and the connections between them. The same computer architecture may be used for organization and realization of different computational architectures. For example, a computer with the parallel architecture can realize sequential computation. At the same time, some types of computational architectures can be realized only by a specific kind of computer architecture. For example, we need a computer with the parallel architecture for performing parallel computation.

1.1.4   Temporal dynamic structure or temporal organization

Usually, we consider only devices (abstract or real) that function in a discrete time, that is, step by step. At the same time, there are theoretical models of computation in continuous time and it is also assumed that analogous computers perform their continuous operations in continuous time.

Problems of time become especially critical when there are many interacting devices and/or we take into account human-computer interaction. Devices in distributed computer architecture have three modes of functioning:

◇ *Synchronous*, when all devices make their step at the same time.
◇ *Synchronized*, when there is a sequence ST of temporal points such that at each point all devices finish some step of computation and/or begin the next step.
◇ *Asynchronous*, when different devices function in their own system time.

Usually it is assumed that there is only one physical time, in which everything functions. However, according to the system theory of time (Burgin, 1997b; Burgin, Liu and Karplus, 2001), each system has its own time. We can easily see this when we take abstract devices. For example, if there are two Turing machines, then their time unit is a step of computation and, as a rule, the steps for different machines are not related to each other. To reduce their functioning to one time, we need to synchronize their operations. Temporal organization of IPS functioning can essentially alter the obtained results (Matherat and Jaekel, 2001).

Thus, we have classified the multitude of existing IPS, although these classifications give only a basis for their further study. Some think that typology and classification of IPS, their models, and other systems are something artificial that has nothing to do with either practice or "real" theory. However, classifications that reflect essential properties of studied systems in cognitive aspects help to compress information, help predict properties of these systems, and are special kinds of scientific laws (Burgin and Kuznetsov, 1994). For example, such classification as the periodic table of chemical elements helped to discover new elements. In the practical sphere, classifications lead to unification and standardization aimed at increasing efficiency and productivity.

1.2   What theory tells us about new directions in information technology

> *She generally gave herself very good advice*
> *(though she very seldom followed it) . . . .*
>
> Lewis Carroll, 1832–1898

With the advent of computers, information technology has been developing and growing very fast. However, the paradox is that it never had enough power.

As Ty Rabe, director of high-performance technical computing solutions at Hewlett-Packard, said, "There are areas in science where the computational power has not reached the stage where it has met the needs. Biology is a good example of that. Biologists couldn't imagine having enough computational power to do what they needed until recently." (cf. Gill, 2002).

This paradoxical situation is the result of the great amount of data and information that scientists have to process to get new knowledge in their area. Numbers that reflect this amount are extremely big. There is an essential difference between small and big numbers.

For example, one of the outstanding mathematicians of the twentieth century, Kolmogorov (1961), suggested that in solving practical problems we have to separate *small*, *medium*, *large*, and *superlarge* numbers.

A number *A* is called *small* if it is possible in practice to process and work with all combinations and systems that are built from *A* elements each of which has two inlets and two outlets.

A number *B* is called *medium* if it is possible to process and work with this *B* but it is impossible to work with all combinations and systems that are built from *B* elements each of which has two or more inlets and two or more outlets.

A number *C* is called *large* if it is impossible to go through a set of size *C* but it is possible to elaborate a system of denotations for these elements.

If even this is impossible, then a number is called *superlarge*.

According to this classification, 3, 4, and 5 are small numbers, 100, 120, and 200 are medium numbers, while the number of visible stars is a large number. Inviting 4 people for a dinner, we can consider all their possible positions at a dinner table. If we come to some place where there are 100 people, we can shake everyone's hands, although it might take too much time. We cannot count the visible stars. However, a catalog of visible stars exists, and we can use it to find information about any one of them.

In a similar way to what has been done by Kolmogorov, the outstanding British mathematician Littlewood (1953) separated all natural numbers into an infinite hierarchy of classes. These classifications of numbers are based on people's counting abilities.

Computers change the borders between classes, but even the most powerful computers cannot erase such distinctions. As a result, we will encounter more and more complex problems that demand working with larger and larger numbers. Thus, we will always need more powerful computers.

To increase the power of computers, we have to understand what we have now and what new directions are suggested for the development of information technology. The results of this book make it possible to evaluate computing power of the new computing schemes. Now, there are several approaches to increasing the power of computers and networks. We may distinguish between chemical, physical, and mathematical directions. The first two are applied to hardware and have an indirect influence software and infware, while the mathematical approach transforms all three components of computers and networks.

The first approach, called molecular computing, has as its most popular branch DNA computing (Cho, 2000). Its main idea is to design molecules that perform computing operations. Engineers and researchers are investigating the field of molecular electronics as a source of new technologies (Overton, 2000). Computers with individual molecules acting as switches could consume far less power. Recent accomplishments by Hewlett-Packard, IBM, ZettaCore, and Molecular Electronics could guide future applications of molecular electronics.

Ari Aviram and Mark Ratner of IBM began the field of molecular electronics by manipulating individual atoms into structures, while Jim Tour and Mark Reed proved that individual molecules can display simple electronic properties. Tour and Reed have since established the startup Molecular Electronics, where Tour is striving to create nanocells, or self-assembled molecules that can be programmed for specific functions. Scientists from UC-Riverside and North Carolina State University are jointly working with porphyrin molecules at their startup, ZettaCore. Porphyrins can store more than 2 bits of data in each molecule and pave the way for faster, more powerful computer devices, claims UC chemist David Bocian. Meanwhile, researchers at Hewlett-Packard Labs have teamed up with UCLA chemists Fraser Stoddart and Jim Heath, who are exploring the possibilities of logic gate-like mechanisms assembled from catenane molecules.

The second direction, quantum computing, is even more popular than the first (cf., for example, Deutsch, 2000; or Seife, 2001). The main idea is to perform computation on the level of atoms and even atomic nuclei, as suggested by Feynman (1982; 1986) and Beniof (1982). Experts write that useful quantum computers are still at least years away. Currently, the most advanced working model can barely factor the number 15. However, if quantum computers can ever be built, they would crack the codes that safeguard the Internet, search databases with incredible speed, and breeze through hosts of other tasks beyond the ken of contemporary computing (Seife, 2001).

Many physical problems have to be solved to make quantum computers a reality. For example, the main feature of quantum objects that makes the quantum computer incredibly powerful is entanglement. However, in 1999, Carlton Caves of the University of New Mexico showed that under the room-temperature conditions large-scale entanglement of atoms is impossible. Last year, MIT physicist Seth Lloyd showed that for some algorithms it is possible to achieve the same results of speedup without entanglement. However, in this case, any quantum computer would need exponentially growing resources (cf. Seife, 2001).

The third direction, the theory of superrecursive algorithms, is based on a new paradigm for computation that changes computational procedure and is closely related to grid computation. Superrecursive algorithms generate and control hypercomputations, that is, computations that cannot be realized by recursive algorithms such as Turing machines, partial recursive functions, and cellular automata. Super-recursive algorithms and their relation to IPS are considered in Chapter 4.

The theory of algorithms shows that both first types of computing, molecular and quantum, can do no more than conventional Turing machines theoretically can do. For example, a quantum computer is only a kind of nondeterministic Turing machine, while a Turing machine with many tapes and heads models DNA and other molecular computers. The theory states that nondeterministic and many-tape Turing machines have the same computing power as the simplest deterministic Turing machine (see

Chapter 2). Thus, DNA and quantum computers will be (when they will be realized) eventually only more efficient.

Superrecursive algorithms (in general) and inductive Turing machines (in particular) go much further, as is demonstrated in Chapters 4 and 5. They surpass conventional computational structures both in computing power and efficiency. Super-recursive algorithms are structural and informational means for the description and realization of hypercomputation.

In business and industry, the main criterion is enterprise efficiency, usually called productivity in economics. Consequently, computers are important not because they make computations with more speed than before but because they can increase productivity. Reaching higher productivity depends on improved procedures. Without proper procedures and necessary skills of the performers, technical devices can even lower productivity. Consequently, methods that develop computer procedures are more important than improvement of hardware or software.

Along the same lines, Minsky in his interview (1998) stated that for the creation of artificial intelligence, software organization is a clue point. Implicitly this contributes to differences between new approaches to computation. From the efficiency perspective, DNA computing is metaphorically like a new model of a car. Quantum computations are like planes at the stage when people did not have them. Super-recursive computations are like rockets, which can take people beyond the "Church–Turing Earth".

These rockets might take us to the moon and other planets if we know how to navigate them. However, we will need new physical ideas for realization of super-recursive algorithms to a full extent. Using our metaphor, we may say that spaceships that will take us to stars are now only in perspective.

If we take grid computation, its real computational power arises from its super-recursive properties. Consequently, this type of computation can overcome limitations imposed on molecular and quantum computers by Turing machines.

Here, it is worth mentioning such new computational model as reflexive Turing machines (Burgin, 1992a). Informally, they are machines that can change their programs by themselves. Genetic algorithms give an example of such an algorithm that can change its program while functioning. In his lecture at the International Congress of Mathematicians (Edinburgh, 1958), Kleene proposed a conjecture that a procedure that can change its program while functioning would be able to go beyond the Church–Turing thesis. However, it was proved that such algorithms have the same computing power as deterministic Turing machines. At the same time, reflexive Turing machines can essentially improve efficiency. Besides, such machines illustrate creative processes facilitated by machines, which is very much on many people's minds. It is noteworthy that Hofstadter (2001) is surprised that a music creation machine can do so well because this violates his own understanding that machines only follow rules and that creativity cannot be described as rule-following.

1.3   The structure of this book

> *"See, boys!" he cried.*
> *"Twenty doors on a side! What a symmetry!*
> *Each side divided into twenty-one equal parts! It's delicious!"*
>
> A *Tangled Tale*, Lewis Carroll, 1832–1898

This book's topic is the future of information technology stemming from a new emerging field in computer science, the theory of superrecursive algorithms, which go beyond the Church–Turing thesis. The majority of computer scientists stand very firmly inside the boundaries imposed by the Church–Turing thesis. Some of them reject any possibility of overcoming the barrier, while others treat superrecursive algorithms as purely abstract constructions that represent a theoretical leap from practical reality.

The attitude of the first group is perfectly explained by a physicist at the Georgia Institute of Technology, Joseph Ford, who quoted Tolstoy:

> I know that most men, including those at ease with problems of greatest complexity, can seldom accept even the simplest and most obvious truth if it be such as would oblige them to admit the falsity of conclusions which they have delighted in explaining to colleagues, which they have proudly taught to others, and which they have woven, thread by thread into the fabric of their life.

With regard to the second group, history of science gives many examples of underestimation of the creative potential of people. One of the brightest cases in this respect was the situation with the great British physicist Rutherford. He made crucial contributions in the discovery of atomic structure, radioactivity of elements, and thermonuclear synthesis. At the same time, when one reporter asked Rutherford when his outstanding discoveries of atom structure and regularities would be used in practice, Rutherford answered, "Never." This opinion of Rutherford was also expressed by his student Kapitsa (cf. Kedrov, 1980).

In a similar way, those who disregard superrecursive algorithms now cannot see that even contemporary computers and networks implicitly possess superrecursivity, opening unimaginable perspectives for future IPS. There is much evidence of the great potential of superrecursive automata and algorithms: they reflect real properties of modern computers and networks better than recursive automata and algorithms, they are can solve many more problems, they are more efficient, and they can better explain many features of the brain and its functioning. All this is explained and proved in Chapter 4. Moreover, in some aspects modern computers possess superrecursive properties even now, while such directions as pervasive computation and grid computing advance these properties and are essentially based on principles of superrecursivity.

The author tried to write this book in a form tentatively interesting to at least three distinct groups of readers. Intelligent laymen can find here an explanation of information processing system functioning, how they are related to algorithms, how they are modeled by mathematical structures, and how these models are used to better understand computers, the Internet, and computation.

Experts in information technology can learn about the models that computer science and mathematics suggest to help them in their work: in designing software, building computers, and developing global and local networks.

Theoretical computer scientists and mathematicians can obtain an introduction to a new branch of both computer science and mathematics, the theory of superrecursive algorithms. These readers will find strict definitions, exact statements, and mathematical proofs.

However, to achieve their personal goal, any reader has to make an intelligent choice of what sections and subsections are relevant to their individual interests, skipping other part or postponing their reading.

The principal goals of Chapter 2 are to systematize conventional models of algorithms, computations, and IPS; to draw a distinction between algorithms and their descriptions; and to show how different types of computations and computer architectures are represented in the theory. Here we consider only algorithms that work with finite words.

Chapter 2 begins with a brief description of the theory of algorithms. We investigate in Section 1 the origin of the term "algorithm" and the development of its meaning in our time. Researchers elaborated many formal and informal definitions of algorithm. Nevertheless, the question "What is an algorithm?" is still open. The works of other researchers and the results of this book indicate that the concept of algorithm is relative with respect to the means and resources of information processing. In this book, algorithms are considered as a kind of procedure and are directly related to IPS as the tool for their investigation and development. However, it is necessary to understand that algorithms and procedures can practically describe the functioning of any system.

The term "algorithm" may be also considered as a linguistic variable in the sense of Zadeh (1973). But to build a theory, it is insufficient to have simply a definition; one needs mathematical models of algorithms. In addition, it is necessary to make a distinction between the informal notion of algorithm and its mathematical models, which help to transform an informal notion into a scientific concept. The informal notion is used in everyday life, in the reasoning of experts, and in methodology and philosophy of computer science, mathematics, and artificial intelligence. At the same time, mathematical models constitute the core of the theory of algorithms.

A brief description of the origin of this theory is given in Section 2.2. Section 2.2 further discusses the Church–Turing thesis, which divides all mathematical models of algorithms into three groups:

◇ *Recursive algorithms*, which are equivalent to Turing machines with respect to their computing power
◇ *Subrecursive algorithms*, which are weaker than Turing machines with respect to their computing power
◇ *Superrecursive algorithms*, which have more computing power than Turing machines

Sections 2.3 and 2.4 discuss two of the most popular models of recursive algorithms, Turing machines and neural networks, as representatives of distinct ap-

proaches to computation and computer architecture. Turing machines model centralized computation, while neural networks simulate distributed computation. Finally,
Section 2.5 gives some application of general mathematical models of algorithms.

Chapter 3 contains a brief exposition and systematization of the theory of subrecursive algorithms. Section 3.1 explains why we need such weaker models. Section
3.2 considers mathematical models of subrecursive algorithms in general, while Sections 3.3 and 3.4 look at two of them in detail: finite automata and recursive functions. Finite automata embody the procedural paradigm, while recursive functions
reflect descriptive and functional paradigms in programming.

The principal goal of Chapter 4 is to systematize and evaluate nonconventional
models of algorithms, computations, and IPS. The main problem is that there are
several approaches in this area, but without a sound mathematical basis, even experts
in computer science cannot make distinctions for different new models of algorithms
and computations. It is important not only to go beyond the Church–Turing thesis
but to do it realistically and to provide possible interpretations and applications. For
example, Turing machines with oracles take computations far beyond the Church–
Turing thesis, but without adequate restrictions on the oracle, they do this beyond
any reason.

The most important superrecursive models (listed in the chronological order) are:
analogue, or continuous time, computation, fuzzy computation, inductive computation, computation with real numbers, interactive and concurrent computation, topological computation, and neural networks with real number parameters, infinite time
computation, and dynamical system computation.

Inability to make distinctions implies misunderstanding and many misconceptions: everything seems the same, although some models are unreasonable and un-
realizable, while others can be realized right now if those who build computers and
develop software begin to understand theoretical achievements. The situation is similar to one when people do not and cannot make distinctions between works of Zi-
olkowsky where theory of space flights was developed and novels of Jules Verne who
suggested to use cannon for space flights.

An analysis of the different models of superrecursive algorithms brings us to the
conclusion that there are good reasons to emphasize inductive Turing machines in
this book. From the theoretical perspective, we have the following reasons to do this:

First, inductive Turing machines are the most natural extension of recursive algorithms and namely, of the most popular model — Turing machine.

Second, the computing and decision power of inductive Turing machines ranges
from the most powerful Turing machines with oracles to the simplest finite automata
(cf. Sections 4.2 and 4.3).

Third, inductive Turing machines have much higher efficiency than recursive algorithms (cf. Chapter 5).

From the practical perspective, we have the following reasons for emphasizing
inductive Turing machines:

First, inductive Turing machines give more adequate models for modern computers in comparison with other superrecursive algorithms (cf. Chapter 4), as well as for
a lot of natural and social processes (cf. Burgin, 1993).

Second, inductive Turing machines provide new constructive structures for the development of computers, networks, and other IPS.

Third, inductive Turing machines explain and allow one to reflect different mental and cognitive phenomena, thus, paving way to artificial intelligence. Pervasive mental processes related to unconscious, associative dynamic processes in memory, and concurrent information processing give examples of such phenomena. Cognition in science and mathematics goes on according to the principles of information processing that are embodied in inductive Turing machines. While cognitive processes on the first level are extensively studied in such area as inductive inference, inductive Turing machines allow one to model, explain, and develop cognitive processes on higher levels, which are relevant to the level of human thinking.

Chapter 4 begins with a brief description of the theory of superrecursive algorithms, which turns out to serve as a base for the development of a new paradigm of computations and provides models for cluster computers and grid computations. We investigate in Section 4.1 the origin of superrecursive algorithms. It is explained that while recursive algorithms gave a correct theoretical representation of computers at the beginning of the "computer era", superrecursive algorithms are more adequate as mathematical models for modern computers. Consequently, superrecursive algorithms serve as a base for the development of a new paradigm of computations. In addition to this, superrecursive algorithms provide for a better theoretical frame for the functioning of huge and dynamic knowledge and databases as well as for computing methods in numerical analysis.

To measure the computing power of inductive Turing machines, which forms the central topic of Chapter 4, we use mathematical constructions like Kleene's arithmetical hierarchy of sets. In this hierarchy, each level is a small part of the next higher level. Conventional Turing machines compute the two first levels of this infinite hierarchy. What is computed by trial-and-error predicates, limit recursive and limit partial recursive functions and obtained by inductive inference is included in the fourth level of the hierarchy. The same is true for the trial-and-error machines recently introduced by Hintikka and Mutanen (1998). At the same time, we prove that it is possible to build a hierarchy of inductive Turing machines that compute the whole arithmetical hierarchy.

Superrecursive algorithms in Chapter 4, as well as recursive algorithms in Chapter 2 and subrecursive algorithms in Chapter 3, are studied from the perspective of three essential aspects of IPS: computability, acceptability, and decidability. These properties describe what computers and their networks can do in principle. However, it is important to know about the efficiency of their functioning. Efficiency, being important per se, also determines what is possible to do in practice. For example, if it is known that the best computer demands hundred years for solving some problem, nobody will use a computer to perform these computations. As a result, this problem will be considered practically unsolvable. Mathematical models of computational efficiency are called measures of complexity of algorithms and computation. Restrictions on complexity imply mathematical formalization for the notion of tractability of problems.

In the concluding section of Chapter 4, a new model for computers and networks is developed and studied. The model consists of a grid array and grid automata. Grid arrays are aimed at computer/network design, description, verification, adaptation, maintenance, and reusability. Grid automata are aimed at the development of a theoretical technique for computer/network studies. In addition to the unification of various models developed for simulation of concurrent processes, the new model allows one to study by mathematical methods and to simulate new kinds of computation, for example, grid computation, and advanced form of IPS like cluster computers.

Chapter 5 contains a study of efficiency of superrecursive algorithms. It is demonstrated that superrecursive algorithms are not only more powerful, solving problems unsolvable for recursive algorithms, but can be much more efficient than recursive algorithms in solving conventional problems. Superrecursive automata allow one to write shorter programs and to use less time to receive the same results as recursive devices.

1.3.1 Some remarks about the theory of algorithms and computation

Although algorithms, both conventional and superrecursive, describe functioning of various types of systems, the central concern of the theory of algorithms is *information processing systems* (IPS) because information processing has become the most important activity. We are interested in the structure of IPS, their functioning, interaction, utilization, and design. We want not only to know how modern IPS are built and work but to look into the future of these systems, how we can improve and develop them.

No science can function properly without evaluation of its results. To understand better the meaning of results presented in this book, we have to evaluate them according to scientific criteria. There are different kinds of evaluation, but one of the most important explains to scientists and society as a whole what is more significant and what is less significant. A quantity of criteria are used for such evaluation, but all of them can be grouped into three classes:

1. *Evaluation directed to the past*. For example, the more time it took the community to solve a problem makes that problem seem more important.

From this point of view, if no one previously tried to solve problem, then its solution is worth almost nothing.

The history of science teaches us that this is not correct. Scientists did outstanding discoveries beginning with their own problem or even without one. For instance, many subatomic particles (positrons, bosons, quarks) were discovered in such a way. Sometimes a discovery shows that the original problem does not have a solution. One of the most important results of the nineteenth century was the construction of non-Euclidean geometries. In particular, this discovery solved the problem that the best mathematicians tried to solve for thousands of years. It demonstrated that it is impossible to prove the fifth postulate of Euclid. However, the initial problem was to prove the fifth postulate. Thus, we can see that in many cases the direction to the past does not give correct estimates of scientific results.

2. *Evaluation directed to the present.* For example, the longer the proof of a mathematical result, the more important the result is considered. Another criterion asserts that the more unexpected some result is, the higher value it has. The latter approach is supported by the statistical information theory of Shannon (1948) that affirms that unexpected results contain much more information than expected ones.

From this point of view, if everybody expects some result, then the result is worth almost nothing.

The history of science teaches us that this is not correct. For example, in 1930s, mathematicians expected that a general model of algorithms would be constructed. It was done. This result is one of the most important achievements of mathematicians in the twentieth century. So, we can see that in many cases the direction to the present also does not give correct estimates of scientific results.

3. *Evaluation directed to the future.* Here the criteria are what influence a given result has on the corresponding field of science (an inner criterion), on science in general (an intermediate criterion), or on the practical life of people, their mentality, nature, etc. (an external criterion). This approach is supported by the general theory of information (Burgin, 1997; 2001), which affirms that the quantity of information and even its value is estimated by its influence. This means that criteria directed into the future are the most efficient and appropriate for science and society, accelerating their development.

At the same time, criteria directed into the future are the most difficult to apply because no one, really, has a clear vision of the future. Nevertheless, to evaluate this book from the point of view of the future will be the most correct one for a reader, since only this approach gives some indication of the future for information technology and computer science. One of the central aims of the author is to facilitate such comprehension of the presented material.

1.4 Notation and basic definitions

Some mathematical concepts, in spite of being basic and extensively used, have different interpretations in different books. In a similar way, different authors use dissimilar notation for the same things, as well as the same notation for distinct things. For this reason, we give here some definitions and notation that is used in this book for basic mathematical concepts.

N is the set of all natural numbers $1, 2, \ldots, n, \ldots$.

N_0 is the set of all whole numbers $0, 1, 2, \ldots, n, \ldots$.

Z is the set of all integer numbers.

Q is the set of all rational numbers.

R is the set of all real numbers.

R^+ is the set of all nonnegative real numbers.

R^{++} is the set of all positive real numbers.

$R_\infty = R \cup \{\infty, -\infty\}$.

ω is the sequence of all natural numbers.

$\emptyset$ is the *empty set*.

$r \in X$ means that r belongs to X or r is a member of X.

$Y \subseteq X$ means that Y is a subset of X, that is, Y is a set such that all elements of Y belong to X.

The union $Y \cup X$ of two sets Y and X is the set that consists of all elements from Y and from X.

The intersection $Y \cap X$ of two sets Y and X is the set that consists of all elements that belong both to Y and to X.

If X is a set, then 2^X is the *power set* of X, which consists of all subsets of X.

If X and Y are sets, then $X \times Y = \{(x, y); x \in X, y \in Y\}$ is the direct or Cartesian product of X and Y; in other words, $X \times Y$ is the set of all pairs (x, y), in which x belongs to X and y belongs to Y.

Y^X is the set of all mappings from X into Y.

$$X^n = \underbrace{X \times X \times \ldots X \times X}_{n};$$

Relations $f(x) \succsim g(x)$ and $g(x) \precsim f(x)$ means that there is a number c such that $f(x) + c \geq g(x)$ for all x.

A fundamental structure of mathematics is *function*. However, functions are special kinds of binary relations between two sets.

A *binary relation* T between sets X and Y is a subset of the direct product $X \times Y$. The set X is called the *domain* of T ($X = \mathrm{D}(T)$) and Y is called the *codomain* of T ($Y = \mathrm{CD}(T)$). The *range* of the relation T is $\mathrm{R}(T) = \{y; \exists x \in X\ ((x, y) \in T)\}$. The *domain of definition* of the relation T is $\mathrm{SD}(T) = \{x; \exists y \in Y\ ((x, y) \in T)\}$.

A *function* or *total function* from X to Y is a binary relation between sets X and Y in which there are no elements from X that are corresponded to more than one element from Y and to any element from X is corresponded some element from Y. Often total functions are also called everywhere defined functions.

A *partial function* f from X to Y is a binary relation in which there are no elements from X which are corresponded to more than one element from Y.

For a partial function f, its domain of definition $\mathrm{SD}(f)$ is the set of all elements for which f is defined.

A function $f : X \to Y$ is *increasing* if $a < b$ implies $f(a) \leq f(b)$ for all a and b from X.

A function $f : X \to Y$ is *decreasing* on X if $a < b$ implies $f(a) \geq f(b)$ for all a and b from X.

A function $f : X \to Y$ is *strictly increasing* on X if $a < b$ implies $f(a) < f(b)$ for all a and b from X.

A function $f: X \to Y$ is *strictly decreasing* on X if $a < b$ implies $f(a) > f(b)$ for all a and b from X.

If A is an algorithm that takes input values from a set X and gives output values from a set Y, then $f_A: X \to Y$ is a partial function defined by A, that is, $f_A(x) = A(x)$ when $A(x)$ is defined, otherwise, $f_A(x)$ is not defined.

If A is a multiple valued algorithm that takes input values from a set X and gives output values from a set Y, then $r_A: X \to Y$ is a binary relation defined by A, that is, $r_A(x) = \{A(x); x \in X\}$ when $A(x)$ is defined, otherwise, $f_A(x)$ is not defined.

A function $f: X \to Y$ is computable when there is a Turing machine T that computes f, that is, $f(x) = f_T(x)$.

A function $f: X \to Y$ is inductively computable when there is an inductive Turing machine M that computes f, that is, $f(x) = f_M(x)$.

In what follows, functions range over numbers and/or words and take numerical and/or word values. Special kinds of general functions are *functionals*, which take numerical and/or word values and have any number of numerical and/or word and/or function variables. Thus, a numerical/word function is a special case of a functional, while a functional is a special case of a general function.

For any set, S, $\chi_S(x)$ is its characteristic function, that is, $\chi_S(x)$ is equal to 1 when $x \in S$ and is equal to 0 when $x \notin S$, and $C_S(x)$ is its partial characteristic function, that is, $C_S(x)$ is equal to 1 when $x \in S$ and is undefined when $\mathrm{x} \notin S$.

A multiset is similar to a set, but can contain indiscernible elements or different copies of the same elements.

A topology in a set X is a system $O(X)$ of subsets of X that are called open subsets and satisfy the following axioms:

T1. $X \in O(X)$ and $\emptyset \in O(X)$.

T2.   For all A, B, if A, $B \in O(X)$, then $A \cap B \in O(X)$.

T3.   For all A_i, $i \in I$, if all $A_i \in O(X)$, then $\bigcup_{i \in I} A_i \in O(X)$.

A set X with a topology in it is called a *topological space*.

In many interesting cases, topology is defined by a metric.

A *metric* in a set X is a mapping $\mathbf{d}: X \times X \to R^+$ that satisfies the following axioms:

M1. $\mathbf{d}(x, y) = 0$ if and only if $x = y$.

M2. $\mathbf{d}(x, y) = \mathbf{d}(y, x)$ for all $x, y \in X$.

M3. $\mathbf{d}(x, y) \leq \mathbf{d}(x, z) + \mathbf{d}(z, y)$ for all $x, y, z \in X$.

A set X with a metric $\mathbf{d}$ is called a *metric space*. The number $\mathbf{d}(x, y)$ is called the distance between x and y in the metric space X.

An *alphabet* or vocabulary A of a *formal language* is a set consisting of some symbols or letters. Vocabulary is an alphabet on a higher level of hierarchy because words of a vocabulary play the same role for building sentences as symbols in an alphabet for building words. Traditionally an alphabet is a set. However, a more consistent point of view is that an alphabet is a multiset (Knuth, 1981), containing an unbounded number of identical copies of each symbol.

A *string* or *word* is a sequence of elements from the alphabet. A^* denotes the set of all finite words in the alphabet A. Usually there is no difference between strings and words. However, having a language, we speak about words of this language and not about its strings.

A *formal language* L is any subset of A^*.

If L and M are formal languages, then their concatenation is the language $LM = \{uw; u \in L \text{ and } w \in M\}$.

The length $l(w)$ of a word w is the number of letters in a word.

ε is the *empty word*.

Λ is the *empty symbol*.

If n and a are natural numbers and $a > 1$, then $\ln_a(n)$ is the length of the representation of n in the number system with the base a. For example, when $a = 2$, then n is represented as finite sequence of 1s and 0s and $\ln_2(n)$ is the length of this sequence.

The logical symbol $\forall$ means "for any".

The logical symbol $\exists$ means "there exists".

If X is a set, then "for almost all element from X" means "for all but for a finite number of them." The logical symbol $\forall\!\!\forall$ is used to denote "for almost all". For example, if $A = \omega$, then almost all elements of A are bigger than 10.

If P and Q are two statements, then P $\rightarrow$ Q means that P implies Q.

2

Recursive Algorithms

In this chapter, we consider the following problems:

◇ What is the general situation with algorithms, their origin, and problems of their representation? Analysis of this situation shows the need for having mathematical models for algorithms and for developing an efficient theory of algorithms (Section 1).

◇ What is the general situation with mathematical models of algorithms, their origin, and existence of the absolute and universal model stated in the Church–Turing thesis (Section 2)?

◇ What is a Turing machine, which is the most popular mathematical model of algorithms? How does this model represent a centralized computer architecture, embodying a symbolic approach to computation, and why is it the main candidate for being an absolute and universal model (Section 3)?

◇ What is a neural network as a complementary mathematical model of algorithms, representing distributed computer architecture? How does it embody a connectionist approach to computation, and become the main candidate for being the model for emergent information processing (Section 4)?

◇ What is useful to know about applications of mathematical models and the theory of algorithms? How does this theory determine our knowledge of programs, computers, and computation (Section 5)?

2.1 What algorithms are and why we need them

People use algorithms all the time, without even knowing it. In many cases, people work, travel, cook, and do many other things according to algorithms. For example, we may speak about algorithms for counting, algorithms for going from New

York to Göttingen or to some other place, algorithms for chip production or for buying some goods, products or food. Algorithms are very important in daily life. Consequently, they have become the main objects of scientific research in such areas as the theory of algorithms.

In addition, all computers, networks, and embedded devices function under the control of algorithms. Our work becomes more and more computerized. We are more and more networked. Embedded devices are integral parts of our daily life and sometimes of ourselves. As a result, our lives becomes entwined in a vast diversity of algorithms. Not to be lost in this diversity, we need to know more about algorithms.

2.1.1 Historical remarks

The word *algorithm* has an interesting historical origin. It derives from the Latin form of the name of the famous medieval mathematician Muhammad ibn Musa al-Khowarizmi. He was born sometime before 800 AD and lived at least until 847. His last name suggests that his birthplace was in Middle Asia, somewhere in the territory of modern Uzbekistan. He was working as a scholar at the House of Wisdom in Baghdad when around the year 825, he wrote his main work *Al-jabr wa'l muqabala* (from which our modern word *algebra* is derived) and a treatise on Hindu-Arabic numerals. The Arabic text of the latter book was lost but its Latin translation, *Algoritmi de numero Indorum*, which means in English *Al-Khowarizmi on the Hindu Art of Reckoning*, introduced to the European mathematics the Hindu place-value system of numerals based on the digits 1, 2, 3, 4, 5, 6, 7, 8, 9, and 0. The first introduction to the Europeans in the use of zero as a place holder in a positional base notation was probably also due to al-Khowarizmi in this work. Various methods for arithmetical calculation in a place-value system of numerals were given in this book. In the twelfth century his works were translated from Arabic into Latin. Methods described by al-Khowarizmi were the first to be called algorithms following the title of the book, which begins with the name of the author. For a long time algorithms meant the rules for people to use in making calculations. Moreover, the term computer was also associated with a human being. As Parsons and Oja (1994) write, "if you look in a dictionary printed anytime before 1940, you might be surprised to find a computer defined as a *person* who performs calculations. Although machines performed calculations too, these machines were related to as calculators, not computers."

This helps us to understand the words from the famous work of Turing (1936). Explaining first how his fundamental model, which was later called a Turing machine, works, Turing writes: "We may now construct a machine to do the work of this computer." Here a computer is a person and not a machine.

Even in a recently published book (Rees, 1997), we can read, "On a chi-chhou day in the fifth month of the first year of the Chih-Ho reign period (July AD 1054), Yang Wei-Te, the Chief Computer of the Calendar – the ancient Chinese counterpart, perhaps, of the English Astronomer Royal – addressed his Emperor . . . " It does not mean that the Emperor had an electronic device for calendar computation. A special person, who is called Computer by the author of that book, performed necessary computations.

Through extensive usage in mathematics, algorithms encompassed many other mathematical rules and action in various fields. For example, the table-filling algorithm is used for minimization of finite automata (Hopcroft, Motwani, and Ullman, 2001). When electronic computers emerged, it was discovered that algorithms determine everything that computers can do. In such a way, the name *Al-Khowarizmi* became imprinted into the very heart of information technology, computer science, and mathematics.

Over time, the meaning of the word *algorithm* has extended more and more (cf., for example, Barbin et al., 1994). Originating in arithmetic, it was explained as the practice of algebra in the eighteenth century. In the nineteenth century, the term came to mean any process of systematic calculation. In the twentieth century, Encyclopaedia Britannica described algorithm as a systematic mathematical procedure that produces – in a finite number of steps – the answer to a question or the solution of a problem.

Now the notion of algorithm has become one of the central concepts of mathematics. It is a cornerstone of the foundations of mathematics, as well as of the whole computational mathematics. All calculations are performed according to algorithms that control and direct those calculations. All computers, simple and programmable calculators function according to algorithms because all computing and calculating programs are algorithms that utilize programming languages for their representation.

Moreover, being an object of mathematical and computer science studies, algorithms are not confined neither to computation nor to mathematics. They are everywhere. Consequently, the term "algorithm" has become a general scientific and technological concept used in a variety of areas. There are algorithms of communication, of production and marketing, of elections and decision making, of writing an essay and organizing a conference. People even speak about algorithms for invention (Altshuller, 1999).

2.1.2 A diversity of definitions

There are different approaches to a defining algorithm. Being informal, the notion of algorithm allows a variety of interpretations. Let us consider some of them.

A popular-mathematics point of view on algorithm is presented by Rogers (1987): *Algorithm is a clerical (that is, deterministic, bookkeeping) procedure which can be applied to any of a certain class of symbolic inputs and which will eventually yield, for each such input, a corresponding output.* Here *procedure* is interpreted as a system of rules that are arranged in a logical order, and each rule describes a specific action. In general, an algorithm is a kind of procedure. *Clerical* or *bookkeeping* means that it is possible to perform according to these rules in a mechanical way, so that a device is actually able to carry out these actions.

Donald Knuth (1971), a well-known computer scientist, defines algorithm as follows: *An algorithm is a finite, definite, effective procedure, with some output.* Here *finite* means that it has a finite description and there must be an end to the work of an algorithm within a reasonable time. *Definite* means that it is precisely definable in clearly understood terms, no "pinch of salt"–type vagaries, or possible ambiguities.

Effective means that some device is actually able to carry out our actions prescribed by an algorithm. Some interpret the condition to give output so that algorithm always gives a result. However, computing practice and theory come with a broader understanding. Accordingly, algorithms are aimed at producing results, but in some cases cannot do this.

More generally, an algorithm is treated as a specific kind of exactly formulated and tractable recipe, method, or technique for doing something. Barbin et al. in their *History of Algorithms* (1999) define algorithm as *a set of step-by-step instructions, to be carried out quite mechanically, so as to achieve some desired result*. It is not clear from this definition whether algorithm *has to be aimed at* achieving some result or it *has to achieve* such a result. In the first case, this definition includes superrecursive algorithms, which are studied in the third and fourth chapters. In the second case, the definition does not include many conventional algorithms because not all of them can always give a result.

According to Schneider and Gersting (1995), *an algorithm is a well-ordered collection of unambiguous and effectively computable operations that when executed produces a result and halts in a finite amount of time*. This definition demands to give results in all cases and consequently, reduces the concept of algorithm to the concept of computation, which we consider later.

For some people, *an algorithm is a detailed sequence of actions to perform to accomplish some task or as a precise sequence of instructions*.

In the Free Online Dictionary of Computing (http://foldoc.doc.ic.ac.uk/) *algorithm* is defined as *a detailed sequence of actions to perform to accomplish some task*.

According to Woodhouse, Johnstone, and McDougall (1984), an algorithm is "a set of instructions for solving a problem." They illustrate this definition with a recipe, directions to a friend's house, and instructions for changing the oil in a car engine. However, according to the general understanding of algorithm in computer science (cf., for example, the definitions of Rogers and of Knuth), this is not, in general, an algorithm but only a procedure.

In a recently published book of Cormen et al. (2001), after asking the question "What are algorithms?" the authors write that "informally, *algorithm is a well-defined computational procedure that takes some value, or set of values, as input and produces some value, or set of values, as output*. An algorithm is thus a sequence of computational steps that transform the input into the output."

If the first part of this definition represents algorithms as a procedure type by a relevant, although vague term *well-defined*, the second part presents some computational process instead of algorithm.

We synthesize the above approaches in the following informal definition:

Definition 2.1.1. An algorithm is an unambiguous (definite) and adequately simple to follow (effective) prescription (organized set of instructions/rules) for deriving necessary results from given inputs (initial conditions).

Here *adequately* means that a performer (a device or person) can adequately achieve these instructions on performing operations or actions. In other words, the performer must have knowledge (may be, implicit) and ability to perform these instructions. This implies that the *notion of algorithm and computability is relative*. In a similar way, Copeland (1999) writes that computability is a relative notion because it is resource dependent. For example, information sufficient for one performer may be essentially insufficient for another one even from the same class of systems or persons. In such a way, an algorithm for utilization of a word processor is good for a computer with such processor, but it is not an algorithm for a calculator. Algorithms of finding the inverse matrix are simple for the majority of mathematicians, but they are in no way algorithms for the majority of population. For them, these algorithms are some mystic procedures invented by "abstruse" mathematicians.

Definition 2.1.1, like most others, implicitly implies that any algorithm *uniquely* determines some process. Computer science has contributed nondeterministic algorithms, including fuzzy algorithms (Zadeh, 1969) and probabilistic algorithms, in which execution of an operation/action is not determined uniquely but has some probability. As examples, we can take nondeterministic and probabilistic Turing machines and finite automata. Here nondeterminism means that there is a definite choice in application of a rule or in execution of an action, allowing an arbitrary choice of input data or/and output result. However, these forms of nondeterminism can be reduced to the choice of a rule or action. In its turn, such a choice is in practice subjugated to deterministic conditions. For instance, when selecting instructions from a list, a heuristic rule is taken, such as "take the first that you find appropriate."

It is possible to find an extensive analysis of the concept of algorithm in (Turing, 1936; Markov, 1951; Kolmogorov, 1953; Knuth, 1971).

Existence of a diversity of definitions for algorithm demonstrates absence of a general agreement on the meaning of the term, and theory experts continue to debate what models of algorithms are adequate. However, experience shows that a diversity of different models is necessary. Some of them are relevant to modern computers, some will be good for tomorrow's computers, while others always will be only mathematical abstractions. However, before we build a model, it is necessary to find out what properties are essential and how to incorporate them.

2.1.3   Properties of algorithms and types of computation

Thus, majority of definitions of algorithm imply that algorithms consist of rules or instructions. As a rule, each instruction is performed in one step. This suggests that algorithms have three features:

1. *Algorithms function in a discrete time.*
2. *All instructions are sufficiently simple.*
3. *Relations between operational steps of algorithm determine topology of computation.*

However, while these properties look very natural, some researchers introduce models of computation with continuous time. An example is given by real number computations in the sense of Shannon (1941) and Moore (1996). In these models instructions look rather simple, while their realization may be very complex. For example, an addition with infinite precision of two transcendental numbers in numerical form is, as a rule, impossible, even though its description in an algebraic form is really simple.

Algorithms for computers generate a diversity of computations with different characteristics. Among the most important of them is the computation topology. This topology separates three classes of processes and corresponding computing architecture are:

1. *Sequential computation.*
2. *Parallel or synchronous computation.*
3. *Concurrent or asynchronous computation.*

In turn, each type has the following subtypes:

1. **Sequential computation** may be
   a) *Acyclic*, with no cycles;
   b) *Cyclic*, organizing computation in one cycle;
   c) *Incyclic*, containing not one but several cycles.
2. **Parallel computation** may be
   a) *Branching*, referring to parallel processing of different data from one package of data;
   b) *Pipeline*, referring to synchronous coprocessing of similar elements from different packages of data (Kogge, 1981);
   c) *Extending pipeline*, which combines properties both of branching and pipeline computations (Burgin, Liu, and Karplus, 2001a).
3. According to the control system for computation, **concurrent computation** may be
   a) *Instruction controlled*, referring to parallel processing of different data from one package of data;
   b) *Data controlled*, referring to synchronous processing of similar elements from different packages of data;
   c) *Agent controlled*, which means that another program controls computation.

While the first two approaches are well known, the third type exists but is not considered to be a separate approach. However, even now the third approach is often used implicitly for organization of computation. One example of agent-controlled computation is the utilization of an interpreter that, taking instructions of the program, transforms them into machine code, and then this code is executed by the computer. An interpreter is the agent controlling the process. A universal Turing machine (cf. Section 2.3) is a theoretical example of agent-controlled computation. The program of this machine is the agent controlling the process. We expect the role of agent-controlled computation to grow in the near future.

Usually, it is assumed that algorithms satisfy specific conditions of nonambiguity, simplicity, and effectiveness of separate operations to be organized for automatic performance. Thus, each operation in an algorithm must be sufficiently clear so that it does not need to be simplified for its performance. Since an algorithm is a collection of rules or instructions, we must know the correct sequence in which to execute the instructions. If the sequence is unclear, we may perform the wrong instruction or we may be uncertain as to which instruction should be performed next. This is especially important for computers. A computer can only execute an algorithm if it knows the exact sequence of steps to perform.

Thus, it is traditionally assumed that algorithm have the following primary characteristics (properties):

1. *An algorithm consists of a finite number of rules.*
2. *The rules constituting an algorithm are unambiguous (definite), simple to follow (effective), and have simple finite description (are constructive).*
3. *An algorithm is applied to some collection of input data and aimed at a solution of some problem.*

This minimal set of properties allows one to consider algorithms from a more general perspective: those that work with real numbers or even with continuous objects, those that do not need to stop to produce a result, and those that use infinite and even continuous time for computation.

2.1.4   Algorithms and their descriptions

Programmers and computer scientists well know that the same algorithm can be represented in a variety of ways. Algorithms are usually represented by texts and can be expressed practically in any language, from natural languages like English or French to programming languages like C^{++}. For example, addition of binary numbers can be represented in many ways: by a Turing machine, by a formal grammar, by a program in C^{++}, in Pascal or in Fortran, by a neural network, or by a finite automaton. Besides, an algorithm can be represented by software or hardware. That is why, as it is stressed by Shore (in Buss et al., 2001), it is essential to understand that algorithm is different from its representation and to make a distinction between algorithms and their descriptions.

In the same way, Cleland (2001) emphasizes that "it is important to distinguish instruction-expressions from instructions." The same instruction may be expressed in many different ways, including in different languages and in different terminology in the same language. Also, some instruction are communicated with other instructions nonverbally, that is, when one computer sends a program to another computer.

This is also true for numbers and their representations. For example, the same rational number may be represented by the following fractions: 1/2, 2/4 , 3/6, as well as by the decimal 0.5. Number five is represented by the Arab (or more exactly, Hindu) numeral 5 in the decimal system, the sequence 101 in the binary number system, and by the symbol V in the Roman number system. There are, at least, three natural ways for separating algorithms from their descriptions such as programs or systems of instructions.

In **the first way**, which we call the *model approach*, we chose some type **D** of descriptions (for example, Turing machines) as a model description, in which there is a one-to-one correspondence between algorithms and their descriptions. Then we introduce an equivalence relation R between different descriptions of algorithms. This relation has to satisfy two axioms:

DA1 Any description of an algorithm is equivalent to some element from the model class D.

DA2 Any two elements from the model class D belong to different equivalence classes.

This approach is used by Moschovakis, who considers the problem of unique representation for algorithms in his paper "What is an Algorithm?" (2001). He makes interesting observations and persuasively demonstrates that machine models of algorithms are only models but not algorithms themselves. His main argument is that there are many models for one and the same algorithm. To remedy this, he defines algorithms as systems of mappings, thus building a new model for algorithms. Moschovakis calls such systems of mappings defined by recursive equations *recursors*. While this indicates progress in mathematically modeling algorithms, it this does not solve the problem of separating algorithms as something invariant from their representations. This type of representation is on a higher level of abstraction than the traditional ones, such as Turing machines or partial recursive functions. Nevertheless, a *recursor* (in the sense of Moschovakis) is only a model for algorithm but not an algorithm itself.

The **second way** to separate algorithms and their descriptions is called the *relational approach* and is based on an equivalence relation R between different descriptions of algorithms. Having such relation, we define algorithm as a class of equivalent descriptions. Equivalence of descriptions can be determined by some natural axioms, describing, for example, the properties of operations:

   Composition Axiom. Composition (sequential, parallel, etc.) of descriptions represents the corresponding composition of algorithms.

   Decomposition Axiom. If a description H defines a sequential composition of algorithms A and B, a description K defines a sequential composition of algorithms C and B, and $A = C$, then H is equivalent to K.

At the same time, the equivalence relation R between descriptions can be formed on the base of computational processes. Namely, two descriptions define the same algorithm if these descriptions generate the same sets of computational processes. This definition of description equivalence depends on our understanding of the processes – different and equal. For example, in some cases it is natural to consider processes on different devices as different, while in other cases it might be better to treat some processes on different devices as equal.

In particular, we have the rule as suggested by Cleland (2001) for instructions:

   Different instruction-expressions, that is, representations of instructions, express the same instruction only if they prescribe the same type of action.

Such a structural definition of algorithm depends on the organization of computational processes. For example, let us consider some Turing machines T and Q. The only difference between T and Q is that Q contains all instructions of T and plus instructions that are never used in computations of the machine Q. Then it is possible to assume that this additional instruction has no influence on computational processes and thus, **T** and **Q** define one and the same algorithm. On the other hand, if a Turing machine in a course of computation always go through all instructions to choose the one to be performed, then the processes are different and consequently, **T** and **Q** define different algorithms.

The **third way** to separate algorithms and their descriptions is the *structural approach* because a specific invariant (structure) is extracted from descriptions. We call this structure an algorithm. Here we understand structures in the sense of (Burgin, 1997). Thus, we come to the following understanding, which separates an algorithm from its descriptions:

Definition 2.1.2. An algorithm is a (finite) structure that contains for some performer (class of performers) exact information (instructions) that allows this performer(s) to pursue a definite goal.

Consequently, algorithms are compressed constructive (that is, giving enough information for realization) representations of processes. In particular, they represent intrinsic structures of computer programs. Hence, an algorithm is an essence that is independent of how it happens to be represented and is similar to mathematical objects. Once the concept of algorithm is so rendered, its broader connotations virtually spell themselves out. As a result, an algorithm appears as to consist of three components: structure, representation (linguistic, mechanical, electronic, and so on), and interpretation.

It is important to understand that not all systems of rules represent algorithms. For example, you want to give a book to your friend John, who often comes to your office. So you decide to take the book to your office (the first rule) and to give it to John when he comes to your office (the second rule). While these simple rules are fine for you, they are much too ambiguous for a computer. In order for a system of rules to be applicable to a computer, it must have certain characteristics. We will specify these characteristics later on in formal definitions of an algorithm. Now we only state that formalized functioning of complex systems (such as people) is mostly described and controlled by more general systems of rules than algorithmic structures. They are called procedures.

Definition 2.1.3. A procedure is a compressed operational representation of a process.

For example, you have a set of instructions for planting a garden where the first step instructs you to remove all large stones from the soil. This instruction may not be effective if there is a ten-ton rock buried just below ground level. So, this is not an algorithm, but only a procedure. However, if you have means to annihilate this rock, this system of rules becomes an algorithm.

It is necessary to remark that the above given definition describes procedure in the theoretical sense. There is also a notion of procedure in the sense of programming.

A *procedure* in a program, or *subroutine*, is a specifically organized sequence of instructions for performing a separate task. This allows the subroutine code to be called from multiple places of the program, even from within itself, in which case the form of computation is called recursive. Most programming languages allow programmers to define subroutines. Subroutines, or procedures in this sense, are specific representations of algorithms by means of programming languages.

Representation for algorithms and procedures fall into three classes.

Automaton representations. Turing machines and finite automata give the most known examples of such representations.

Instruction representations. Formal grammars, rules for inference, and Post productions give the most known examples of such representations.

Equation representations. Here is an example of a well-known recursive equation:

$$\text{Fact}(n) = \begin{cases} 1 & \text{when } n = 1, \\ n \cdot \text{Fact}(n-1) & \text{when } n > 1. \end{cases}$$

The fixed point of this recursive equation defines a program for computation of the factorial $n!$.

Algorithms are connected to procedures in a general sense, being special cases of procedures. If we consider algorithms as rigid procedures, then there are also soft procedures, which have recently become very popular in the field of soft computing.

To discern algorithms from procedures, it is assumed that algorithms satisfy specific conditions:

1. *Operational decomposition* means that there is a system of effective basic operations that are performed in a simple way with some basic constructive objects, while all other operations can be reduced to the basic operations.
2. *Purposefulness* means that execution of algorithms is aimed at some purpose.
3. *Discreteness* means that operations are performed in a discrete time, that is, step by step with each step separated from the others.

Some experts demand additional conditions that are not always satisfied both in the theory of algorithms and practice of computation:

4. *Substantial finiteness* means that all objects of the algorithm operation and the number of objects involved in that operation at each step are finite.
5. *Operational finiteness* means that the number of algorithm operations and operations themselves are finite.
6. *Temporal finiteness* means that the result of the algorithm functioning/execution is obtained in a finite time.
7. *Demonstrativeness* means that the algorithm provides explicit information when it obtains the necessary result.
8. *Definability* means that given a relevant input, the algorithm always obtains the result.

Like algorithms, procedures (in a general sense) are also different from their descriptions.

Definition 2.1.4. A representation or description of a procedure or algorithm is a symbolic materialization of this procedure/algorithm as a structure.

According to this definition, algorithms and procedures are similar to mathematical objects because, as it is demonstrated in Burgin (1998), all mathematical objects are structures. This explains why algorithms in a strict sense appeared in mathematics, were named after a mathematician, and have been developed into a powerful and advanced mathematical theory – the theory of algorithms and computation. However, in this theory there is no distinction between algorithms and their descriptions. In what follows, we follow this tradition to make comprehension easier.

Processes represented by algorithms vary, implying the corresponding classification of algorithms.

Transformation algorithms describe how to transform some input into a definite output, for example, how to calculate $123 + 321$.

Performance algorithms describe some activity or functioning, for example, algorithms of how a car engine functions or algorithms of human–computer communication. Performance algorithms also describe how mental activity is organized.

Construction algorithms describe how to build objects.

Decision algorithms are specific cases of construction algorithms describing decision-making.

Algorithms are also grouped by objects that are involved in the processes they describe. We have

1. *Material algorithms*, which work with material objects (for example, algorithm of a vending machine or algorithms of pay phone functioning);
2. *Symbolic algorithms*, which work with symbols (for example, computational algorithms are symbolic algorithms that control computing processes); and
3. *Mixed algorithms*, which work both with material and symbolic objects (for example, algorithms that control production processes or algorithms of robot functioning).

All mathematical models of algorithms are symbolic algorithms. Thus, in computer science and mathematics, only symbolic algorithms, that is, algorithms with symbolic input and output, are studied. Computer algorithms, that is, such algorithms that are or may be performed by computers, form an important class, contributing to the concept of *computation* in two ways. According to an engineer, computation is any thing computer can do. On the one hand, this restricts computation to computers that exist at a given time. Each new program extends the scope of computation. On the other hand, not everything that computers do is computation. For example, interaction with users or with other computers, sending e-mails, and connecting to the Internet are not computations. Mathematical approach reduces dependence on computers. According to a mathematician, computation is a sequence of symbolic transformations that are performed according to some algorithm. From the mathematical point of view, computers function under the control of algorithms, which are embodied in computer programs. So, to understand and explore the possibilities of computers and their boundaries, we have to study algorithms.

2.2 Mathematical models of algorithms and why we need them: History and methodology

> *Of course the first thing to do*
> *was to make a grand survey of the country*
> *she was going to travel through.*
>
> Lewis Carroll, 1832–1898

We begin this section with an informal overview of formal mathematical models of algorithms. With a diversity of such models, we need to understand this "algorithmic universe" from a general perspective of the whole picture. In their aggregate and latitude, models of algorithms constitute a whole world of ideas and techniques. However, many do not make a distinction between algorithms and their mathematical models, resulting in misunderstanding and misconceptions, especially, when professionals in computer technology consider computer science. That is why we give here an explanation of relations between algorithms and their mathematical models.

The main difference is that given some relevant input an algorithm determines a computational process. At the same time, a mathematical model needs some further specification to become an algorithm. Only after such specification is given, it is possible to provide some relevant input and begin computation.

For instance, in a Turing machine (cf. Section 2.3) as a model for algorithms and computation, the alphabet has the form $\{a_1, \ldots, a_n\}$ and the rules of functioning are

$$q_h a_i \rightarrow a_j q_k,$$
$$q_h a_i \rightarrow R a_j q_k,$$
$$q_h a_i \rightarrow L a_j q_k.$$

In a Turing machine as a particular algorithm that checks if some input x is an even or odd number, the alphabet is $\{1, 0, B\}$ and the rules of functioning are (Minsky, 1967):

$$q_0 0 \rightarrow R 0 q_0,$$
$$q_0 1 \rightarrow R 0 q_1,$$
$$q_0 B \rightarrow 0 q_0,$$
$$q_1 0 \rightarrow R 0 q_1,$$
$$q_1 1 \rightarrow R 0 q_0,$$
$$q_1 B \rightarrow 1 q_0.$$

2.2.1   Methodological considerations

In the previous Section 2.1, we have considered algorithms informally, deriving their basic properties. We see that an informal notion of algorithms is comparatively vague, flexible, and easy to treat. Consequently, it is insufficient for exact studies. In contrast, mathematical models are precise, rigid, and formal. As a result, they capture, as a rule, only some features of informal notions, but are suitable for theoretical investigation. This is why we need mathematical models for algorithms. Such situation has always emerged when mathematics acquired its basic notions from the real world. For example, the notion of number was turned into an exact concept and has been developing through the ages: from natural numbers to rational and integer numbers to real to complex numbers to hypernumbers and transfinite numbers. This notion gave birth to a series of mathematical concepts: groups and algebras, fields and topological spaces, order relations and measures. The same is true for the notion of algorithm. An exact concept of algorithm has been introduced in a form of a mathematical model of algorithm.

Being rather practical, the theory of algorithms is a typical mathematical theory with a quantity of theorems and proofs. However, the main achievement of this theory has been elaboration of an exact mathematical model of algorithm. The first models were constructed in mathematics less than seventy years ago – in thirties of the twentieth century – in connection with its intrinsic dilemmas of finding solutions to some mathematical problems. Some of the first models of algorithms also included a formal device (abstract automaton) for realization of algorithmic scheme. Examples of such models are Turing machines and neural networks, which are considered later in this chapter. These constructions give also more or less relevant models for computers. Other models of algorithms gave only a description of rules for computation. Examples of such models are recursive functions and formal grammars, which are considered in the next chapter.

Some think that mathematical models of algorithms and abstract automata were constructed prior to the advent of digital computers. This is not true. The first digital computer was designed by the British scientist Babbage in the nineteenth century. However, his computer was a mechanical device. Electronic computers really appeared after their mathematical models had been designed. Consequently, computer science did not exist at that time, but its foundations inevitably emerged before the first electronic computers were built.

Creation of mathematical models of algorithms was caused by the following situation. Many algorithms had been elaborated and used in mathematics and beyond. Each field of mathematics has its specific algorithms. For example, we have algorithms of adding and subtracting integer numbers and fractions in arithmetic. Many algorithms have been developed in geometry for comparison of different geometrical figures, such as triangles and segments. For instance, we know how to find whether two triangles are congruent or when one segment is larger than another segment. However, mathematicians were not able to find algorithms for solving some important mathematical problems and suggested a hypothesis that it is impossible to find such algorithms. To make a mathematical statement out of a hypothesis, it is necessary to prove it. But to prove, mathematicians need mathematical structures because informal notions are not relevant for this purpose. Consequently, necessity in mathematical models of algorithms became very urgent.

2.2.2   A beautiful diversity of mathematical models

Aiming at solving different problems, mathematicians have suggested a diversity of exact mathematical models for a general notion of algorithm. The first models that were *recursive functions* introduced by Gödel (1934), ordinary *Turing machines* constructed by Turing (1936) and in a less explicit form by Post (1936), *recursive functions* (Church, 1936), *partial recursive functions* Kleene, 1936), and λ-*calculus* built by Church (1932/33). Creating λ-calculus, Church was developing a logical theory of functions and suggested a formalization of the notion of computability by means of λ-definability. Later Kleene (1936a) demonstrated how λ-definability is related to the concept of recursive function and Turing (1937) showed how λ-definability is related to the concept of Turing machine. It is interesting to know that the theory of Frege (1893) actually contains λ-calculus. So there were chances to develop a theory of algorithms and computability in the nineteenth century. However, at that time the mathematical community did not feel a need in such a theory and probably, would not accept it if somebody created it.

In his 1934 Princeton lectures, Gödel proposed a precise characterization, based on an idea of Herbrand, of the notion of *recursive function*. This construction is equivalent to the current notion of *general recursive function* (cf., for example, Davis, 1982).

As stated in (Barbin et al., 1999), the concept of recursive function is based on double recursion. A function defined by double recursion appeared in the works of Ackermann in 1920. Hilbert presented this function in 1925 in a lecture, so as to prove the continuum hypothesis (Guillame, 1978), and Ackermann studied it in 1928 (Heijenoort, 1967).

Gödel suspected that all effectively computable functions are general recursive, but was not convinced of this (nor of the first version of the Church's thesis of 1934 that stated that effective computability is equivalent to λ-computability) until he read Turing's 1936 paper. The most influential in this area paper of Alan Turing (1936) was aimed at investigating the famous Hilbert's *Entscheidungsproblem*. The *Entscheidungsproblem* (posed in 1922), or 'decision problem', is to find a constructive procedure by which, in a finite number of steps, it can be tested whether any given formal expression can be deduced from a given system of axioms. Turing examined the intuitive idea of computation and argued in a convincing manner that, if irrelevant aspects are omitted, we are led to what is now called Turing machine and Turing computability. Then he applied Turing machine to prove that the *Entscheidungsproblem* had no solution. Less well known is the fact that Post, in his 1936 paper, carried out an analysis, analogous to Turing's but much less elaborate, which reduced the intuitive idea of computability to a precise mathematical form (later proved equivalent to Turing computability).

Then other mathematical models of algorithms were suggested. They include a variety of Turing machines: *multihead, multitape Turing machines, Turing machines with n-dimensional tapes, nondeterministic, probabilistic* (Leeuw, Moore, Shannon, and Shapiro, 1956), *alternating* (Chandra and Stockmeyer, 1976; Kozen, 1976), *reflexive* (Burgin, 1992a) *Turing machines, Turing machines with oracles* (Turing, 1939), *Las Vegas Turing machines* (Hopcroft et al., 2001), etc.).

Other popular mathematical models of algorithms are

◇ *Neural networks*, the simplest case of which appeared in the work of McCulloch and Pitts (1943) and which, like Turing machines, have several types – *fixed-weights, unsupervised, supervised, feedforward*, and *recurrent neural networks*;

◇ *Von Neumann automata* (von Neumann, 1949) and general *cellular automata*;

◇ *Kolmogorov algorithms* (Kolmogorov, 1953);

◇ *Finite automata* (in the simplest form usually attributed to McCulloch and Pitts (1943), while in the developed form as a formal construction introduced by Mealy (1953), Kleene (1956), and Moore (1956)), which, like Turing machines, have several forms – *automata without memory, autonomous automata, automata without output or accepting automata, deterministic, nondeterministic* (Rabin and Scott, 1959), *probabilistic automata*, etc.;

◇ *Minsky machines* (Minsky, 1967);

◇ *Storage-modification machines* or simply, *Shönhage machines* (Shönhage, 1980);

◇ *Random Access Machines* (RAM), which were introduced by Shepherdson and Sturgis (1963) and also have modifications – *Random Access Machines with the Stored Program* (RASP) introduced by Elgot and Robinson (1964); *Parallel Random Access Machines* (PRAM);

◇ *Petri nets* introduced by Petri (1962), which like Turing machines have several forms – *ordinary* and *ordinary with restrictions* (Hack, 1975), *regular, free, colored* (Zervos, 1977), *self-modifying* (Valk, 1978) *Petri nets*, etc.;

◇ *Vector machines* introduced by Pratt, Rabin, and Stockmeyer (1974);

◇ *Array machines* introduced by Leeuwen and Wiedermann (1985);

◇ *Multidimensional structured model of computation and computing systems* developed by Burgin and Karasik (1975; 1976; 1980; 1982);

◇ *Systolic arrays* (Kung and Leiserson, 1978);

◇ *Hardware modification machines* introduced by Dymond and Cook (1989);

◇ *Post productions* (Post, 1943);

◇ *Normal Markov algorithms* (Markov, 1954);

◇ *Formal grammars*, which, like Turing machines, have many forms – *regular, context-free, context-sensitive, phrase-structure*, etc. (Chomsky, 1956; Bakus, 1959; Naur, 1960).

We see that an important peculiarity of the exact concept of algorithm, or a model of algorithm, is that it exists in various forms. In addition, we come to a situation when we distinguish algorithms, their descriptions, and their models. The following example demonstrates differences between these objects. Turing machine as a mathematical structure is a model of algorithm. A specific Turing machine T, for example, a Turing machine that adds two numbers, is a description of an algorithm, in our case, of an algorithm of addition. Finally, an algorithm of addition is a corresponding to the machine T structure. Taking an algorithm of addition with a description in the form of a Turing machine T, we can give other descriptions of the same algorithm: by programs in Lisp, in C^+ or in Pascal, by a formal grammar, by a random access machine, and by a neural network.

Each model of algorithm defines a class of algorithms. For example, a model as a Turing machine corresponds to the class of all Turing machines. This class contains many subclasses: deterministic, nondeterministic, alternating, reflexive, and other Turing machines. In turn, classes of deterministic, nondeterministic Turing machines contain subclasses of Turing machines with one linear tape and one head, with one linear tape and two heads, with two linear tapes and two heads, with one two-dimensional tape and one head and so on. In its turn, each of these classes contain subclasses of Turing machines with the alphabet $\{1, 0\}$, with the alphabet $\{a, b\}$, with the alphabet $\{a_1, \dots, a_i\}$ and so on. Turing machines from these classes solve different problems: comparing and transforming words; adding, subtracting, and multiplying numbers; building or accepting languages, etc.

Realization of an algorithm is a computational process. Consequently, mathematical models of algorithms give formalize the concepts of computation and computable function. Some of these models (such as recursive functions or Post productions) give only rules for computing. Others (such as Turing machines or finite automata) also present a description of an abstract computing device or *abstract automaton*, which functions according to given rules and computes what is predetermined.

2.2.3   Algorithms and abstract automata

To form an algorithm, a system of rules must have a description how these rules are applied. This description consists of metarules for an algorithm, given in a form of some abstract machine. For this reason, metarules often have the form of an abstract automaton.

Definition 2.2.1. An *abstract automaton* is an abstract information processing system (IPS).

Abstract automata and algorithms work with symbols, words, and other symbolic configurations, transforming them into other configurations of the same type. For example, words are transformed into words. While words and strings are linear configurations, there are many useful abstract automata and algorithms work with such configurations as graphs, vectors, arrays etc. (cf., for example, Kolmogorov, 1953; Rabin, 1969; Pratt, Rabin, and Stockmeyer, 1974; Leeuwen and Wiedermann, 1985; Burgin and Karasik, 1975; 1982).

As an IPS, an abstract automaton consists of an abstract information processing device (*abstract hardware*), an algorithm/program of its functioning (*abstract software*), and description/specification of information which is processed (*abstract infware*).

In a general case, the hardware of an abstract automaton, as well as of any other IPS, consists of three main parts: abstract *input device*, abstract *information processor*, and abstract *output device*, which are presented in Figure 2.1. For example, Turing machines (cf. Section 2.3) can have special input and output tapes, or the same tape works as an input/output device and a part of the processor. Neural networks (cf. Section 2.4) also contain these parts: the input device that comprises all input neurons, the output device that consists of all output neurons, and the information processor that includes all hidden neurons. In some cases, input and output neurons are combined into one group of visible neurons.

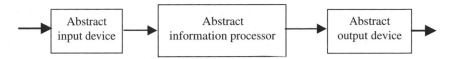

Figure 2.1. The structure of an abstract automaton.

Accordingly, we have a *structural classification* of automata (Fisher, 1965):

1. Abstract automata without input devices. They are called *generators*.
2. Abstract automata without output devices. They are called *acceptors*.
3. Abstract automata with both input and output devices. They are called *transducers*.

A finite state transducer, for example, is a finite state machine with a read-only input and a write-only output. The input and output cannot be reread or changed. Decision tables (Humby, 1973) represent the simplest case of transducers that have no memory. Basically any abstract automaton can be considered as a transducer.

At the same time, any transducer can work in the mode of an acceptor or generator. For example, very often Turing machines are considered as acceptors (cf. Hopcroft et al., 2001), although they produce some output, which is written in their tapes at the time when they stop.

Even such abstract automata as neural networks have the same triadic structure presented in Figure 2.1. The input device consists of input neurons; output neurons form the output device; and hidden neurons form the processor of such IPS (cf. Section 2.4). However, it is more relevant to consider that all neurons constitute the processor, while input and output devices are built into this processor.

The general idea of an abstract automaton implies three main modes of processing input data by an abstract or real IPS as follows:

1. The *computing* mode occurs when an automaton produces (computes or outputs) some words (its output or configuration) as a result of its activity.
2. The *deciding* mode occurs when an automaton, given a word/configuration u and a set X of words/configurations, indicates (decides) whether this word/configuration belongs to X or not.

3. The *accepting* mode occurs when an automaton, given a word/configuration u, either accepts this word/configuration or rejects it.

In addition, we have two partial deciding modes:

4. The *positive deciding* mode occurs when an automaton, given a word/configuration u and a set X of words/configurations, indicates (decides) whether this word/configuration belongs to X.
5. The *deciding* mode occurs when an automaton, given a word/configuration u and a set X of words/configurations, indicates (decides) whether this word/configuration does not belong to X.

Sometimes (cf. Section 4.2), the positive deciding mode is called semidecidable.

These types not only reflect the principal modes of computer functioning, but also define the main utilization modes for algorithms and programs. We have several kinds of each mode. For example, there is *acceptance by a state* and *acceptance by a result*.

Definition 2.2.2. An automaton A accepts a word u by a result if A gives some result, or gives a definite result (e.g., the symbol 1 or the word *yes*), when u is its input.

For example, a Turing machine T accepts a word u if and only if: T gives some output when u is its input, and after producing the output T halts. At the same time, it is possible to demand that a Turing machine T accepts a word u if and only if the produced result is equal to 1. It is possible to show that both ways of acceptance are equivalent.

To define acceptance by a state, we need to choose some states of the automaton as final or accepting.

Definition 2.2.3. An automaton A accepts a word u by a state if A comes to a final or accepting state when u is its input.

Remark 2.2.1. For finite automata (Trahtenbrot and Barzdin, 1970) and for pushdown automata (Hopcroft et al., 2001), it is proved that acceptance by a result is functionally equivalent to acceptance by a state.

Remark 2.2.2. For many classes of algorithms or abstract automata, acceptance of a word u means that the automaton/algorithm that works with the input u comes to an inner state that is an accepting state for this algorithm or automaton. Finite automata give an example of such a class. However, for such algorithms that produce output, the acceptance assumption means that whenever an algorithm comes to an inner accepting state it produces some chosen result (e.g., the number 1) as its output. In such a way, this algorithm informs that it has reached an inner accepting state.

Another way of defining an accepting state is to consider a state of some component of an abstract automaton. For example, pushdown automata accept words not only by an accepting inner state, but also by an empty stack, that is, by a definite state of their stack, which is an external state for these automata.

Definition 2.2.4. An automaton A accepts a word u by a component state if a chosen component of A comes to a final or accepting state when u is the input of A.

For a Turing machine, such accepting component is its control device (cf. Section 2.3), while for push down automaton such accepting component is either its stack or its control device (Hopcroft et al., 2001).

All these forms of acceptance are static. At the same time, there are such forms that depend on the behavior of the automaton. For instance, a finite automaton A accepts infinite strings when A comes to an accepting state infinitely many times. Such automata are called Büchi automata (Büchi, 1960). Another example is inductive Turing machine (cf. Chapter 4). It produces a result or accepts a word when its output stabilizes.

The first abstract automaton was Turing machine created by Turing (1936). A general concept of an abstract automaton was introduced by von Nemann (1951). Finite automata were formalized, modified and studied by Mealy (1953), Kleene (1956), and Moore (1956). Potentially infinite automata were formalized, modified and studied by Church (1957).

An important distinction exists between deterministic and nondeterministic algorithms. At first, such condition as complete determination of each step of an algorithm was considered as necessary for a general model of algorithm. For a long time, all models were strictly deterministic. However, necessity to reflect real situations and model computational processes influenced introduction of nondeterministic algorithms (Rabin and Scott, 1959).

Thus, we may have an impression that the extensive diversity of models results in a similar diversity for the concepts of algorithm, computation and computable function. Nevertheless, mathematicians and computer scientists found that the algorithmic reality is well organized. They have found a unification law for this reality, which was called the Church–Turing thesis. It is considered in Section 2.2.5 because to give a mathematically correct formulation of this thesis, we need some additional concepts.

2.2.4   Computational power of algorithms: comparison and evaluation

Having such a diversity of models for algorithms, we need to compare them. To do this, we introduce special concepts: *computing power*, *accepting power*, *decision power*, and *equivalence* of algorithms and their classes.

Definition 2.2.5. The *computing power* of an algorithm A is *less than or equal to* (is *weaker than or equivalent to*) the *computing power* of an algorithm B when the algorithm B can compute everything that A can compute. Naturally, the algorithm B is *stronger than or equivalent to* (has *more than or the same computing power*) the algorithm A.

Let us take as an example algorithms that solve some algorithmic problem P for a class of algorithms **A**. We assume that any result of such an algorithm is a solution to P. Such problem P may be: to define whether a given algorithm from **A** gives the result for some data; to define whether a given algorithm from **A** gives the result for all possible data; or to define whether a given algorithm from **A** has more computing power than another given algorithm from **A**.

Proposition 2.2.1. If an algorithm A solves the problem P strictly for the class A_1, an algorithm B solves the problem P strictly for the class A_2, and $A_1 \subset A_2$, then the computing power of B is larger than the computing power of A.

Definition 2.2.6. Two algorithms A and B are *functionally equivalent* (or simply, *equivalent*) if the algorithm B can compute everything that A can compute and the algorithm A can compute everything that B can compute.

Taking into account different modes of algorithm functioning, we come to a more general concept.

Definition 2.2.7. Two algorithms are called *functionally equivalent with respect to computability* (acceptability, positive decidability, negative decidability or decidability) if they define in the corresponding mode the same function f_A or relation r_A.

Example 2.2.1. In the theory of finite automata, functional equivalence means that two finite automata accept the same language (Hopcroft et al., 2001). This relation is used frequently to obtain different properties of finite automata. The same is true for the theory of pushdown automata.

Algorithms that work with finite words in some alphabet X are the most popular in the theory of algorithms. As a rule, only finite alphabets are utilized. For example, natural numbers in the decimal form are represented by words in the alphabet $X = \{0, 1, 2, 3, 4, 5, 6, 7, 8, 9\}$, while in binary form they are represented by words in the alphabet $X = \{0, 1\}$. The words in X may represent natural numbers or other entities, but we have a natural procedure to enumerate all such words. This makes it possible, when it is necessary, to assume that algorithms can work with natural numbers. In such a way, through enumeration of words, any algorithm A defines a partial function $f_A : N \to N$ (cf., (Burgin, 1985)). However, there are many reasons to consider algorithms that work with infinite words (Vardi and Wolper, 1994) or with such infinite objects as real numbers (Blum et al., 1998).

Remark 2.2.3. Many algorithms (cf., for example, Krinitsky, 1977 or Burgin, 1985) work with more general entities than words. As an example, it is possible to consider as input, working, and output data such configurations that were utilized by Kolmogorov (1953) in his analysis of the concept of algorithm and construction of the most general definition. Configurations are sets of symbols connected by relations and may be treated as multidimensional words or hypertexts. Hypertext technology is now very important in information processing both for people and computers (Barrett, 1988; Nielsen, 1990; Landow, 1992). Other configuration examples are discrete graphs.

While in practice we usually compare individual algorithms, for theory it is even more important that we compare power of different classes of algorithms.

Definition 2.2.8. A class of algorithms A has *less or equal computing power* than (is *weaker than or equivalent to*) a class of algorithms B when algorithms from B can compute everything that algorithms from A can compute. Naturally, the class of algorithms B is *stronger than or equivalent to* (has *more or equal computing power than*) the class of algorithms A.

Remark 2.2.4. Any mathematical or programming model of algorithms defines some class of algorithms. Thus, when we compare classes corresponding to models, we can compare power of these models.

For example, the class of all finite automata is weaker than the class of all Turing machines. It means that Turing machines can compute everything that can compute finite automata, and in addition, certain functions that finite automata cannot compute.

Remark 2.2.5. Definitions 2.2.5, 2.2.6, and 2.2.8 are not exact because they contain such a term as *everything*, which is very imprecise. To develop a mathematical theory of algorithms and automata, we need completely formal constructions. Formalization is achieved in two different ways. The first one is to relate to an algorithm the function that it computes. Such formalization is done in Definitions 2.2.7, 2.2.10, and 2.2.11. The second approach is based on introduction of sets or languages defined by an automaton/algorithm and consideration of three modes of automaton functioning: *computation, acceptance,* and *decision*. Such formalization is done in Definition 2.2.9.

As it is known, any set of words form a *formal language*. This allows us to consider as a formal language full output of algorithms, that is, the set L_A of all words that an algorithm A computes, accepts or decides.

Given an automaton/algorithm A, we say that A:

computes a set X_A (a formal language L_A) if X_A (correspondingly, L_A) consists of all outputs of A;
accepts a set X_A (a formal language L_A) if X_A (correspondingly, L_A) consists of all elements (words) accepted by A;
decides a set X_A (a formal language L_A) if A, given a word/configuration u, indicates (decides) whether this word/configuration belongs to X_A (to L_A) or not.

This set L_A is called the *computation* (*acceptance* or *decision*) *language of the algorithm/automaton A*.

Remark 2.2.6. Usually, when the mode of computation is fixed, L_A is called simply the language of the automaton/algorithm A. For example, in (Hopcroft et al., 2001) only the accepting mode is treated. This makes possible to speak about languages of finite automata, push down automata and Turing machines.

This allows us to compare computing, accepting, and deciding power of algorithms.

Definition 2.2.9. A class of algorithms A has *less or equal computing (accepting or decision) power* than (is *weaker than or equivalent to*) a class of algorithms B when algorithms from B can compute (accept or decide, correspondingly) any language that algorithms from A can compute (accept or decide). Naturally, the class of algorithms B is *stronger than or equivalent to* (has *more or equal computing (accepting or decision) power than*) the class of algorithms A.

Instead of using sets or languages, it is possible to use functions for comparison of computing power.

Definition 2.2.10. A class of algorithms A has *less or equal functional computing power* than (is *functionally weaker than or equivalent to*) a class of algorithms B when algorithms from B can compute any function that algorithms from A can compute.

Definition 2.2.11. Two classes of algorithms are *functionally equivalent* (or simply, *equivalent*) if they compute the same class of functions, or relations for nondeterministic algorithms.

Remark 2.2.7. Equivalence of algorithms is stronger than functional equivalence because more algorithms are glued together by equivalence than by functional equivalence. Really, when two algorithms compute the same function, then they compute the same language. However, it is possible to compute the same language by computing different functions. For example, let us take the alphabet $\{a, b\}$ and two automata A and B. The first one gives as its output the word that is its input. It computes the identity function $f(x) = x$. The automaton B gives the following outputs: $B(\varepsilon) = \varepsilon$, $B(a) = b$, and $B(b) = a$. As a result, A and B are equivalent, but not functionally equivalent.

2.2.5 The Church–Turing thesis

> *Our engraved knowledge may bite into our thinking*
> *certain errors that become well-nigh ineradicable.*
>
> Rogers MacVeagh and Thomas Costain, *Joshua*

In spite of all differences, there is a unity in the system of algorithms. While new models of algorithm appeared, it was proved that any of them could not compute more functions than the simplest Turing machine with one one-dimensional tape. Consequently, the class of each of the mathematical models of algorithm is either functionally weaker or equivalent to the class of all Turing machines (or equivalently, to the class of all partial recursive functions).

This situation influenced the emergence of the famous *Church–Turing thesis*, or better called the *Church–Turing conjecture*, in different forms. Here we consider a Turing's version that states that *the informal notion of algorithm is equivalent to the concept of a Turing machine* and a Church's version that asserts that *any computable function is a partial recursive function*. In other words, an algorithmic form of the Church–Turing thesis states:

Any problem that can be solved by an algorithm can be solved by some Turing machine and any algorithmic computation can be done by some Turing machine.

Saying that a problem can be solved by an algorithm, we assume potentially infinite, or unbounded, computability and skip, as it is traditionally done in the Church–Turing thesis, the problem of efficiency because conventional Turing machines, being very simple, are very inefficient. Consequently, we consider problems of computability in general and tractability or practical computability separately, making at first emphasis on computability and computing power.

All models of algorithms that are functionally equivalent to the class of all Turing machines are called *recursive algorithms*, while classes of such algorithms are called Turing complete.

Those kinds of algorithms that are functionally weaker than Turing machines are called *subrecursive algorithms*. Finite and stack automata with one stack, recursive and primitive recursive functions give examples of subrecursive algorithms.

It is possible to give other forms of the Church–Turing thesis.

An *automata form* of the thesis states:

Any computation performed by an abstract automaton can be performed by some Turing machine.

A *prescription form* of the thesis states:

Any computation performed according to some algorithmic/constructive rules can be performed by some Turing machine.

A (physical) *machine form* of the thesis states:

Any computation performed by a computing machine (by a computer) can be performed by some Turing machine.

For many years all attempts to find mathematical models of algorithms that were stronger than Turing machines have been fruitless. For example, Kolmogorov algorithms were developed aiming directly at this goal but appeared to be equivalent to the class of all Turing machines (Kolmogorov and Uspensky, 1958). The same equivalence has been proved for many other models of algorithms. That is why the majority of mathematicians and computer scientists believe that the Church–Turing thesis is true. More accurate researchers consider this conjecture as a law of the theory of algorithms and complexity theory, as well as in the methodological context of computer science. It is similar to the laws of nature that might be supported by more and more new evidence or refuted by a counter-example but cannot be proved. At the same time, many logicians assume that the thesis is an axiom that does not need any proof. Few believe that it is possible to prove this thesis utilizing some evident axioms.

The Church–Turing thesis is extensively utilized in the theory of algorithms, as well as in the methodological context of computer science. It has become almost an axiom. However, it has always been only a plausible conjecture like any law of physics or biology. It is impossible to prove such a conjecture completely.

As Nelson writes (1987), "Although Church–Turing thesis has been central to the theory of effective decidability for fifty years, the question of its epistemological status is still an open one." Nelson's view, is prompted by a naturalistic attitude toward such questions in mathematics, is that the thesis is an empirical statement of cognitive science, which is open to confirmation, amendment, or discard, and which, on the current evidence, appears to be true.

Thus, we can only add some supportive evidence to the thesis or refute it. At the same time, inside mathematics, we can prove or refute this conjecture as some statement about properties of mathematical models of algorithms if we choose an adequate context. This possibility is considered below.

The Church–Turing thesis is central in computer science and is implicitly one of the cornerstones of mathematics as it separates provable propositions from those that are not provable. So, it not a surprise that there is an extensive literature on the Church–Turing thesis. The domineering opinion is that the thesis is true. It is supported by numerous arguments and examples (cf., Turing, 1936; Kolmogorov and Uspensky, 1958; Rogers, 1987). Many developments in artificial intelligence and cognitive psychology provide strong empirical support for the Church–Turing thesis.

At the same time, there are researchers who suggest arguments against validity of the Church–Turing thesis. For example, Kalmar (1959) raised intuitionistic objections, while Lucas and Benacerraf discussed objections to mechanism based on theorems of Gödel that indirectly threaten the Church–Turing thesis.

Another aspect of discussion related to the Church–Turing thesis is its ontological status. The traditional and most popular opinion is that validity of the thesis can be settled only empirically. Nevertheless, some mathematicians support the opposite point of view. For example, Shoenfield (1967) suggests that Church's thesis might be proved from evidently true axioms about the notion of effective computability. Stahl (1981) considers the prospects for proving Church's thesis from a formal definition of effective computability, and argues that this possibility cannot be excluded automatically. Mendelson (1990) challenges the standard conception of the Church–Turing thesis as an unprovable thesis. To support his point of view, he asserts that Turing's justification for his definition of Turing machine (1936) is as clear a proof as he have seen in mathematics, and it is a proof in spite of the fact that it involves the intuitive notion of effective computability. At the same time, Thomas (1971) discusses the arguments which Kleene, Turing and Church have written in favor of Church's thesis, demonstrating that these arguments are not convincing. This discussion shows that the concept of mathematical proof is still vague and is not the same as the concept of a formal proof.

There are methodological arguments in support of the idea that the thesis cannot be proved by conventional logical technique from evidently true and decidable axioms about the notion of effective computability. Rigorous mathematical proofs are done in formal mathematical systems. As it is demonstrated (cf., for example, Smullian, 1962), such systems are equivalent to Turing machines as they are built by means of Post productions (Post, 1943). Thus, as Turing machines can model proofs in formal systems, it is possible to assume that proofs are performed by Turing machines.

As a result, we come to a circle in argumentation because we have to prove a statement about Turing machines by means of Turing machines. However, it is possible to overcome this limitation. If we use a sufficiently general context or more powerful methods of deduction such as provided by superrecursive algorithms, we can prove some versions of the Church–Turing thesis. For example, in (Burgin and Borodyanskii, 1991) a very general computational scheme is defined. Then it is proved that a natural axiom determines the class of all inductive Turing machines (cf. Chapter 4). Addition of the axiom of computability in a finite number of steps characterizes the class of all Turing machines. In a similar way, it is possible to find general computational schemes that definitely include algorithms. Then axioms can separate recursive algorithms in this class.

Different modifications of the Church–Turing thesis have been suggested. For example, Gandy (1980) presents a new version of the thesis:

Whatever can be calculated by a (discrete deterministic) machine is computable.

Here a machine is taken to consist of a (possibly infinite) set S of states and a transition function "$\rightarrow$" that satisfies four rather complicated principles. States are hereditarily (finite and nonempty) sets built up from an infinite collection of labels. Each of the four principles implies that some aspect of the machine is finite.

Yao (2003) considers the Extended, or Polynomial Time, Church–Turing thesis:

Any function computable by a computing machine (hardware device) in time $T(n)$ for input of size n can be computed by some Turing machine in time $(T(n))^k$ for some fixed k.

Tucker and Zukker (2002) introduce a Generalized Church–Turing thesis:

Computability of functions on many-sorted algebras can be formalized by the theory of partial recursive functions on many-sorted algebras.

Copeland, in the *Stanford Encyclopedia of Philosophy*, gives a detailed exposition of the history, meaning, and role of the Church–Turing thesis.

Thus, for a long time, Turing machines and other equivalent computational structures were treated as the absolute boundary for algorithmic computations. Discovery of new more powerful constructions extended possibilities of algorithms. As a result, treating computational power of algorithms, we separate three main classes: *recursive*, *subrecursive*, and *superrecursive automata* and *algorithms*. Each type of recursive automata or algorithms form a class in which it is possible to compute exactly the same functions that are computable by Turing machines. Examples of recursive algorithms are partial recursive functions, RAM, von Neumann automata, Kolmogorov machines, and Minsky machines. Each type of subrecursive automata or algorithms forms a class that has less computational power than all Turing machines. Examples of subrecursive algorithms are finite automata, primitive recursive functions and recursive functions. Each type of superrecursive automata or algorithms forms a class that has more computational power than all Turing machines. Examples of superrecursive algorithms are inductive and limit Turing machines, limit partial recursive functions and limit recursive functions.

In addition to everything that is written and said concerning the Church–Turing thesis, it is necessary to understand that all mathematical constructions embodying the informal notion of algorithm are only models of algorithm. Consequently, what is proved for these models has to be verified for real computers. In our case, we need: (1) to test whether recursive algorithms give an adequate representation for modern computers and networks, and (2) to find whether it is possible to build such computers that go beyond the recursive schema. We will see in Chapter 4 that the answer to the first question is negative, while the second problem has a positive solution.

This brings us a new vision of the Church–Turing thesis. While in general this thesis has been disproved by invention of different classes of superrecursive algorithms, under definite reasonable conditions that restrict computational processes, it would be possible to prove validity of the thesis.

2.3 Centralized computation: Turing machines

> "Begin at the beginning,"
> *the King said gravely,*
> "and go on until you come to the end; then stop."
>
> Lewis Carroll, 1832–1898

A Turing machine is an ideal device, which is in many aspects similar to a modern computer. At first, we consider the structure and then describe functioning of Turing machine. To make the reader's understanding easier, we consider, at first the simplest Turing machine with one tape.

2.3.1 A simple Turing machine

The structure of Turing machine, as an automaton, consists of three main components, which we can call *hardware*, *software*, and *infware*. We begin with the infware, that is, with description and specification of information that is processed by Turing machine. Infware of a computer consists of information, or more exactly, of data that are processed by the computer. A Turing machine T is an abstract automaton, which works with symbolic information. Consequently, formal languages with which T works constitute its infware. Usually, these languages are divided into three categories: input, output, and working language(s). In contrast to the languages that we use in our everyday life (such as English, German or French), Turing machines use formal languages.

A *formal language* **L** consists of three parts: the alphabet A of **L**, which is a finite set of symbols; the set A^* of all words in A, which are finite strings of symbols; and of the subset L of the set A^*. Elements from L are called the words of the language **L**. The set A^* is often represented by generating rules R. Because a formal language is an arbitrary subset of A^*, it is possible to consider the languages of the Turing machine T as one language $\mathbf{L}(T)$, which consists of three parts:

$\mathbf{L} = (\mathbf{L_I}, \mathbf{L_W}, \mathbf{L_O})$ where $\mathbf{L_I}$ is the *input* language, $\mathbf{L_W}$ is the *working* language, and $\mathbf{L_O}$ is the *output* language of **T**. Each of them has the following structure $\mathbf{L_X} = (A_X, R_X, L_X)$ where A_X is the *alphabet*, R_X is the set of *generating rules*, and L_X is the *set of all words* of $\mathbf{L_X}$. Usually the generating rules for formal languages consist of one operation, which is called concatenation and combines two words into one. For example, if x and y are words, then xy is the concatenation of x and y. Taking the alphabet $A = \{1, 0\}$ with two words $x = 1001$ and $y = 001$ in this alphabet, we have 1001001 as the result of concatenation. A^* is also a formal language and it includes the empty word ε that contains no symbols.

Now let us look at the *hardware* or *device of Turing machine*. What is the hardware of a computer? It consists of all devices (processor, system of memory, display, keyboard, etc.) that constitute the computer. In a similar way, a Turing machine T has three abstract devices: a finite automaton A, which we may the controller of T and which controls performance of the Turing machine T; a *processor* or *operating device* h, which is traditionally called the *head* of the Turing machine T; and the *memory* L, which is traditionally called the *tape* or tapes of the Turing machine T. These devices are presented in Figures 2.2–2.4, which give the structure of the simplest kind of a Turing machine.

The *tape* L is divided into different but uniform cells. Each cell can contain a symbol from an alphabet of the Turing machine T or be empty. Each cell from a linear two-sided tape has two neighbors left and right. In a one-sided tape, the beginning cell has only one neighbor.

The head h performs information processing in T. However, in comparison to computers, this operational device perform very simple operations. There are three types of such operations:

1. Reading a symbol from the cell, in which the head is situated.
2. Writing a symbol to the cell, in which the head is situated.
3. Going to the next cell.

In the theory of algorithms and computation, these operations are usually considered as elementary, that is, indivisible entities. However, more detailed analysis displays their inner structure. For instance, the operation *reading a symbol* consists of three stages: at first, the head begins to recognize or identify the symbol, proceeds to make a decision whether to accept or reject this symbol, and then performs the corresponding action (acceptance or rejection). The operation *writing a symbol* also consists of three stages: at first, the head begins to check whether the cell is empty or not, proceeds to choose the necessary symbol for writing in the first case and substitution in the second case, and then performs the corresponding action (writing or rewriting). The operation *going to the next cell* depends on the metarules that explain

**Turing Machine
With the Head as a Static Agent**

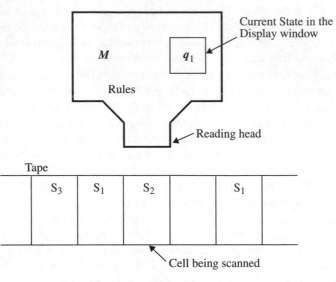

Figure 2.2.

**Turing Machine
With the Head as a Dynamic Agent**

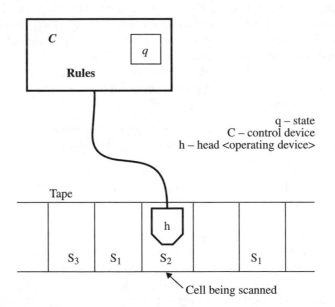

Figure 2.3.

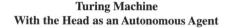

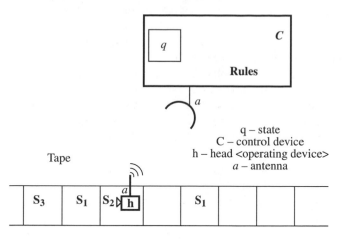

Figure 2.4.

where the head can go. When T has a linear tape, then the head has two options: go to the right or to the left cell. When T has a two-dimensional tape, the head traditionally has four options: to go to the right, up, down or to the left. With n-dimensional tape the number of option becomes 2^n, where n is any natural number.

The control automaton A has the *state configuration or space* $S = (q_0, Q, F)$ where Q is *the set of states or the state space* of both A and T, q_0 is an element from Q that is called the *start or initial state*, and F is a subset of Q that is called the set of *final* (in some cases, *accepting*) states of T. The automaton A regulates the state of T, performing specific operations. Such operations are grouped into three types:

1. Input operations, the first of which is the operation of starting the Turing machine T.
2. Output operations that include stopping the Turing machine T.
3. Computational operations that change the full state of the Turing machine T. This full state includes the state of A, the symbol that is written in the cell where the head is situated, and the position of the head.

For each operation, there are specific rules. Some of them are written explicitly in the program of the Turing machine T, while others are given implicitly as metarules. An example of the first kind is given by the rules of changing the state of T. An example of the second kind is given by the rules of starting or stopping the Turing machine T.

Connections between the automaton A and the head h may be organized in a different way. The head h may be rigidly connected to A. In this case, the tape L moves when it is necessary to observe the next cell. This structure is presented in Figure 2.2. The connection between A and h may be flexible, allowing h to move from one cell to another. This structure is presented in Figure 2.3. One more option is that h exists autonomously from A, only sending to A the information about the content of cells and receiving from A the instructions of what to do next. This structure is presented in Figure 2.4.

In such a way, the control device A regulates behavior of the operating device h and the tape. When h is a static agent (Figure 2.2), A sends a direct command to h if it is necessary to change a symbol in the observed cell and to the tape if it is necessary to move to the adjacent cell. When h is a dynamic agent (Figure 2.3), A also sends a direct command to h if it is necessary to change a symbol in the observed cell and/or to move to the adjacent cell. When h is an autonomous agent (Figure 2.4), A sends information on the current state to the head h and in turn h decides what to do using the rules of T.

As we know, different programs constitute software on a computer. Programs tell our computer what to do and what not to do. *Software of a Turing machine* is a single program in a form of simple rules. Using traditional notation, we present these rules in the following form:

$$q_h a_i \to a_j q_k,$$
$$q_h a_i \to R q_k,$$
$$q_h a_i \to L q_k.$$

Each rule directs one step of computation of the corresponding Turing machine. The first rule means that if the state of the Turing machine is q_h and the head h observes in the cell the symbol a_i, then the state of the Turing machine becomes q_k and the head h writes the symbol a_j in the cell where it is situated. The second rule means that if the state of the Turing machine is q_h and the head h observes in the cell the symbol a_i, then the state of the Turing machine becomes q_h and the head h moves to the next cell to the right from the cell where it is situated. The third rule means that if the state of the Turing machine is q_h and the head h observes in the cell the symbol a_i, then the state of the Turing machine becomes q_h and the head h moves to the next cell to the left from the cell where it is situated.

Other representations of rules allow Turing machine to rewrite a symbol on the tape and to make a move in one step.

Besides, it is possible to define functioning of Turing machine not by rules, but by a transition function δ, which is defined by the corresponding rules (Hopcroft et al., 2001).

$$\delta(q_h, a_i) = (q_k, a_j, D).$$

For a linear tape, the direction D is either R, which means a move to the right, or L, which means a move to the left, or S, which means no move.

Thus, rules of a Turing machine T define the transition function δ of T and describe changes of A, h, and L. In particular, they determine the transition function δ_A of the automaton A.

An important concept for Turing machine is its instantaneous description (ID). ID reflects a given Turing machine T between two consecutive steps of computation. ID includes: the state of the control device A, the position of the head, and the word(s) written on the tape(s) of the Turing machine T.

2.3.2 Types and basic properties of Turing machines

Turing machines come in different types and kinds.

Definition 2.3.1. A Turing machine is called *deterministic* if it has no two or more rules with the same left part. Otherwise, a Turing machine is called *nondeterministic*.

In other words, a deterministic Turing machine has at most one rule for continuation at each step of computation, while a nondeterministic Turing machine allows in some cases many successions.

Let us consider functioning of deterministic and nondeterministic Turing machines in more detail.

A deterministic Turing machine T with one tape begins its functioning from the start position, where the control device A of T is in the state q_0, the head h is in a cell of the tape, and a word, which may be empty, is written onto the tape. Usually it is assumed that the head h observes the first symbol of the word written in the tape, but this does not restrict the computing and accepting power of Turing machines. At the beginning, the control device A finds the rule the left part of which corresponds to the start position, that is, the left part of the rule must be equal to $q_0 a_i$ if the head h observes the symbol a_i. Then the machine T performs actions according to this rule, going to a new position. Then A finds the rule the left part of which corresponds to the new position, that is, the left part of the rule must be equal to $q_k a_j$ if the head h observes the symbol a_i and A is in the state q_k. Then T performs actions according to this rule, coming into a new position and so on. Because each time there is no more than one rule with the necessary left part, each action of the machine T is determined uniquely. If T comes to a position which is not final and for which there is no corresponding rule for continuation, T halts without giving a result in the computing mode and does not accept the initial word in the accepting mode.

If T comes to a position in which the state of A is final, T halts. In the computing mode, T gives the word that is written in the tape as its result. In the accepting mode, T accepts the word that was written in the tape at the beginning of its functioning. If T never halts, then it gives no result in the computing mode and does not accept the initial word in the accepting mode.

For a nondeterministic Turing machine T, it is possible that at some positions, there is more than one rule with the necessary left part. In this case, it is assumed that T performs all possible actions. Thus, rules generate several processes in T. If one of these processes brings T to a final state, then in the accepting mode, T accepts the word that was written in the tape at the beginning of its functioning. In all other cases, T does not accept the initial word. From a different perspective, a nondeterministic accepting Turing machine T 'guesses' what choice to make at any step. So, the input is accepted if T can choose the right path from the start.

In the computing mode of T, there are four options for defining its result. One option is to take as the result all final words in the tapes for all processes that terminate in a final state. Another option is to order all processes in T and to take as the result the final word in the tapes for the least process that terminates in a final state. One more possibility is to take as the result all final words in the tapes for all shortest processes that terminate in a final state.

The third option is to define a result of computation for a nondeterministic Turing machine T by considering this machine as a set-valued algorithm in the sense of (Burgin, 1984). That is, nondeterminism of T results in a possibility for T sometimes to choose what will be the next step of computation and go along several (occasionally infinitely many) paths of computation. In this case, the result of the computation is the set of all words that may be written in the tapes of T if T does not have output tapes or in the output tapes of if T has output tapes when T stops after performing computation along one of the possible paths.

The fourth option is to take as the result all final words in the tapes for all shortest processes that terminate in a final state. In this case, T also defines a set-valued algorithm.

However, it is necessary to remark that accepting modes are, as a rule, studied for nondeterministic Turing machines. This is justified by the two following results.

Theorem 2.3.1. *A set X is acceptable by a deterministic Turing machine if and only if X is computable by some deterministic Turing machine.*

Theorem 2.3.2. *For every nondeterministic Turing machine, there is a deterministic Turing machine that accepts exactly the same set of words.*

It is possible to find a proof of this result in (Hopcroft et al., 2001).

Theorem 2.3.2 shows that classes of all deterministic and all nondeterministic Turing machines have the same accepting power.

Turing machine can have multiple tapes and heads. The structure of a Turing machine with three tapes and three heads is given in the picture 4.1. However, the quantity of tapes does not add to the computing and accepting power of Turing machines.

Theorem 2.3.3. *For every Turing machine with m tapes and n heads, there is a Turing machine with one tape and one head that computes (accepts) exactly the same function (the same set of words).*

For accepting Turing machines, it is possible to find a proof of this result in (Hopcroft et al., 2001). For computing Turing machines, the proof is similar.

Theorem 2.3.3 shows that the class of all Turing machines with multiple tapes and heads has the same computing and accepting power as the class of all Turing machines with one tape and one head.

Even more, it is possible to prove a similar result for Turing machines with an infinite number of tapes that are enumerated by natural numbers.

Theorem 2.3.4. *For every Turing machine with the countable quantity of tapes and/or heads, there is a Turing machine with one tape and one head that computes (accepts) exactly the same function (the same set).*

The proof is similar to the proof of Theorem 2.3.2 because there are only finite number of finite parts of tapes at any given moment of computation.

All tapes that are considered above are one-dimensional. However, Turing machine can have multidimensional tapes. This can cause changes in several possible moves of the head. For example, the head can move in four directions (left, right, up, and down) in a two-dimensional tape (cf. Figure 2.5).

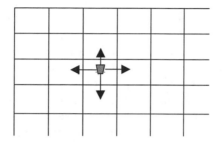

Figure 2.5. Two-dimensional tape with a head and directions for possible moves.

Theorem 2.3.5. *For every Turing machine with a multidimensional tape, there is a Turing machine with one linear tape that computes/accepts exactly the same function.*

It is possible to find a proof of this result in (Hartmanis and Stearns, 1965).

Theorem 2.3.5 shows that the class of all Turing machines with multidimensional tapes has the same computing and accepting power as the class of all Turing machines with one linear tape and one head. At the same time, it is proved that many tapes or multidimensional tapes increase efficiency of Turing machines, for example by reducing the time for different computations.

2.3.3   Universal Turing machines and operations with Turing machines

As written in (Blum et al., 1997), "A pillar of the classical theory of computation is the existence of a universal Turing machine, a machine that can compute any (recursively, M.B.) computable function. This theoretical construct foretold and provides a foundation for the modern general-purpose computer."

As our aim is to go beyond recursive computations, we begin with a definition of an automaton or algorithm that is universal in some class $\mathbf{K}$ of automata/algorithms that work with inputs from a set X. As before, we assume that X is the set of all words in some alphabet. It is useful to build and study universal automata and algorithms in different classes. The construction process begins with some description $\mathbf{c} \colon \mathbf{K} \to X$ of all automata or algorithms in $\mathbf{K}$.

Definition 2.3.2. An automaton or algorithm U is *universal* for the class **K** if given a description $\mathbf{c}(A)$ of an automaton or algorithm A from **K** and some input data x for it, U simulates A, working with the same input x, and gives the same result as A.

Definition 2.3.3. If U is universal for the class **K** and belongs to **K**, then U is called a *universal automaton or algorithm* in **K**.

For instance, a universal Turing machine is any Turing machine that is universal for the class of all Turing machines.

A description $\mathbf{c}(T)$ of a Turing machine T given to a universal Turing machine U plays the role of a program for a computer. Thus, each stored program in a computer that executes this program is, in this sense, a separate automaton.

To build a universal Turing machine, we need, at first, to codify all Turing machines. Theorems 2.3.1–2.3.3 show that it is sufficient to consider only Turing machines with one linear tape. That is, we construct a function $\mathbf{c}\colon \mathbf{T} \to \mathbf{\Sigma}^+$ where **T** the class of all Turing machines with an alphabet $\mathbf{\Sigma}$ and $\mathbf{\Sigma}^+$ is the set of all nonempty words in this alphabet. Because any alphabet can be coded in words of the two element alphabet $\{1, 0\}$, we take from the beginning that $\mathbf{\Sigma} = \{1, 0\}$.

At first, we enumerate all binary strings so that each string corresponds to one integer, and each integer corresponds to one string. The most natural way to do this, at first, to order all words by length, and then to lexicographically ordered words of equal length. A lexicographical order means that two words u and v are compared symbol by symbol from the left to the right. The first time they have different symbols determines their order. If these symbols are 1 in u and 0 in v, then u is larger than v. After all words are ordered, a natural number is corresponded to each word, preserving this order. In such a way, (becomes the first word, 0 becomes the second, 1 becomes the third, 00 becomes the fourth, 01 becomes the fifth, and so on.

Now we can codify Turing machines, building a function $\mathbf{c}\colon \mathbf{T} \to \mathbf{\Sigma}^+$. To represent a Turing machine T as a binary string, we at first assign integers to the states, tape symbols, and directions L, R, and S. If T has the states $q_1, q_2, \ldots, q_k$ for some number k, then the start state is always q_1 and we correspond to each q_i the string 0^i. To the tape symbols 0, 1, and the blank symbol B, which denotes an empty cell, we correspond 0, 00, and 000, respectively. Then we denote the direction L as D_1, direction R as D_2, and direction S, which means no move at all, as D_3, and correspond D_1, D_2, and D_3 words 0, 00, and 000, respectively.

Once we have established an integer to represent each state, symbol, and direction, we can encode the transition rules or, equivalently, the transition function δ. When we have a transition rule $q_i, a_j \to q_k, a_l, D_m$, for some natural numbers i, j, k, l, and m, we code this rule using the string $0^i 10^j 10^k 10^l 10^m$. As each number i, j, k, l, and m is greater than or equal to one, there are no occurrences of two or more consecutive 1's within the code for one transition rule. Having codes for all transition rules, we write them as one word in which they are separated by couples of 1's. For example, we have the code $0^i 10^j 10^k 10^l 10^m 110^h 10^t 10^r 10^q 10^p$ for two transition rules $q_i, a_j \to q_k, a_l, D_m$ and $q_h, a_t \to q_r, a_q, D_p$. As a Turing machine is completely described by its states, initial state, tape symbols, and transition rules, in such a way, we obtain a complete coding $\mathbf{c}(T)$ of T by a binary word. This coding is used for many purposes. For example, the coding **c** of a Turing machine allows us to give the instantaneous description (ID) of T. An instantaneous description of a Turing machine T represents the complete state of T after some step of computation. This description includes the word(s) written in the tape of T, the state of the control device of T, and the position of the head H of T. In its turn, the coding and instantaneous descriptions are used for building a universal Turing machine.

Since for each Turing machine, it is possible to assign natural numbers to its states and tape symbols in many different orders, there exist more than one encoding of the typical Turing machine. Nevertheless, any such encoding may be used by a universal Turing machine for simulating one and the same Turing machine. The only difference may be in efficiency of simulation, as some encodings give better efficiency in comparison with others.

Taking a function $\mathbf{c}\colon \mathbf{T} \to \Sigma^{+}$, we obtain the following result.

Theorem 2.3.6. *For any class* **T** *of all Turing machines with the same working alphabet* Σ *and alphabet* Q *of states, there exists a universal Turing machine.*

It is possible to find a proof of this result, for example, in (Minsky, 1967). In addition, this theorem is a corollary of Theorem [4.3.11.],which is proved in Section 4.3 for universal inductive Turing machines.

Building computers, networks, and software systems, we often combine different devices and programs. For instance, to develop software systems subprograms are frequently used and now the component-based software development becomes more and more popular. Such combination of devices and programs is represented in the theory by operations with automata and algorithms. Studying properties of such operations, we learn better how to build more advanced computers, networks, and their software.

In this section, we consider operations with Turing machines.

Any operations with Turing machines that are performed by other Turing machines do not allow one to go beyond the space **TM** of all Turing machines. This is the consequence of the famous Church–Turing thesis. Let us consider exact definitions and results.

Definition 2.3.4. An n-place algorithmic operation on a space **AL** of algorithms is an algorithm H that gives rules for combining any system of algorithms $A_1, \ldots, A_n$ from **AL** into an algorithm $H(A_1, \ldots, A_n)$ from **AL**.

As an external operation on the space of all Turing machines, the algorithm H works in the following manner. Given some input x_0, the algorithm H selects some Turing machine A_i after which this input is sent and the machine A_i begins to process x_0. The algorithm H controls this processing, utilizing its rules/criteria. When the state and/or the memory content of A_i satisfy these criteria, H selects another Turing machine A_j, takes a definite part x_1 of the memory content of A_i, and sends this x_1 to A_j as its input. Then A_j begins to process x_1 under the control of H and the process continues until one of the A_i or the algorithm H comes to a final state.

As an internal operation, the algorithm H can change the state of any of the Turing machines A_i, as well as direct the head of this machine to a definite cell.

Let us consider algorithmic operations in the space of recursive algorithms, that is, algorithms that are functionally equivalent to Turing machines.

Theorem 2.3.7. *If an algorithmic operation H with Turing machines T_i is performed by a Turing machine, then the algorithm $H(T_1, \ldots, T_n)$ is a equivalent to a Turing machine.*

To prove this statement, we use induction on the number of rules in the algorithm H, demonstrating that a specially built Turing machine can model the whole functioning process of the algorithm $H(T_1, \ldots, T_n)$ functioning. This machine works in a similar way to the universal Turing machine.

Thus, we see that application of recursive algorithms to recursive algorithms does not extend the class of algorithms although it is a nonlinear operation on the space of all recursive algorithms. It shows that in computability spaces of all recursive algorithms the nonlinear phenomena do not appear.

Remark 2.3.1. There are algorithmic operations on Turing machines transforming them in superrecursive algorithms. For instance, a simple algorithmic operation can make an inductive Turing machine out of a conventional Turing machine.

2.4 Distributed computation: Neural networks and cellular automata

> *"Ah well! It means much the same thing,"*
> *said the Duchess, "and the moral of that is –*
> *'Take care of the sense,*
> *And the sounds will take care of themselves.' "*
>
> Lewis Carroll, 1832–1898

Turing machines, due to their variety, can model several computing architectures and realize many computational paradigms. However they do not encompass the whole diversity of computing strategies. The reason is that their creation was based on analysis of the behavior of a person who performs some computations (Turing, 1936), that is, Turing machine is a *behavioristic model* of intellectual aspects of people.

Another way to achieve the same goal of creating a machine that assists people in their intellectual activity is to model the structure of the human mechanism for such activity. This mechanism is the brain. Copying its structure, we get another model of abstract automata – abstract neural networks. They were introduced by McCulloch and Pitts (1943) to model the most important features of the brain. Thus, three kinds of neural networks are studied: *natural*, such as the brain, and *artificial*, which are subdivided into *abstract* and *physical artificial neural networks*. As the brain is treated as the mechanism for intelligence, the artificial neural network is a *structural model* of intellectual aspects of people.

Consequently, an *artificial neural network* is an interconnected assembly of simple processing elements, which are called *units*, *nodes* or *neurons* and whose functionality is based on a similarity to the biological neuron. That is why we begin this section with a description of the human brain and its comparison to the main theoretical computational model, Turing machine. Then we describe the structure, properties and functions of artificial neuron and neural networks. Artificial neural networks represent computer architecture complimentary to Turing machines. They embody distributed computation. And like Turing machines, artificial neural networks have many types.

There are also other models of distributed computing architecture: cellular automata, which has become very popular; systolic arrays; iterated arrays; Petri nets; and grid automata. The first three models are considered in this section, while grid automata are studied in Chapter 4.

2.4.1 The brain and the Turing machine

We are not going to discuss here the whole structure of the brain. It is possible to find it in more or less detail in many books. Our goal is to demonstrate what features of the brain gave birth to such model of algorithms and computation as abstract neural networks.

It is known that the brain contains many billions of very special kinds of cells, which are called *nerve cells* or *biological neurons*. An estimation is that there are $10^{11\pm1}$ neurons in the human brain. This number is approximately equal to the number of stars in our galaxy. Brain neurons are wired up in the 3-dimensional space into a very complicated intercommunicating network. Real brains are also orders of magnitude more complex than any artificial neural network so far considered.

Typically each neuron is physically connected to tens of thousands of others, from which it receives and/or to which it sends information by specific signals. Signals are of two kinds – electrical and chemical. Electrical signals transmit information inside neurons, while communication between neurons is realized by chemical signals. These connections are not merely *on* or *off* – the connections have varying *strength* which allows the influence of a given neuron on one of its neighbors to be either very strong, very weak (perhaps even no influence) or anything in between. Furthermore, many aspects of brain function, particularly the learning process, are closely associated with the adjustment of these connection strengths. Brain activity is then represented by particular patterns of firing activity amongst this network of neurons. It is supposed that this simultaneous cooperative behavior of very many simple processing units is at the root of the enormous sophistication and computational power of the brain.

It is interesting to know that some researchers reject the role of neurons in the brain that is ascribed to them by standard theories. For example, Pribram explains that according to the latest experimental data information processing in the brain is performed not by neurons but by dendrites; that is, dendrites perform those operations that are traditionally related to neurons (cf. Pribram, 2002).

According to Pribram's holographic theory of the brain, both time and spectral information are simultaneously stored in the brain. However, similar to the Heisenberg's uncertainty principle, there is a limit with which both spectral and time values can be concurrently determined in any measurement (Pribram, 1991). This uncertainty describes a fundamental minimum defined in 1946 by Gabor (the inventor of the hologram) as a quantum of information. Dendritic microprocessing is conceived (by Pribram) to take advantage of the existing uncertainty relation to achieve optimal information processing. The brain, as a whole, operates as a "dissipative structure" that continually self-organizes to minimize its uncertainty.

Although neurons in the brain have different forms, they have the same structure. A biological neuron consists of three parts:

◇   The *cell body*, the central part, contains the nucleus.
◇   The *axon*, which is a fiber (often very long), carries impulses from the cell body to other neurons.
◇   The *dendrites*, shorter fibers, receive impulses and carry them toward the cell body.

An important role in information processing in the brain belongs to the gap between the dendrite of one neuron and the axon of the next. Such a gap is called a *synapse*. A neural signal crosses the synapse by means of neurotransmitters. Namely, the electrical propagation along the transmitting neuron releases neurotransmitters, which diffuse across the gap and trigger receptors on the next, accepting neuron, inducing a new electrical signal.

A nerve is a bundle of elongated axons belonging to hundreds or thousands of neurons.

Signal transmission between neurons begins with electrical pulses (*action-potentials* or "spike" trains), traveling along the axon. These pulses come to synaptic terminals of the axon, where they initiate the release of a small amount of chemical substance or *neurotransmitter*. This substance travels across the synaptic cleft and is then received at post-synaptic *receptor* sites of the dendrites, situated on the other side of the synapse. This initiates a change in the dendritic membrane potential. Such *post-synaptic-potential* (PSP) change may serve to increase (*hyperpolarize*) or decrease (*depolarize*) the polarization of the post-synaptic membrane. In the former case, the PSP tends to inhibit generation of pulses in the receiving neuron, while in the latter, it tends to excite the generation of pulses. The size and type of PSP produced depends on various factors such as the geometry of the synapse and the type of neurotransmitter. Over 50 different synapses have been identified. Each PSP will travel along its dendrite and spread over the cell body. The receiving neuron sums or integrates the effects of thousands of input PSPs over its dendritic tree and over time. The integrated potential eventually reaches the base of the axon (*axon-hillock*) and if it exceeds a threshold, the cell 'fires' and generates an action potential or spike, which starts to travel along its axon. This then initiates the whole sequence of events again in neurons contained in its pathway. As complicated as the biological neuron is, it can be simulated in computer science with a simplified model of artificial neuron.

In spite of having a huge diversity of forms, there are three types of biological neurons (Atkinson et al., 1990). The *sensory* neurons transmit impulses from the sense organs (receptors), such as eyes, ears, and skin. The *motor* neurons send signals from the central nervous system to the effectors, which are the muscles and glands of the body, enabling us to make movements. The *connector* neurons or *interneurons* receive signals from the sensory neurons and send impulses to other interneurons or to motor neurons.

It is necessary to remark that besides neurons the brain contains other constituents, including glia cells and blood vessels that bring physical resources for functioning of the brain. It is assumed that the brain is the mechanism for human intelligence. At the same time, artificial means of simulating intelligence are based on algorithms and automata developed by people. The basic model for algorithms is Turing machine. As we can see the brain and Turing machine have very different structures. The brain is a very complicated network of relatively simple processing units with a high degree of interconnection. In contrast to this, Turing machine usually consists of three parts: the control device, the operating device (the head), and a uniform memory. Functioning of these devices is organized by (eventually a very sophisticated) program in the form of simple instructions.

Turing machines model human behavior, ignoring imitation of the structure of the human control organ, the brain. As the majority of computers are organized in the same fashion as Turing machines, we can compare the situation with aviation. At first, people, in their attempts to fly, tried to imitate birds. When this did not give positive results, people invented other flying schemes: planes, balloons, and rockets. All are built in a dissimilar way in comparison with birds, although some structural similarities exist; for example, planes and birds have wings and tail. Parallel to this, the brain, Turing machines, and traditional computers have similar components: memory and input/output devices.

However, people still have not abandoned their efforts to create mechanisms that fly like the birds do. In the field of artificial intelligence, people also try to create artificial means that imitate the brain in its structure and functioning. Such means are called artificial neuron networks and neurons.

2.4.2   A model of a neuron

An *artificial neuron*, a building block of an artificial neural network, is rather loosely based on the brain's nerve cell, or natural neuron. Simple forms of an artificial neuron are called the Threshold Logic Unit (TLU). Artificial neurons may be realized as a program on a computer, as some electronic device or to be a formal construction, which is called abstract neuron. The simplest form of abstract neurons, Boolean neurons, was originally proposed by McCulloch and Pitts (1943). It was the first mathematical model of brain neurons. Boolean neurons have only two options for input and output: 1 and 0.

Later much more complicated models of neurons have been considered. They can work with arbitrary natural, rational, and even real numbers (Siegelman, 1999), realizing a variety of arithmetical and logical functions.

The contemporary structure of an artificial neuron is presented in Figure 2.6.

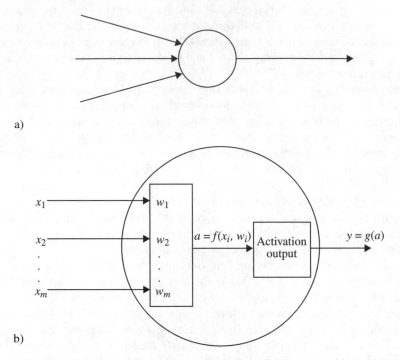

a)

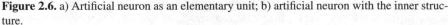

b)

Figure 2.6. a) Artificial neuron as an elementary unit; b) artificial neuron with the inner structure.

Neurons receive inputs via weighted links from other neurons. In Figure 2.6, all left arrows show *input connections*, or *inlets*, of the neuron, the right arrow represent the *output connection*, or *outlet* of the neuron. The neuron inputs are processed according to the neuron *activation function*. Signals are then passed on to other neurons if the neuron receives a sufficiently strong input signal from the other neurons to which it is connected.

In Figure 2.6b, $\{x_i; \ i \ = \ 1, 2, \ldots, m\}$ is a set of m inputs, and $\{w_j; \ j \ = \ 1, 2, \ldots, n\}$ is the set of n internal parameters, *weights* of the neuron. As a rule, $n = m$. The *activation a* is a function of the inputs and parameters, $a = f(x_i, w_i)$. Usually, $f(x_i, w_i)$ has the form of a linear combination $f(x_i, w_i) = x_1 w_1 + \ldots + x_m w_m = \sum_{i=1}^{m} x_i w_i$. The output y is a function of the activation, $y = g(a)$. It is usually defined by the activation-output relation by means of a *threshold parameter* θ. In this case, we have:

$$g(x) = \begin{cases} 1 & \text{when } a \geq \theta, \\ 0 & \text{when } a < \theta. \end{cases}$$

The *threshold function* $g(x)$ is sometimes called a *step-function* or *hard-limiter*.

We can take $y = h(x_i, w_i)$ for some function h, eliminating a from the description. However, splitting the neuron functionality can assist the understanding of what 'ingredients' have been used in the artificial neuron make-up, how the neuron functions, and how to build necessary neurons. In addition, the function $f(x_i, w_i)$ models the integrating process in the body of a biological neuron, while the function $g(x)$ represents functioning of synaptic terminals of the axon of this neuron. Artificial or more exactly, abstract neuron is a finite automaton (cf. Section 3.3) without inner states, that is, it has the input-output or action-reaction type.

However, being realized as a physical device, an artificial neuron functions in the physical space-time manifold. Taking time into consideration, we come to the conclusion that it is more reasonable to define four inner states of a neuron. In one state, the neuron gives no output and does not accept information. In the second state, it gives no output but accepts information. In the third state, it does not accept information, but gives an output, or as say experts, it *fires*. In the fourth state, it accepts information and gives an output. Usually it is assumed that a neuron always accepts information. This results in a possibility to consider only two states of a neuron: *firing* and *silent*.

Usually the following *inner criteria* for classification of artificial neurons are considered:

◇ According to the type of the processed information, there are:
   - *Boolean neurons*, processing binary data;
   - *rational number analogue neurons*, used with rational numbers;
   - *real number analogue neurons*, working with real numbers.
◇ According to the form of the function $f(x_i, w_i)$, there are:
   - *linear neurons*,
   - *quadratic neurons*,
   - *cubic neurons*,
   - *sigma-pi neurons*, and so on.
◇ According to the form of the activation-output relation, there are:
   - *activation linear neurons*,
   - *hard-limiter neurons*,
   - *sigmoidal neurons*.
◇ According to the form of the dynamics of the node, there are:
   - *deterministic neurons*,
   - *stochastic neurons*,
   - *nondeterministic neurons*.

In addition to this, the role of a neuron in the network to which it belongs implies the *network criterion*. According to this criterion, there are three different types:

◇ *Input* neurons receive encoded information from the external environment.
◇ *Output* neurons send signals out to the external environment in the form of an encoded answer to the problem presented in the input.

◇ *Hidden* neurons, which may not be accessible by the environment, allow inter-mediate calculation between input and output neurons.

There is a natural correspondence between artificial and biological neurons. Input neurons correspond to sensory biological neurons. Output neurons correspond to motor biological neurons. Hidden neurons correspond to connector biological neurons.

In some nets, the same neurons can send and receive signals from the external environment. Then all neurons in such a net are divided into *visible* and *hidden* neurons. Visible neurons are connected to the external environment, while hidden neurons may not be accessible by the environment.

2.4.3 Artificial neural networks

An *artificial neural network* is an interconnected assembly of artificial neurons. The processing ability of the network is stored in the connections between different neurons, as well as in the activation function and inter-unit connection strengths, or *weights*. Neural network is a form of multiprocessor computer system. In many cases, it has the following properties:

◇ relatively simple processing units;
◇ a high degree of interconnection;
◇ a relatively simple form of information exchange;
◇ adaptive interaction between elements.

Neural networks represent the connectionist approach to computation when the process is based mostly on transmission of data. The connectionist approach is considered as complementary, or sometimes, as contrasting to the symbolic computation based on transformation of symbols.

Neuron does not have memory. However, neural networks allow one to organize some kind of memory, being capable of storing and retrieving data from this memory. There are two formats for saving data. Permanent information is usually designed into the weights of the neurons. These weights play the role of the long-term memory of people or secondary memory of computers: magnetic disks and tapes. Experience of a network written into the weights of neurons is obtained by a process of *adaptation* to, or *learning* from, a set of training patterns. This is a static storage by the structure and/or properties of a system.

Dynamic storage is utilized for temporary data. In this case, a part of a network is used only for preserving information. If we take a neuron, which can be in two states: firing and silent, then it is possible to interpret a silent neuron as containing the symbol '0', while a firing neuron is considered as containing the symbol '1'. When a neuron can fire several kinds of output, for example any rational number, then we can store in it more than two symbols. To preserve the firing state, a neuron can be initiated to go into a loop, until it is stopped. Since the output of the neuron feeds back to itself, there is a self-sustaining loop that keeps the neuron firing even when the top input is no longer active. Activating the lower input suppresses the looped input, and the node stops firing. The stored binary bit is continuously accessible by looking at the output. This configuration is called a latch. The corresponding schemes are presented in Figure 2.7. A unit of memory in Figure 2.7a consists of one neuron N with two weights $w_1 = w_2 = 1$ and the output function $g(a) = 1$ when $a = xw_1 + vw_2 = 1$ and $g(a) = 0$ when $a = xw_1 + vw_2$ is equal to 0 or to 2. The input x is used to change the state of N, while the output x is used to read the stored symbol. A unit of memory in Figure 2.7b consists of two neurons: N_1 and N_2. The neuron N_1 is almost the same as the neuron N in Figure 2.7a, that is, it has the same weights and the same output function. The difference is that the second neuron N_2 is used for reading the stored data: when the input $z = 1$, the neuron N_2 fires its output y. Having another neuron only for reading information, makes possible to retrieve information without exerting any influence on the storing neuron N_2. This is especially important when computing occurs on the molecular or quantum level.

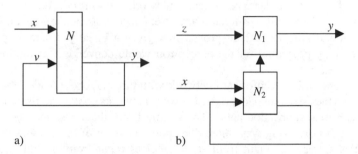

Figure 2.7. a) Artificial neuron N as an element of a short-term memory; b) A two-neuron unit of a short-term memory.

The part of a network that used only for dynamical storage plays the role of the short-term memory of people or primary memory of computers: random access memory (RAM) and read-only memory (ROM).

While dynamic memory works perfectly in this model, a biological neuron would not behave quite this way. The difference is in temporal characteristics. After firing, a biological neuron has to rest for a thousandth of a second before it can fire again. To get around this limitation, it is possible to link several neurons together in a duty-cycle chain that substitutes separate neurons in the Figure 2.4.2 and achieves the same result in storing data. Existence of such memory is supported by the experimental evidence that some patterns in the brain are preserved in a dynamical fashion (Suppes and Han, 2000; Suppes, Han, Epelboim and Lu, 1999; 1999a).

2.4.4   Types and basic properties of neural networks

Neural networks can solve many problems. At the beginning of formal treatment of artificial neural networks, McCulloch and Pitts (1943) considered simple networks with Boolean neurons that transformed binary symbols and had fixed thresholds. For example, a Boolean neuron that realizes the function $x \vee y$ has two inlets and two weights $w_1 = w_2 = 1$. Its output function $g(x)$ gives the following results: $g(a) = 0$ when $a = xw_1 + yw_2 = 0$ and $g(a) = 1$ when $a = xw_1 + vw_2$ is equal to 1 or to 2. McCulloch and Pitts (1943) obtained the following result.

Theorem 2.4.1. *Neural networks with Boolean neurons can realize any Boolean function.*

Later learning potential of neural networks was considered by Hebb (1949). He described a tentative mechanism for learning in real brains. Namely, to learn, neuron weights in a form of synaptic strengths change so as to reinforce any simultaneous reciprocity of activity levels between the pre-synaptic and post-synaptic neurons. In artificial neural networks, this means augmenting the weights of a node so that they can reflect the correlation between the input and output. Networks that learn from their experience in a training environment are based on this "Hebb rule."

One of the most popular type of neural networks was introduced by Rosenblatt (1958) and called *perceptron*. It can learn to connect or associate a given input to a random output unit. Rosenblatt proved that given a linearly separable problem, a simple training procedure for the perceptron would converge if a solution to the problem existed.

The perceptron is a feedforward network with three layers, in which the middle layer is called the association layer. Feedforward networks contain no feedback loops in their inter-node connection paths. The simplest forbidden case may be formally described in the following way: there does not exist a set of three nodes A, B and C, such that C receives input from B, B receives input from A, and A receives input from C. As a result, a general feedforward net is separated into a series of distinct layers (cf. Figure 2.8), which consist of three types of neurons: *input nodes* or *neurons*, which have no inlets; *output nodes* or *neurons* that does not have outlets; and *hidden neurons* or *nodes* that have both inlets and outlets.

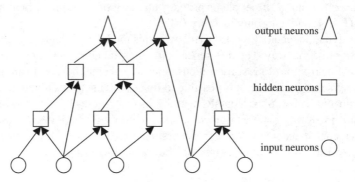

Figure 2.8. A feedforward network.

Layered networks are naturally represented by such graphs as trees or forests (Berge, 1973) and by chains of named sets (Burgin, 1997, Ch. 9).

The dynamics of feedforward networks is naturally straightforward. That is, at the beginning, the input neurons get their inputs from the environment. Then input neurons produce their output that goes to the next level. In the same way, subsequent layers evaluate the outputs of the previous layers and give their results to the next layer until the output layer is reached. In a more general case, it is necessary to ensure that all the inputs to a given node are valid before its evaluation. This whole process is referred to as a *forward pass*.

In 60s neural networks became very popular. Some researchers claimed that very soon neural networks would be able to perform the majority of intelligent actions and imitate intelligent behavior of people. However, enthusiasm for neural networks was dampened to some extent by the results of Minsky and Papert (1969). They demonstrated that perceptrons cannot solve many interesting problems (those that are not linearly separable), and they held out little hope for the training of multilayer systems that might deal successfully with some of these. However, in the same book the authors discussed a new more powerful scheme for neural network functioning. Thus, implicitly it was a beginning of the next stage in the neural network theory development.

Later, it was shown that it was possible to find exact solutions for problems that are not linearly separable by means of nonrecurrent (feedforward) networks. Namely, Werbos in 1974 elaborated algorithm for the credit assignment problem. This algorithm realized the method called "back error propagation" or simply *backpropagation*. But in spite of all interest to these problems, several years passed before this approach was popularized. The result was that back-propagation networks were rediscovered by Parker in 1982. Then discovered again and made popular by Rumelhart, Hinton and Williams (1986). In essence, a back-propagation neural network is an advanced perceptron with multiple layers, a different threshold function in the artificial neuron, and a more robust and capable learning rule. Today back-propagation networks are, probably, the best known and widely applied class of the neural networks.

It is necessary to mention other paradigms and directions for neural networks. Amari (A. Shun-Ichi) established a mathematical theory for a learning basis (error-correction method) dealing with adaptive pattern classification (1967). Klopf (1972) developed a basis for learning in artificial neurons based on a biological principle for neuronal learning called *heterostasis*. Fukushima (F. Kunihiko) developed (1975) a stepwise trained multilayered neural network, called the *Cognitron*, for interpretation of handwritten characters. Kohonen (1982; 1984) investigated nets that used topological feature maps. Grossberg developed Adaptive Resonance Theory (ART) (1987; 1988) and founded a school of thought that explores resonating algorithms.

More sophisticated and thus, more powerful than feedforward networks are *recurrent networks*. They have feedback loops in their inter-node connection paths. As a result, the distinction between input and output nodes becomes superficial and a new classification of neurons is introduced. There is a set of *visible nodes*, which interact with the environment, and, as in the feedforward case, a number of *hidden nodes*, which do not (cf. Figure 2.9).

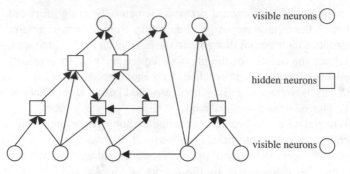

Figure 2.9. A recurrent network.

Recurrent networks have two fundamental modes of operation – *synchronous* or *parallel* update, and *asynchronous* update. In the former, all the nodes evaluate their new output together, and the net is globally synchronized. In the latter case, at each step a node is selected at random to be updated. Both synchronous and asynchronous modes can be thought of as extreme instances of the more general case where there is a probability p that any node gets updated. In the case of synchronous dynamics and with deterministic nodes, the network is a *deterministic finite automaton* with a characteristic state space, where the state of the network is just the vector of node outputs.

Comparison of abstract neural networks with other models of automata and computation gives us the following results.

Theorem 2.4.2. (Minsky, 1967). *Neural networks with Boolean neurons can simulate any finite automaton.*

Recently, much more powerful abstract neural networks have been constructed.

Theorem 2.4.3. (Siegelman and Sontag, 1991). *Neural networks with rational number neurons are equivalent to, that is, have the same computing power as, Turing machines.*

However, realized as physical devices, neural networks, like conventional computers, can work only with a finite subset of rational numbers. Moreover, now as a rule, the majority of artificial neural networks are simulated on conventional computers. However, the result of Theorem 2.4.3 is true for conventional neural networks.

Theorem 2.4.4. (Hyotyniemi, 1996). *Neural networks can simulate arbitrary Turing machines.*

To prove this result, the author uses recurrent neural networks of the second order. They are called perceptron networks because their elements are not neurons but perceptrons, which, as we know, consist of neurons.

The main applications of neural networks are pattern recognition (recognition of speakers in communications; three-dimensional object recognition; handwritten word recognition; and facial recognition), diagnosis and analysis (diagnosis of hepatitis; texture analysis; customer research; industrial process control; recovery of telecommunications from faulty software; data validation; undersea mine detection; interpretation of Chinese words with multiple meaning); and prediction or forecasting (sales forecasting, risk management, target marketing).

All these tasks are performed better when neural networks learn how to do it. Learning is an essential potential of neural networks. They learn by adaptively updating the weights of their nodes. The weights are updated according to the information extracted from new training patterns. Usually, the optimal weights are obtained by optimizing (minimizing or maximizing) certain "energy" functions. For example, a popular criterion in supervised learning is to minimize the least-squares-error between the teacher value and the actual output value.

According to their learning modes, neural networks are commonly categorized in terms of their corresponding training algorithms: *fixed-weights networks*, *unsupervised networks*, and *supervised networks*. There is no learning mode for the fixed-weight networks. Supervised learning networks have been the mainstream of neural model development. Their training data consist of pairs of input/output training patterns, which supervise learning and teach the network. For unsupervised learning networks, the training set consists of input training patterns only. The network learns to adapt based on the experiences collected through the previous training patterns.

There are three ways of how a neural network receives information from the environment. One is to clamp a subset of the visible units and keep the clamp on for the entire test, while the net reaches equilibrium. The clamp can be freely reselected at each trial from the entire collection of visible units. Alternatively, it is possible to initialize the state of the net from an input vector, and then let the net operate completely freely, with *all* nodes partaking in the update process. The third way is to give the net new input data from time to time. This approach actually includes two others as the first methods can be simulated by repetition of the same input data, while the second mode only reduces the sequence of inputs to the only one input.

Neural information processing is usually separated into two phases: the *learning* or *training phase* and the *retrieving* or *working phase*. In the training phase, a training data set is used to determine the weight parameters that define the resulting neural network. This trained neural network is used later in the working phase to process real test patterns and yield the network output, for instance, classification results.

A more comprehensive scheme includes the third phase, which is called the *evaluating phase*. In the evaluating phase, the network gets a feedback from the environment (in particular, from a user). This feedback informs the network on the quality of the output. The low quality may result in the repetition of the first phase.

There is many misconceptions related to neural networks. One of them is the belief that Turing machines and other similar models (for example, RAM or PRAM) work with symbols, while neural networks process signals. Turing computations as symbolic are contrasted with neural net computations as subsymbolic.

In reality, even the simplest abstract neural networks (cf., for example, (McCulloch and Pitts, 1943)) simulate logical functions. It means that they work with symbols 1 and 0. Modern abstract neural networks work even with more advanced symbols that represent rational and real numbers (cf., for example, (Siegelman, 1999)).

At the same time, Turing machines can compute everything that they are able to compute, using only two symbols: 1 and 0 (Theorem 2.3.4). That is why now the majority of neural networks are simulated on conventional computers. When a Turing machine, realized as a physical device or a conventional computer works with these symbols, they are usually represented by the corresponding physical states: existence of a signal means 1 and absence of a signal means 0. Such interpretation makes computation of Turing machine or of a conventional computer equivalent to a process that only changes signals. It means that, in some sense, Turing machines and computers also works only with signals.

Besides, there is a claim that neural networks do not have set of rules or programs, while their complex behavior emerges from interconnections. First, the weights w_i together with the activation function $a = f(x_i, w_i)$ and the output function $g(a)$ of each neuron constitute a program for the neuron functioning. Second, as Machlin and Stout (1991) demonstrate, even a simple Turing machine can realize very complex and unpredictable, chaotic behavior. Learning is not a privilege of neural networks. Reflexive Turing machines can learn from their experience in problem solving and modify their programs (Burgin, 1992).

2.4.5   Related models of autonomous distributed computation

There are other models of autonomous distributed computation, which are closely related to neural networks: cellular automata, systolic arrays, Petri nets, and grid automata.

The idea and construction of cellular automata and systolic arrays stems from the works of von Neumann in automata theory (1951; 1966). Von Neumann in the late 1940s and early 1950s attempted to design such a system that could construct any automaton from a proper set of encoded instructions, so that it would make a copy of itself as a special case. It was assumed that such machine would operate in a very simplified environment, giving a chance to see just what was involved in reproduction. For the building blocks of this automaton, von Neumann decided on identical chips placed in a rigid two-dimensional array and connected to their four nearest neighbors. The chips in this array, which was later called a systolic array, changed states synchronously in discrete time-steps. The state of each chip for the next time-step was determined from its own current state and those of its four neighbors, using a set of transition rules specified by the automaton rules. In practice, this automaton was simulated by a single large computer that modeled the work of the many small chips.

It is necessary to remark that when von Neumann introduced his automata, they were considered as constructing devices in contrast to Turing machines as computing devices. However, as this construction always takes place in a symbolic domain, von Neumann automata are devices that actually perform symbolic computations.

Later cellular automata became very popular. Some of them are even named after von Neumann. Thus, an infinite string of finite automata, which is a one-dimensional cellular automaton, is called a Neumann automaton (Trahtenbrot, 1974). Von Neumann's self-reproducing automata have been greatly simplified. For example, Codd (1968) reduced the number of states needed for each chip from 29 to 8.

Definition 2.4.1. A *cellular automaton CA* is a system of identical finite automata, or *cells*, which form a net and interact with one another.

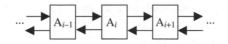

Figure 2.10. A one-dimensional cellular automaton.

A cellular automaton *CA* is described by the following net characteristics:

1. The *space organization* of the cells. Most often the system of the *CA* is organized as a simple rectangular grid (mostly it is a one-dimensional string of cells and a two- or three-dimensional grid of cells), but other arrangements, such as a honcycomb or Fibonacci trees, are sometimes used.
2. The *topology* of the *CA*, which is determined by the type of the cell neighborhood. Such neighborhood of a cell is the set of other cells that this cell interacts with. In a grid, these are normally the cells physically closest to the cell in question.

   For example, if each cell has only two neighbors (right and left), it defines linear topology. Such cellular automata are called linear or one-dimensional. It is possible to consider linear automata with the neighborhood of some radius $r > 1$ (cf. Crutchfield and Mitchell, 1995). When there are four cells (upper, below, right, and left), the *CA* has two-dimensional rectangular topology. Such cellular automata are called linear or one-dimensional.
3. The *dynamics* of a cellular automaton, which determines by what rules cells exchange information with each other.

A separate cell in *CA* is a finite automaton, which is described in detail in Section 3.3. It is usually assumed that all automata in *CA* are the same.

Traditionally, only rectangular organization of the cells and their neighborhoods has been considered for cellular automata. Recently, researchers have begun studies of cellular automata in the hyperbolic plane or on a Fibonacci tree (Margenstern, 2000; Margenstern and Morita, 2001; Margenstern, 2003). It is proved that such automata are more efficient than traditional cellular automata in the Euclidean plane. This higher efficiency is a result of a better topology in cellular automata in the hyperbolic plane.

Now cellular automata are so popular that they are used as models almost in every imaginable field. Some researchers suggest that the whole world is a kind of a cellular automaton (cf., for example, (Talbot, 2001) or (Wolfram, 2002)).

Nobili and Pesavento (1994) introduce and study generalized cellular automata, which can change not only the state of its neighbors, but also the program (the transition function) of these neighbors. The authors give a description of a planar cellular automaton, which is called a universal constructor and which, if properly programmed, generates any sort of a planar cellular automaton.

One more model of distributed computation is Petri net, or place-transition net. They were introduced by Petri in 1962 as a tool for studies of concurrency, synchronization, forking, blocking, and interaction of components in different systems. They provide a convenient framework for correctly and faithfully describing systems with many interacting parts. At the same time, Petri nets serve in many cases as a convenient simulation model.

Definition 2.4.2. A Petri net A is a system $(P, T, \textbf{in}, \textbf{out})$ that consists of two sets: the set of *places* $P = \{p_1, p_2, p_3, \ldots, p_n\}$ and the set of *transitions* $T = \{t_1, t_2, t_3, \ldots, \ldots t_m\}$, for which $P \cup T \neq \emptyset$, $P \cap T = \emptyset$; as well as of two functions: the *input function* **in**: $(P \times T) \to N_0$ that defines directed arcs from places to transitions; and the *output function* **out**: $(P \times T) \to N_0$ that defines directed arcs from transitions to places, where N_0 is the set of all whole numbers (nonnegative integers).

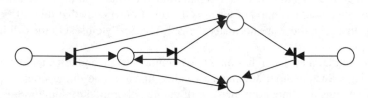

Figure 2.11. The graph of a Petri net, where $\bigcirc$ is a place and $|$ is a transition.

Graphically, places are represented by circles and transitions are indicated by horizontal or vertical bars. If $\textbf{in}(p_i, t_j) = k$, where $k > 1$ is an integer, a directed arc from place p_i to transition t_j is drawn with the label k. This number can be also interpreted as the number of arcs connecting the pair (p_i, t_j). If $k = 1$, we include an unlabeled arc, and if $k = 0$, then no arc is drawn. Petri nets are bipartite directed graphs, whose nodes are divided into two disjoint sets called *places* and *transitions*. Directed arcs in the graph connect places to transitions (called input arcs, which go from input places) and transitions to places (called output arcs, which go to output places). Places usually represent states or resources in the system, while transitions model the activities of the system. In such a way, Petri nets provide an efficient and mathematically rigorous modeling framework for discrete event concurrent dynamically systems.

Formally functioning of a net begins with an initial marking of the net. Any marking is a state of a Petri net and is represented by the number of tokens in places. Traditionally tokens are represented by black dots. Movement of tokens between places changes marking and describes the evolution of the system and is accomplished by the firing of the enabled transitions. Each change occurs through transition firings. A firing of a transition is an atomic action in which one or more tokens are removed from the input place of the transition and one or more tokens are added to each of the output place of the transition

As in other models of algorithms, time is absent in the initial definition of Petri net. Real systems, including IPS, that are modeled by Petri nets are functioning in time. So, the concept of time needs to be incorporated in the definition. A convenient way to do this is to correspond time intervals to states (markings) of the net, assuming that the net or, at least, a separate place preserves its state for some time. Events when time is changing are transaction firings. In such a way, timed Petri nets are usually built. However, it is possible to relate time intervals to firings and time shifts to states. In both cases, we have a system time of a Petri net in the sense of (Burgin, 2002a).

Another related model is systolic array (Kung and Leiserson, 1978).

Definition 2.4.3. A *systolic array* is an arrangement of processors in an array (often rectangular) where data flows synchronously across the array between neighbors, usually with different data flowing in different directions.

Systolic array is called so by analogy with the regular pumping of blood by the heart. For example, it can function in the following manner. Each processor at each step takes in data from one or more neighbors (e.g., from the left and up processors), processes it and, in the next step, outputs results in the opposite direction (to the right and down processors). Systolic arrays realize systolic algorithms. An example of a systolic algorithm might be matrix multiplication. One matrix is fed in a row at a time from the top of the array and is passed down the array, the other matrix is fed in a column at a time from the left hand side of the array and passes from left to right. Dummy values are then passed in until each processor has seen an entire row and an entire column. At this point, the result of the multiplication is stored in the array and can now be given as the output a row or a column at a time, flowing down or across the array.

One more model of distributed computation is iterative array (Hennie, 1961).

Definition 2.4.4. An *n-dimensional iterative array* is a system that consists of identical finite automata indexed by n-tuples of the nonnegative integers, which interact with their neighbors.

Two automata are considered neighbors if their indices are identical in all but one component and differ by one in that position. Inputs and outputs for the array are given to and taken from the machine indexed by $(0, 0, \ldots, 0)$. Except for the machine at the origin, the state of any machine in the array at time $t + 1$ depends only upon the states of it and its neighbors at time t, and its state at time 0 is a particular quiescent state.

All constructions of neural networks, Petri nets, systolic arrays, iterative arrays, and cellular automata are synthesized in the concept of grid automaton (Section 4.4). This new model for computation and communication allows us to combine in one structure centralized and distributed computer architecture.

2.5 Applications

> *She generally gave herself very good advice*
> *(though she very seldom followed it).* . . .
>
> Lewis Carroll, 1832–1898

There are so many applications of the theory of recursive algorithms that it will take many volumes to describe all of them. For example, Turing machines are used to evaluate what computers can do and what they can't do, to develop complexity theory for computations, to study problems of computational efficiency and so on. Some researchers even use Turing machines and their generalizations to build a theory of everything (cf., for example, (Schmidhuber, 2000)).

Neural networks are used for pattern recognition (recognition of speakers in communications; three-dimensional object recognition; handwritten word recognition; and facial recognition), diagnosis and analysis (diagnosis of hepatitis; texture analysis; customer research; industrial process control; recovery of telecommunications from faulty software; data validation; undersea mine detection; interpretation of Chinese words that have multiple meaning), prediction or forecasting (sales forecasting, risk management, target marketing).

Examples of systolic array applications are: solving linear systems of equations, sorting and searching, and matrix multiplication. Systolic arrays have high efficiency for these operations. For example, a systolic array processor requires only about n time units to multiply $n \times n$ matrices. On such a machine, the time complexity for matrix multiplication is thus linear rather than cubic as for sequential machines.

Cellular automata are used as models in a variety of imaginable fields. Some researchers suggest that the whole world is a kind of a cellular automaton. This approach is inspired by the so-called game of Life, which is usually attributed to John Conway. In its turn, the game of Life led Langton (1989) to creation of a new discipline known as Artificial Life (cf. Talbot, 2001).

The game of Life is realized by a two-dimensional cellular automaton in which each cell can have only two states: bright or dark, on or off, "alive" or "dead," and has eight neighbors. There are three rules for functioning of this automaton:

1. If exactly two of a cell's neighbors are alive at the clock tick ending one interval, the cell will remain in its current state (alive or dead) during the next interval of time;
2. If exactly three of a cell's neighbors are alive, the cell will be alive during the next interval of time regardless of its current state; and
3. In all other cases, that is, if less than two or more than three of the neighbors are alive, the cell will be dead during the next interval of time.

The game starts with some initial configuration of bright cells and then, with each tick of the clock, the configuration changes as the rules are applied.

As suggested by Peter Cochrane, head of research at British Telecom, "It may turn out that it is sufficient to regard all of life as no more than patterns of order that can replicate and reproduce" (cf. Talbot, 2001). Langton is even more decisive, "Life isn't just *like* a computation, in the sense of being a property of the organization rather than the molecules. Life literally is a computation" (cf. Waldrop, 1992).

Here we consider in more detail some applications of Turing machines that help us to compare recursive and superrecursive algorithms and are related to the study of computer programs. This is very important as there are many problems with computer programs. Alan Cooper, widely considered the "father" of Visual Basic and now the head of Cooper Interaction Design, says the main problem with much of the software available currently is that it is poorly thought out because its programmers did not take into account what users really want; programs may be more and more powerful, but they frustrate users to no end (Vizard, 2001). The theory of algorithms can help us to evaluate needs of users and to design better software.

At first, let us consider the problem of program correctness. We all want to use correct programs. However, as states the proverb, *to err is human*. Thus, the vast majority of written programs contain errors. So, we need to get rid of these errors. Practitioners and researchers suggested three main approaches to this problem:

1) Find all errors by testing the program on a computer and correct them.
2) Find all errors in the process of proving the program's correctness and correct them.
3) Build a software system that finds all errors in the process of proving the program's correctness and corrects them.

The first approach is good only for small, simple programs. For complex programs, the number of computational processes that these programs generate is so big that it possible to test only a small part of them.

Thus, we come to the second and third approaches and ask the question whether it is possible to find a procedure or to write a program that allows us to debug all computer programs. The theory of algorithms gives a negative answer to this question under the assumption of the Church–Turing thesis. This thesis, in particular, equates programs with Turing machines and allows us to reduce all questions concerning computer programs to questions related to Turing machines.

From this perspective, one kind of program mistake is that the program never halts because according to the Church–Turing thesis to give a result at all, not to speak of the correct result, the program has to halt. So, we may ask if we can write a program that, at least, determines whether a given program halts after it begins calculations with a given input.

Moreover, it is possible to consider only such programs that do not give a result only if they do not halt. Really, let us consider a program P that stops in some state q but does not give a result. Such q is called a dead state for P. Then we can transform this program so that the new program Q gives the same results as P but coming to the state q, it never stops. To get such a program Q from P, we need to add an infinite cycle that begins from the state q. After we do the same transformation for all dead states of P, we get the program R that does not give a result only if it does not halt.

The reduction of computer programs to the class of all Turing machines converts this question to the question whether there is a Turing machine D that determines whether a given Turing machine T halts after T begins calculations with a given input. This is the famous *halting problem*.

Let us consider the class $\mathbf{T}$ of all Turing machines that work with words in an alphabet A. The set of all such words is denoted by A^*.

Theorem 2.5.1. *The halting problem is undecidable in the class* $\mathbf{T}$, *that is, there is no such Turing machine D that determines whether a given Turing machine T halts after T begins calculations with a given input.*

Proof. At the beginning, we code all Turing machines by words in A. There are standard methods for building such coding function $c\colon \mathbf{T} \to A^*$. Such coding is described in Sections 2.3 and 4.3. The value of this function $c(T)$ is the "code" or "description" of a Turing machine T.

Let us assume that such a Turing machine D exists. Applied to a pair $(c(T), w)$, it gives 1 as the result when T gives a result for the input w, and D gives 0 as the result when T does not give a result for the input w. Then we take two simple Turing machines C and A_C. Here C is a checking Turing machine such that it checks whether a given word w is equal to $c(T)$ for some Turing machine T and then if this is true it converts w to the pair $(c(T), w)$. The Turing machine A_C gives the result 1 for the input 0 and gives no results for all other possible inputs. It is easy to build such Turing machines by standard methods (cf., for example, (Rogers, 1987) or (Minsky, 1967)).

Taking these two Turing machines and an arbitrary Turing machine T, we build a new Turing machine M as the sequential composition $M = C \circ D \circ A_C$. Sequential composition means that the output of each Turing machine in the composition goes as input to each next Turing machine. It is easy to build such composition of Turing machines by standard methods (cf., for example, (Rogers, 1987) or (Minsky, 1967)). The structure of M is presented in the Figure 2.12.

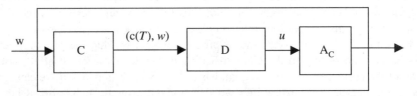

Figure 2.12. The structure of $M = C \circ D \circ A_C$. Here u is some output of D.

Now let us find what happens when the Turing machine M receives the word $w = c(M)$ as its input. This word goes first to the Turing machine C, which produces the pair $(c(M), w)$. This pair goes to the Turing machine D as its input. Now we have two options for M: M halts for the input w or does not. In the first case, the output of D is 1, which goes to A_C as an input. According to its rules, A_C does not halt, which means that M does not halt for w as its input. This contradicts our assumption that M halts. So, M does not halt for w and the output of D is 0, which goes to A_C as input. According to its rules, A_C halts and produces 1 as its output. Consequently, this means that M also halts for w as its input. This contradicts our assumption that M does not halt and shows that whatever case we assume for M, we come to a contradiction. This contradiction shows that the Turing machine D cannot exist, and thus, Theorem 2.5.1 is proved, stating that the halting problem is undecidable.   □

In spite of the undecidability of the halting problem for Turing machines and programs, we can ask other questions about programs. For instance, we can if a given program produce any result or if all its result belong to some given set X. The theory of algorithms shows that all such general questions about programs are undecidable. This follows directly from the, so-called, Rice theorem (Rice, 1951). Here for completeness, we prove informally one version of this theorem.

We assume, as before, that programs and Turing machines are equivalent and call a property of programs or Turing machines *functional* if it separates all Turing machines into two classes according to the following rule. If two Turing machines T and Q are equivalent, meaning that they produce the same output language, then T and Q belong to the same class. This means that functional properties of Turing machines correspond to properties of languages accepted or computed by these machines.

Remark 2.5.1. There are properties of programs or Turing machines that are not functional.

Example 2.5.1. Let D be a description of a Turing machine and B be a text in a programming language. A Turing machine T has the property P_D if its description coincides with D. A program p has the property P_B if its description coincides with B.

Example 2.5.2. Let a be a symbol. A Turing machine T (program p) has the property P_a if its description (the program p) begins with a.

Example 2.5.3. A Turing machine T (program p) has the property P_a if its description (the program p) contains a.

Definition 2.5.1. A property of Turing machines is called *nontrivial* if some Turing machines have this property and other Turing machines do not have it.

Definition 2.5.2. A property P of Turing machines is called *functional* if for any two Turing machines A and B, $A_f = B_f$ implies $P(A) = P(B)$.

Theorem 2.5.2. *Any nontrivial functional property P of Turing machines is undecidable in the class* **T**, *that is, there is no such Turing machine D that determines whether a given Turing machine T has property P or not.*

Proof. For the property P in question we have two options: Turing machines that give no output have this property or do not have it. As P is a functional property, all Turing machines that give no output either have this property or they do not have it. At first, we consider the case when all Turing machines that give no output have the property P.

As the property P is nontrivial, there is a Turing machine M that does not have the property P.

Let us assume that there is such Turing machine D that determines whether a given Turing machine T has the property P or not.

To prove Theorem 2.5.2, we reduce the initial problem to the halting problem. The goal of problem reduction is to use knowledge about one problem to infer something about a different problem. If you have a problem P that you know is undecidable and you want to show that a problem Q is undecidable, then it is sufficient to demonstrate that being able to solve the problem Q allows you to solve the problem P.

For such reduction, we build a system { $A_{T,w}$; where T is some Turing machine from **T** and w is an arbitrary word in A } of Turing machines $A_{T,w}$. The machine $A_{T,w}$ consists of copies of the machines T and M and a finite automaton G with two inputs. One input come from the outside, while the second is the output of T. The automaton G can be in two states: closed and open. Initially G is closed until it received some input from T, which makes it open. When G is closed, it gives no output. When G is open, it gives as its output the word that comes to G from M as its output. The structure of the machine $A_{T,w}$ is presented in the Figure 2.13.

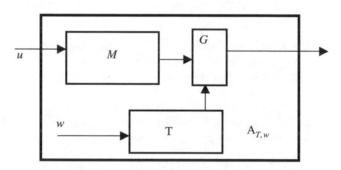

Figure 2.13. The structure of the Turing machine $A_{T,w}$.

Functioning of the machine $A_{T,w}$ is determined by the following rules. At the beginning the word w is given as the input for T. Then an arbitrary word u is given as the input of the whole $A_{T,w}$. Consequently, $A_{T,w}$ gives no result when the Turing machine T does not halt after T begins calculations with the input w. Otherwise, the output of T coincides with the output of M. Thus, if we know whether $A_{T,w}$ has property P or not, we also know whether $A_{T,w}$ gives some result or not because when T halts the machine $A_{T,w}$ is equivalent to the machine M, while P is a functional property. At the same time, if we know whether $A_{T,w}$ gives some result or not, we also know whether the machine T halts after it begins calculations with the input w. Because the latter problem is undecidable, the initial problem is also undecidable.

To complete the proof, we need to consider the second case when all Turing machines that give no output do not have the property P.

However, not to have the property P is also some property Q. As P is a functional property, so is Q. As P is a nontrivial property, so is Q. Besides, P is decidable if and only if Q is decidable. All Turing machines that give no output have the property Q. As it is demonstrated the property Q is undecidable. Thus, the property P is also undecidable.

Theorem 2.5.2 is proved. □

This result shows that the majority of algorithm or program properties that are defined by their output are undecidable for recursive algorithms. This implies that the majority of program properties related to their output are undecidable by means of modern computers when they work in the recursive mode. We cannot write a program that debugs all other programs or discerns those programs that give some necessary result x or some dangerous result y. Similar results concerning limits of software estimation are considered by Lewis (2001), using the theory of algorithmic complexity.

As we will see later, superrecursive algorithms can solve many of such problems for ordinary programs and contemporary computers.

At the same time, it is necessary to remark that there are nontrivial properties of programs that are decidable by conventional Turing machines and even by finite automata. For instance, all properties from Examples 2.5.2 and 2.5.3 are decidable in the class of all finite automata, while the property from Example 2.5.1 is decidable in the class of Turing machines.

The proof of Theorem 2.5.2 gives us another interesting result that places the halting problem in the hierarchy of all problems related to Turing machines.

Definition 2.5.3. The measure of undecidability of a problem X is less than the measure of undecidability of a problem Y if undecidability of the problem X implies undecidability of the problem Y.

In other words, a problem X has lower measure of undecidability than a problem Y if X is reducible (in a given class of algorithms) to Y. There are different types of reducibility: Turing or T-reducibility, tt-reducibility, m-reducibility, and 1-reducibility (Rogers, 1987). Here we consider Turing reducibility, that is, reducibility by means of arbitrary Turing machines.

Theorem 2.5.3. *The halting problem has minimal measure of undecidability in the class of all problems of determining nontrivial functional properties P of Turing machines.*

3

Subrecursive Algorithms

> *"What's one and one and one and one*
> *and one and one and one and*
> *one and one and one?,"*
> *"I don't know," said Alice. "I lost count."*
> *"She can't do addition," said the Red Queen.*
>
> Lewis Carroll, 1832–1898

As many believe, the class of recursive algorithms appears to be a natural class, independent of any specific model of computation. As a result, computer scientists look into it for an answer to the basic question of decidability and computability about their dependence on formalism. The Church–Turing thesis is an assertion of belief that the classical formalisms completely capture our intuitive notion of computable function, giving one a great deal of confidence in the theoretical foundations of computer science. Nevertheless, much weaker than recursive algorithms are subrecursive algorithms, which are studied in the theory of algorithms and computation. They are successfully used in practice of computer industry and programming.

In this chapter, we consider the following problems:

◇ What is the general situation with subrecursive algorithms (Section 1)?
◇ What mathematical models are used for subrecursive algorithms (Section 2)?
◇ What is a finite automaton (the most popular procedural model of subrecursive algorithms)? What kinds of finite automata have been studied and what properties do they have (Section 3)?
◇ What are recursive and partial recursive functions (the most popular functional models of subrecursive algorithms) and what properties do they have (Section 4)?

3.1 What subrecursive algorithms are and why we need them

> *Every device to its function.*
>
> A proverb

Taking a class **R** of recursive algorithms, such as classes of all Turing machines, all formal grammars or all RAM, researchers often consider some specific subclasses of these classes. For example, they study Turing machines that perform only polynomial in time computations or finite automata as the simplest model of algorithms. In general, those algorithms that belong to a proper subclass of **R** are called *subrecursive*.

Having a developed theory of recursive algorithms, we ask: Why do we need to consider weaker algorithms if recursive algorithms can simulate any subrecursive algorithm? Our experience gives us at least three reasons for doing this.

First, many programs and computing devices do not have the full power of recursive algorithms. For example, any adder in a computer or programs for multiplication and division belong to much smaller classes of automata and algorithms than the class of all recursive algorithms. Using Turing machines for modeling these devices is unreasonable and even misleading. A model of a system has to be as simple as possible. To achieve this goal, researchers introduced the concept of finite automaton.

Moreover, in some cases computations demand so much computer time that nobody will perform such computations as it will be unreasonable or even impossible to do this when, for example, computations demand billions of years. To reflect such situations in theory, researchers claimed that realistic computations are only those that demand polynomial (in the length of the input) time for computation. In such cases, subrecursive algorithms provide more adequate models of real computing devices and programs.

Second, in general, we can know more about relevant classes of subrecursive algorithms than about a class of all recursive algorithms. It is possible to formulate the following property:

Thesis of Cognitive Complementarity. The more powerful a model of algorithms is taken, the less knowledge we can get about this model.

For example, for finite automata, almost all algorithmic problems (such as: "Is the language of a given automaton empty?" or "Is a given word w accepted by a given automaton A?") are decidable. For pushdown automata with one stack or context free grammars, which are more powerful than finite automata, the quantity of decidable problems becomes much less. For Turing machines, which provide the most tractable model for recursive algorithms, almost all algorithmic problems are undecidable.

Third, in addition to better decidability of algorithmic problems, subrecursive algorithms possess other good properties. In particular, they are simpler when you work with them (build automata, transform them, test them, etc.). For example, building a compiler with a context free grammar instead of a general phrase-structure grammars allows one to avoid many complications.

Besides, as Cleland (2001) writes, from a preanalytic, intuitive standpoint, a procedure is effective if correctly following it reliably yields a definite output. The Euclidean algorithm, for example, is ordinarily thought to be effective for computing the greatest common divisor of two integers because correctly applying it to two integers invariably yields their greatest common divisor. Recursive algorithms do not give a definite output for all inputs. So, we need to restrict recursive algorithms to achieve a total computability for a given domain.

Historically, subrecursive algorithms were introduced in the following situation. Researchers first tried to design a formal mathematical model that completely encompassed all kinds of computations as phenomena. As a result, they created different models of recursive algorithms. Then they discovered that lots of systems have algorithmic nature or, at least, express some algorithmic properties. Examples of systems of the first kind are various automata like vending machines used by people. The brain gives an example of the system of the second kind. An attempt to apply to such system general recursive models did not give positive results. In comparison with models, such systems were either much simpler or had a very different structure. Consequently, general models of recursive algorithms provided too inexact and weakly tractable approximation to real systems. This brought researchers to creation of new models. Neural networks, finite automata, context free grammars and other models appeared. Investigation of model properties demonstrated that they are subrecursive algorithms, giving new supportive evidence to the validity of the Church–Turing thesis.

Subrecursive algorithms are built in different ways. However, with respect to recursive algorithms, there are three approaches to the construction of subrecursive algorithms:

◇ by *excluding some components* of a recursive model, for example, finite automata do not have memory;

◇ by *restricting* resources that are used in operation, for example, polynomial time computations allow only such Turing machines that run in an amount of time that is polynomial in the size of the input (Cobham, 1964) or memory of push down automata (cf., for example, Hopcroft, Motwani, and Ullman, 2001) is more restricted than memory of Turing machines;

◇ by *restricting rules* for operation, for example, primitive recursive functions and general recursive functions cannot use minimization operation (cf., Section 2.4); or context free grammars have only transformation rules in which the head of the rule is a variable (cf., for example, Hopcroft, Motwani, and Ullman, 2001).

However, originally many classes of subrecursive algorithms appeared without any reference to recursive algorithms and only later they have been related to the general structure. Independently of recursive algorithms, McCulloch and Pitts (1943) built artificial neural networks or Petri (1962) constructed Petri nets, another class of subrecursive algorithms.

3.2 Mathematical models of subrecursive algorithms and why we need them

> *When schemes are laid in advance,*
> *it is surprising how often the circumstances*
> *fit in with them.*
>
> William Osler, 1849–1919

Subrecursive algorithms like recursive algorithms have many different models and types. We consider them from this perspective.

3.2.1 A variety of mathematical models for subrecursive algorithms

The first models of subrecursive algorithms were *primitive recursive functions* and *recursive functions* introduced by Gödel in 1931 and 1934, respectively. The latter concept is equivalent to the current notion of *general recursive function*. However, when these functions were built, the goal was to develop the most general model of algorithms. So soon in 1936 recursive algorithms appeared: Kleene described partial recursive functions, while Turing built Turing machines. Thus, the first mathematical models for recursive algorithms (Turing machines and partial recursive functions) had been introduced before the actual development of the theory of subrecursive algorithms began.

The next type of subrecursive algorithms were artificial neurons and their networks built by McCulloch and Pitts (1943). However, active development of the theory of subrecursive algorithms began in the fifties of the twentieth century when the formal construction of finite automaton was introduced and studied by Mealy (1953), Kleene (1956), and Moore (1956)),

The most popular of mathematical models of subrecursive algorithms are:

◇ *logic circuits*, also called *combinatorial machines*, which are systems of logic elements each of which realizes a Boolean function (their origin can be traced to the works of Shannon (1938; 1949) and Riordan and Shannon (1942));
◇ *finite automata*, also called *sequential machines* (in the simplest form they are usually attributed to McCulloch and Pitts (1943), and, while in the developed form as a formal construction, they were introduced by Mealy (1953), Kleene (1956), and Moore (1956)). Like Turing machines, finite automata have several forms: *automata without memory, autonomous automata, automata without output or accepting automata or simply, acceptors, automata without input or generators, deterministic, nondeterministic* (Rabin and Scott, 1959), *probabilistic automata*, etc.;
◇ *iterative arrays* (Hennie, 1961);
◇ *neural networks* with Boolean neurons (McCulloch and Pitts, 1943) and with integer number neurons;
◇ a variety of *Petri nets* (Petri, 1962);
◇ various *formal grammars*: *regular, context-free, context-sensitive*, etc. (Chomsky, 1956; Backus, 1959; Naur, 1960);
◇ *pushdown automata* with one stack: *general* (Oettinger, 1961) and *deterministic* (Fisher, 1963; Schutzenberger, 1963);
◇ *recursive functions* (Gödel, 1934);
◇ *primitive recursive functions* (Gödel, 1931);
◇ *n-tape one-way acceptors* (Rabin and Scott, 1959), which recognize sets of n-tuples of strings via computations in which at each step one of the n input tapes is advanced one square and a new machine state is entered; they can be deterministic and nondeterministic;
◇ *counter machines* which can be deterministic and nondeterministic (Evey, 1963; Fischer, 1963)

◇ *generalized sequential machines* (Ginsburg, 1962);
◇ *hardware modification machines* (Dymod and Cook, 1980; 1989), which may be considered as variably connected sequential machines
◇ *multi-tape finite-state transducers* (Ginsburg and Spanier, 1964);
◇ *deterministic n-tape automata* with tapes that are blank except for a single end symbol (Elgot and Rutledge, 1964);
◇ *aggregates* (Dymod and Cook, 1980; 1989), which may be considered as combinatorial machines with cycles;
◇ *linear-bounded automata* (Myhill, 1960), which are one-tape Turing machines given only enough tape to hold the input string, what is equivalent to saying that the amount of information storage permitted is a linear function of the length of the input;
◇ *output restricted Turing machines*: 1) *monotone Turing machines*, which have the same structure as ordinary Turing machines with an output tape, but monotone Turing machines can only write into this tape without a possibility to change what is written (an important direction in the theory of Kolmogorov complexity and inductive inference is based on monotone Turing machines, cf., for example, (Li and Vitanyi, 1997)); 2) *enumerable output machines* (Schmidhuber, 2000), which have the same structure as monotone Turing machines, but enumerable output machines, in contrast to monotone Turing machines, can edit their previous output without decreasing it lexicographically (as a result, the computational power of enumerable output machines lies in between those of monotone Turing machines and ordinary Turing machines).
◇ *resource-restricted Turing machines*: Turing machines that perform only time-bounded or space-bounded computations, with a bounded number of head reversions and so on; taking different classes of functions as boundaries for computations, many different classes of subrecursive algorithms are studied; the most popular of them are: logarithmic time or space computations (LOG-TIME, NLOG-TIME, LOG-SPACE, and NLOG-SPACE), deterministic polynomial time or space computations (**P** or P-TIME and P-SPACE), nondeterministic polynomial time or space computations (**NP** or NP-TIME and NP-SPACE), and exponential time or space computations (E-TIME and E-SPACE).

According to (Fischer, 1965), the first person to consider time-restricted Turing machines was Yamada (1960; 1962). He was especially interested in a class of strictly increasing functions from the nonnegative integers into the nonnegative integer which were, in his terminology, *real-time computable*. Myhill (1960) is considered as the first researcher who studied tape-restricted Turing machines.

According to (Aho, Hopcroft, and Ullman, 1976), general classes of functions and corresponding classes of Turing machines restricted by the time of computation were first studied by Hartmanis and Stearns (1965). However, nine years earlier, there was a paper of Trahtenbrot (1956), which was unknown to Western readers and in which time and space complexity and their corresponding classes of Turing machines were already studied. Independently, Zeitin studied time complexity, making a presentation on this topic in 1956 (cf. Yanovskaya, 1959).

3.2.2   Why we need both recursive and subrecursive algorithms as mathematical models for computation

Subrecursive algorithms have proved themselves useful in many aspects. For example, finite automata are used for description and modeling many computing devices and parts of a computer. Moreover, resources that people can use for computation are always limited. So, any computer is in a definite sense a finite automaton. Only it is not an abstract but a material automaton. Being finite, any computer can be modeled by some abstract finite automaton. As a result, we come to a natural question: Why do we need more powerful algorithmic structures, for example, Turing machines, if in practice we have nothing more than finite automata?

There are, at least, three reasons for considering such classes as Turing machines.

The **first reason** is perfectly explained by Hopcroft, Motwani, and Ullman (2001). They write:

"In fact one could argue that a computer with 128 megabytes of main memory and 30 gigabyte disk, has "only" $256^{30\,128\,000\,000}$ states, and is thus a finite automaton.

However, treating computers as finite automata (or treating brains as finite automata, which is where the finite automata idea originated), is unproductive. The number of states involved is so large, and the limits are so unclear, that you don't draw any useful conclusions. In fact, there is every reason to believe that, if we wanted to, we could expand the set of states of a computer arbitrarily."

In a similar manner, we can always extend the time of computation.

Thus, we come to the conclusion that the first reason for utilization of Turing machines and other potentially infinite models is *indeterminacy* of possible time of computation and memory utilization in computers. One would like to suggest that it might be feasible to take some very big numbers and consider them as boundaries for time and memory. Natural arguments show that such boundaries exist. For example, it is possible to find limits to the state expansion because according to contemporary physics, the whole universe consists of a finite number, say N, of subatomic particles. Consequently, no computer can have more than N elements. Consequently, if these elements can have only two states, the numbers of states of the whole computer will be not larger than 2^N. Even if these elements can have k states, the numbers of states of the whole computer will be not larger than k^N. It means that that the number of states is always bounded. However, given any particular computer, we can always extend its memory and thus the number of states.

However, having these boundaries does not solve the problem of modeling computers exclusively by finite automata. It is due to the fact that these models become overcomplicated and consequently, intractable when the number of their elements grows above some limit. Thus, the **second reason** to use Turing machines is that they can reduce complexity of modeling in many times. It looks strange that an infinite, at least, potentially infinite, model can be simpler than a finite model. But it will not be a surprise if we understand that finite automata may have a very irregular structure, while in a Turing machine such irregularity is reduced only to a small control device, while the memory of the machine is a uniform tape or several tapes. Uniform structures are essentially less complex in comparison with nonuniform structures. It is possible to see this from the everyday experience, to derive it as a consequence of the theory of Kolmogorov complexity (Li and Vitanyi, 1997), or explicate it from the principle of asymptotic homogeneity (Bratalsky and Burgin, 1986).

The **third reason** why Turing machines are so good for theoretical studies is the fact that there is a universal Turing machine (cf. Section 2.3). It is such a machine that can simulate any other Turing machine. In some sense, it is the most powerful Turing machine that serves as milestone for many problems. For example, existence of a universal Turing machine provides a base for the development of such important field as the theory of algorithmic complexity (cf. Chapter 5 and (Li and Vitaniy, 1997)) and duality of complexity measures of algorithms and automata (Burgin, 1982). At the same time, in the class of finite automata and in many other classes of subrecursive algorithms the most powerful or universal automaton does not exist. For other classes where universal algorithms exist this fact can be deduced from the existence of universal Turing machines. Thus, Turing machines give a universal framework for an investigating subrecursive algorithms. For example, as the authors of neural networks, McCulloch and Pitts, stated, their model was essentially influenced by Turing machine.

3.3   Procedural programming as know-how: Finite automata and finite-state machines

> *"It's too late to correct it," said the Red Queen;*
> *"when you have once said a thing,*
> *that fixes it, and you must take the consequences."*
>
> Lewis Carroll, 1832–1898

3.3.1   Structure of a finite automaton

According to the traditional approach (cf., for example, Hopcroft, et al., 2001), there are three forms of representation of finite automata or, what is the same, sequential machines (Savage, 1976): *analytical, dynamic,* and *table.* We begin with the analytical form.

A finite automaton $\mathbf{A}$ consists of three structures, that is, $\mathbf{A} = (L, S, \delta)$:

◇ The *linguistic structure* $L = (\Sigma, Q, \Omega)$ where Σ is a finite set of *input symbols*, Q is a finite *set of states*, and Ω is a finite set of *output symbols* of the automaton $\mathbf{A}$;

◇ The *state structure* $S = (Q, q_0, F)$ where q_0 is an element from Q that is called the *initial* or *start state* and F is a subset of Q that is called the set of *final* (in some cases, *accepting*) states of the automaton $\mathbf{A}$;

◇ The *action structure*, which is traditionally called the *transition function*, or more exactly, *transition relation* of the automaton $\mathbf{A}$

$$\delta \colon \Sigma \times Q \to Q \times \Omega.$$

The state structure is realized by the memory unit M and the action structure is realized by the logic unit L of the automaton **A**. The scheme of the automaton **A** is given in Figure 3.1.

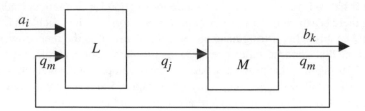

Figure 3.1. The structure of a sequential machine or finite automaton **A**.

Remark 3.3.1. Sometimes the initial state and final states are not included in the definition of finite automaton (cf., for example, Savage, 1976). In this case, automata with the initial state and final states form a special subclass of all automata and are called *adjusted finite automata* (Trahtenbrot and Barzdin, 1970).

Remark 3.3.2. Sometimes instead of one initial state, the set of initial states is included in the definition of finite automaton (Balcazar, Diaz and Gabarro, 1988). This model is closer to real automata. However, it is possible to prove that from the theoretical point of view finite automata with several initial states, as well as several final states are equivalent to finite automata with one initial and one final states. In other words, both classes generate or accept the same class of languages, namely, regular languages.

The *dynamic form* of representation of a finite automaton **A** is different from its analytic form only in its transition function representation. The function or relation δ is given in the form of a transition diagram.

The *table form* of representation of a finite automaton **A** is also different from its analytic form only in its transition function representation. The function or relation δ is given in the form of a table.

Example 3.3.1. **A** $= (L, S, \delta)$ is an automaton without output. The linguistic structure L consists of the set $\Sigma = \{1, 0\}$ of input symbols and the set $Q = \{q_0, q_1, q_2\}$ of states. The state structure S consists of the set Q, the start state q_0 and the set $F = \{q_2\}$ of accepting states.

The *action structure* or *transition function* $\delta\colon \Sigma \times Q \to Q$ in *analytical form* is given by a formula or by the set of triples:

$$\delta = \big\{(q_0, 1; q_1), (q_0, 0; q_0), (q_1, 1; q_2), (q_2, 1; q_0), (q_1, 0; q_1), (q_2, 0; q_2)\big\}$$

or

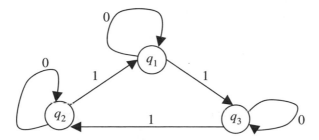

Figure 3.2. The transition diagram of the automaton **A** from Example 3.3.1. Here circles are states and arrows are transitions marked by the corresponding input symbols.

Input Output	1	0
q_0	q_1	q_0
q_1	q_2	q_1
q_2	q_0	q_2

Figure 3.3. The table for the transition function of the automaton **A** from Example 3.3.1.

$$\delta(q_i, k) = q_j \text{ where } k \in \{1, 0\}, i, j \in \{0, 1, 2\} \text{ and } j = i + k \pmod 2.$$

The *dynamic form* of representation of the automaton **A** is given in Figure 3.2, while the *table form* of representation of the automaton **A** is given in Figure 3.3.

When the automaton **A** works, some word w (for example, $w = abc$) is given to **A** letter by letter. When the automaton **A** consumes some letter (for example, a), its state and the output change according to the transition function or relation δ.

However, when we model real systems, we must not forget about time. Taking time into consideration, we see that the transition relation $\delta(q, a)$, even when it is a function, does not define uniquely functioning of the automaton **A**. There are three conventional ways to define information processing in a timed automaton.

Let a word $a(1)a(2)\ldots a(n)$ be given to **A** as its input. Then the rules of the first type are defined (Trahtenbrot and Barzdin, 1970) by the following formulas:

$$q(1) = q_0,$$
$$q(t + 1) = \delta_1(q(t), a(t)),$$
$$b(t) = \delta_2(q(t), a(t)).$$

Here $\delta(q(t), a(t)) = \big(\delta_1(q(t), a(t)), \delta_2(q(t), a(t))\big)$, while $a(t)$ belongs to Σ, $q(t)$ belongs to Q, and $b(t)$ belongs to Ω.

The rules of the second type are defined (Mealy, 1955) by the following formulas:

$$q(1) = q_0,$$
$$q(t + 1) = \delta_1(q(t), a(t + 1)),$$
$$b(t + 1) = \delta_2(q(t), a(t + 1)).$$

Many authors call an automaton **A** with the rules of the second type a *Mealy automaton*.

The rules of the third type (Moore, 1956) are defined by the following formulas:

$$q(1) = q_0,$$
$$q(t + 1) = \delta_1(q(t), a(t)),$$
$$b(t) = \delta_2(q(t)).$$

In this case, the output depends only on the inner state, while the automaton **A** with the rules of the third type is called a *Moore automaton*.

As in the case of Turing machines, it is traditionally assumed that finite automata work with finite words.

In this case, *the result of computation* of **A** is defined in the following way. When a word w is given to **A** as its input, all letters are consumed, and **A** eventually comes to a state in F, producing a word u, this word is taken as the output. If **A** does not come to the state in F, then **A** gives no output.

The result of acceptance of **A** is defined in the following way. When a word w is given to **A** as its input, all letters are consumed, and **A** eventually comes to a state in F, the word w is accepted. If **A** does not come to the state in F, then the word w is rejected.

There are other approaches to the definition of the result of automaton functioning. Some are considered in the next section.

It is worth knowing that finite automata that work with infinite words were considered very early (Burks and Wright, 1953). The reason is explained by Vardi and Volper (1994), who write that when we deal with concurrent or nonterminating processes (like those that are supported by operating systems) there is a need to reason about infinite computations. Thus, instead of considering the first and the last states of a program that realizes finite computations, we need to deal with infinite sequences of states that the program, such as operating system, goes through.

Finite automata do not have inner structure or, at least, this structure is implicit. There are different ways of realization of such a structure. One of them is utilization of Boolean circuits (cf. Figure 5.3.1). As it is proved, Boolean circuits can model arbitrary finite automata (Minsky, 1967). Consequently, it is possible to consider Boolean circuits as the inner structure of finite automata. Moreover, Boolean circuits can simulate functioning of a Turing machine (cf., for example, Balcazar, et al., 1988).

For some time, Boolean circuits were very popular due to their usage as models for computer circuits and components. If computer is built on transistors, then Boolean elements are good models of these transistors. Minimization of Boolean functions related to Boolean circuits was utilized for optimization of computer hardware.

However, implementation of LSI and VLSI as the base for computer hardware changed the situation. Although it is possible to represent VLSI by a Boolean circuit, such representation would be inefficient as it would contain a huge amount of Boolean elements. As a result, Boolean circuits became inept as models of computer hardware.

Now Boolean circuits are used mostly in the theory of computational complexity estimating the size of algorithms and for investigation of different complexity classes.

3.3.2 Types and basic properties of finite automata

Traditionally, three types of finite automata (FA) are considered: deterministic, nondeterministic, and nondeterministic finite automata with ε-transitions.

Definition 3.3.1. A *deterministic finite automaton* (DFA) A is a finite automaton in which both components $\delta_1(q, a)$ and $\delta_2(q, a)$ of the transition relation $\delta(q, a)$ are functions, that is, each is defined for all pairs (q, a) with q from Q and a from Σ.

This means that A performs a transition from a state q to a state p if and only if it is in the state q and it is given the input a from Σ for which $\delta_1(q, a) = p$. This transition is uniquely determined by the input and state of A. In addition, A gives output b when $\delta_2(q, a) = b$.

Definition 3.3.2. A *nondeterministic finite automaton* (NFA) is a finite automaton in which the $\delta(q, a)$ is not necessarily a function but can be an arbitrary binary relation between the Cartesian products $Q \times \Sigma$ and $Q \times \Omega$.

This means that A does not necessarily work for some inputs from Σ and its transition and output are not always uniquely determined by the input and state of A that is, it can be possible to make several transitions and give different outputs for the same pair (q, a).

Definition 3.3.3. A *nondeterministic finite automaton with ε-transitions* (ε-NFA) is a finite automaton in which the $\delta(q, a)$ is an arbitrary binary relation between the Cartesian product $Q \times \Sigma_\varepsilon$ and $Q \times \Omega$ where ε is a symbol of the empty word and $\Sigma_\varepsilon = \Sigma \cup \{\varepsilon\}$.

However, theoretical studies mostly involve more restricted types of finite automata.

Definition 3.3.4. A finite automaton A is called an *automaton without output* if it gives no output, that is, the set Ω is empty and its transition relation has the form $\delta \colon \Sigma \times Q \to Q$.

Such automata can work only in the accepting mode. Many popular textbooks restrict their exposition to automata without output (cf., for example, (Hopcroft et al., 2001) or (Davis and Weyuker, 1983)).

As in the general case, three types of automata without output are considered.

Definition 3.3.5. A *deterministic finite automaton without output* (usually it is denoted by the abbreviation DFA) A is a finite automaton in which the transition relation $\delta(q, a)$ is a function, that is, it is defined for all pairs (q, a) with q from Q and a from Σ.

This means that A performs a transition from a state q to a state p if and only if it is in the state q and it is given the input a from Σ for which $\delta(q, a) = p$. This transition is uniquely determined by the input and state of A.

Definition 3.3.6. A *nondeterministic finite automaton without output* (usually it is denoted by the abbreviation NFA) is a finite automaton in which the $\delta(q, a)$ is not necessarily a function but can be an arbitrary binary relation between the Cartesian products $Q \times \Sigma$ and the set Q.

This means that A does not necessarily work for some inputs from Σ and its transition is not always uniquely determined by the input and state of A, that is, it can be possible to make several transitions for the same pair (q, a).

Definition 3.3.7. A *nondeterministic finite automaton with ε-transitions and without output* (usually it is denoted by the abbreviation ε-NFA) is a finite automaton in which $\delta(q, a)$ is an arbitrary binary relation between the Cartesian product $Q \times \Sigma_\varepsilon$ and Q where ε is a symbol of the empty word and $\Sigma_\varepsilon = \Sigma \cup \{\varepsilon\}$.

In addition to the features of a nondeterministic finite automaton, ε-transitions allow such an automaton to make transitions even without input.

A deterministic finite automaton works in a strictly sequential mode. At the same time, a nondeterministic finite automaton works in an independent branching mode. It means that functioning of such automaton may be considered either as emergence at each point of indeterminacy and existence of independent processes in one automaton or as appearance new identical automata at each point of indeterminacy, which continue to function independently of all others. An intermediate mode of functioning is realized by Petri nets (Petri, 1962; Hack, 1975). These nets are considered as a generalization of finite automata, to allow for the occurrence of several actions (state transitions) autonomously. Petri nets and their generalizations are extensively used for modeling different asynchronous processes, including computational processes. It is proved (Peterson, 1981) that all languages of ordinary Petri nets are context sensitive. This implies that these nets are subrecursive algorithms. However, simple generalizations of Petri nets allow one to model arbitrary Turing machines and thus, to achieve the recursive level (Agerwala, 1974).

As nondeterministic finite automata work differently in comparison with deterministic finite automata, we need to define the result of their computation and acceptance.

The *result of computation* of a nondeterministic finite automaton A is defined in the following way. When a word w is given to A as its input, all letters are consumed, and A eventually comes to some state in F, producing a word u, this word is taken as one from the complete output of A. Usually one input defines several paths of computation. Thus, in a general case, the result of A is a set of words and A is a multivalued algorithm in the sense of (Burgin, 1993a). If beginning with the input word w, A does not come to any state in F, then A gives no output.

The result of acceptance of a nondeterministic finite automaton A is traditionally defined in the following way (cf. Hopcroft et al., 2001). When a word w is given to A as its input, all letters are consumed, and there is such a path that A comes to some state in F, the word w is accepted. If A does not come to any state in F, then the word w is rejected.

A variety of other rules for acceptance for a nondeterministic finite automaton A are considered by Peichl and Vollmer (2001). While any input defines a path of transition for a deterministic automaton, in a nondeterministic case, we have several paths, which form a tree. The terminal nodes of these trees are called leaves. Each such node is a state of the automaton A. This allows us to attach symbols 1 and 0 to these leaves. When the leaf is a final state of A, then we correspond 1 to this leaf. Otherwise, we correspond 0 to the leaf.

In such a way, any tree generated by A, generates in its turn some set of 0's and 1's. When the paths that belong to this tree are generated in specific order, then we correspond to the tree a word in binary symbols 1 and 0. This allows us to take some class T of sets in the first case and of words in the second case and to make this class accepting. That is, a word w is accepted by A if and only if the set (the word) that is corresponded to the leaves of the tree generated by w belongs to T. For example, the class T consists of all sets of binary symbols that have even number of 1's. For the standard rule of acceptance of a nondeterministic finite automaton A, the class T consists of all sets or words of binary symbols that contain, at least, one symbol 1.

Finite automata with such acceptance rule are called (Peichl and Vollmer, 2001) *leaf automata*. As it is proved (cf. Theorem 3.3.5), the acceptance power of leaf automata is much higher than the acceptance power of standard nondeterministic automata. However, our main interest here is standard deterministic and nondeterministic automata.

Theorem 3.3.1. (Rabin and Scott, 1959). *Both classes of nondeterministic finite automata with the standard rule for acceptance and deterministic finite automata have the same accepting power.*

Theorem 3.3.2. *Both classes nondeterministic finite automata and nondeterministic finite automata with ε-transitions have the same accepting power.*

It is possible to find a proof of this result in (Hopcroft et al., 2001).

Theorem 3.3.3. *The class of all finite automata without output is equivalent to the class of all finite automata with output, that is, they generate the same class of formal languages.*

It is possible to find a proof of this result in (Trahtenbrot and Barzdin, 1970).

There are also other types of finite automata.

Definition 3.3.8. A finite automaton A is called an *automaton without memory* if it has no inner states Q.

Logic elements or, in another terminology, Boolean neurons, each of which realizes a Boolean function, are automata without memory.

Theorem 3.3.4. (Minsky, 1967). *For any deterministic finite automaton, there is an
equivalent logical circuit with logical elements that realize Boolean functions (not
x) and (x ∧ y) or an equivalent neural network with Boolean neurons.*

Definition 3.3.9. A finite automaton A is called an *autonomous automaton* or *automaton without input* if it has no input, that is, the set Σ is empty and its transition
relation has the form $\delta : Q \to Q \times \Omega$.

In conventional automata models, time is either ignored or is present implicitly
in form of a system time, which is determined by steps of computation. However,
real systems are functioning in physical time. To have more exact models of real systems than give conventional automata, it is necessary to introduce time into the rules
of automaton functioning. This problem is considered in many works. Alur and Dill
(1994) distinguish qualitative and quantitative temporal reasoning. The first type has
been studied in great detail. It is based on the assumption that any functioning of a
system can be completely modeled as a sequence of states or system events, called
an *execution trace* (or just *trace*). The *behavior* of the system is a set of such execution sequences. When the systems do not have too big a number of different states,
as many have, conventional finite automata give an appropriate model, leading to
effective constructions and decision procedures for automatically manipulating and
analyzing system behavior. In particular, the universal acceptance of finite automata
as the canonical model of finite-state computation can be attributed to the robustness
of the model and the appeal of its developed theory.

Although such abstraction away from quantitative time has had many advantages,
it is ultimately counterproductive when reasoning about systems that interact with
physical processes. The correct functioning of the control system of airplanes, cars
and toasters depends crucially upon *real-time* considerations. To reflect these properties, finite automata are modified for this task and a theory of *timed* finite automata
is developed (Alur and Dill, 1994).

There are three ways to represent time in automata models. One alternative,
which leads to the *discrete-time model*, requires the time sequence to be a monotonically increasing sequence of integers. This model is appropriate for certain kinds
of synchronous digital circuits, where signal changes are considered to have changed
exactly when a clock signal arrives. One of the advantages of this model is that it can
be transformed easily into an ordinary formal language. Each timed execution trace
can be expanded into a trace where time increases by exactly one unit at each step, by
inserting a special *silent* event as many times as it is necessary between events in the
original execution trace. Once this transformation has been performed, the time of
each event is the same as its position, so the time sequence can be discarded, leaving
an ordinary string. Hence, discrete time behaviors can be manipulated using ordinary finite automata. Of course, in physical processes events do not always happen
at integer-valued moments of time. The discrete-time model requires that continuous

time is approximated by taking some fixed quantum of time *a priori*, which limits the accuracy with which physical systems can be modeled.

The *fictitious-clock model* is similar to the discrete time model, except that it only requires the sequence of integer moments of time to be nondecreasing. The interpretation of a timed execution trace in this model is that events occur in the specified order at real-valued moments of time, but only the (integer) readings of the actual moments of time with respect to a digital clock are recorded in the trace. This model is also easily transformed into a conventional formal language (Alur and Dill, 1994). It is conceptually simple to manipulate these behaviors using finite automata, but the compensating disadvantage is that it represents time only in an approximate sense.

The third approach is a *dense-time model*, which is developed by Alur and Dill (1994) and in which time is a dense set, because it is a more natural model for physical processes, which are functioning in continuous time. In this model, the moments of time of events are real numbers, which increase monotonically without bound. Moreover, it is presupposed that an automaton can have several clocks, registering times with different scales. This correlates with the system theory of time (Burgin, 1997b; 2002a), according to which one system can have several times. Dealing with dense time in a finite-automata framework is more difficult than the other two cases, because it is not obvious how to transform a set of dense-time traces into an ordinary formal language.

In addition, timed automata provide an efficient model for human-computer interaction, in which time is often a critical parameter (Burgin, Liu, and Karplus, 2001).

3.3.3 Languages and automata

Finite automata are usually compared by and utilized for producing, that is, accepting, deciding or computing, some languages.

Definition 3.3.10. A *formal language L* is any set of words in some alphabet *A*.

There are three main ways to build a formal language L. The first two forms give dynamic representation of languages, while the third form provides a static language description. A formal language may be defined:

1. *By an automaton*: for example, a language can be built by finite automata, pushdown automata or Turing machines in the mode of computation or acceptance. In the first case, *L* consists of all words that are computed and in the second case, accepted by the corresponding automaton. This is a **system representation of a language**.

2. *By a system of rules*: for example, a language can be built using formal grammars or Post productions. This is a **linguistic representation**.

3. *By a formula*: for example, a language can be defined by regular expressions. This is an **analytic representation**.

Regular expressions over an alphabet *A* are defined by induction:

The base of induction:

1. Ø is a regular expression
2. ε is also a regular expression
3. For any single letter a from A, a is a regular expression

The step of induction: If U and V are regular expressions, then:

4. $U + V$ is regular expression, which is called the union of the expressions U and V;
5. UV is regular expression, which is called the concatenation of the expressions U and V;
6. U^* is regular expression, which is called the Kleene closure of U;
7. (U) is regular expression, assuming that parenthesis are not allowed to be in A.

Example 3.3.2. If $\Sigma = \{a, b, 1\}$, then a, $a + b$, $1a$, 1^*, $(a + b)^*$, and $(1a)$ are all regular expressions.

Regular expressions over an alphabet A define languages in the alphabet A. This is an analytic representation of languages.

Definition 3.3.11. A *regular language* L is any set of all words in some alphabet A that are obtained by the following inductive rules:

The base of induction:

1. The regular expression Ø defines the empty language $L(\emptyset) = \emptyset$.
2. The regular expression ε defines the empty language $L(\varepsilon) = \{\varepsilon\}$, which consists of the empty word.
3. The regular expression a defines the empty language $L(a) = \{a\}$, which consists of the single word a.

The step of induction: If U and V are regular expressions define languages $L(U)$ and $L(V)$, correspondingly, then:

4. $U + V$ defines the union of languages of languages $L(U)$ and $L(V)$ by the formula $L(U + V) = L(U) \cup L(V)$;
5. UV defines the concatenation of languages $L(U)$ and $L(V)$ by the formula $L(UV) = L(U)L(V)$;
6. U^* defines the Kleene closure of the language $L(U)$ by the formula $L(U)^* = \bigcup_{n=1}^* L(U)^n$ where $L(U)^n = L(U) \ldots L(U)$;
7. (U) defines the same language as U.

Theorem 3.3.5 (Kleene, 1956). a) *All languages that correspond to regular expressions (which are called regular languages) are languages of finite automata.*
b) *Finite automata accept only regular languages.*

So, the class of regular languages is the same as the classes of languages accepted by finite automata. This theorem shows equivalence between system and analytic representations for the given class of languages.

There are many languages that are not regular. For example, the language $L = \{ww;\ w$ is an arbitrary word$\}$ is not regular. Theorem 3.3.5 states that finite automata, both deterministic and nondeterministic, can accept only regular languages. At the same time, *leaf automata*, that is, nondeterministic finite automata with a leaf acceptance condition (cf. Section 3.3.2), have much higher acceptance power.

Theorem 3.3.5. *For any formal language L, there is a leaf automaton A that accepts L.*

Proof. For simplicity, we give a proof only for the alphabet $\{1, 0\}$. A general case is considered in similar way. It only demands more space.

Let us consider the nondeterministic finite automaton **A** with the following rules: $\delta(q_0, 1) = q_0$, $\delta(q_0, 0) = q_0$, $\delta(q_0, 1) = q_1$, $\delta(q_0, 0) = q_2$, $\delta(q_1, 1) = q_1$, $\delta(q_1, 0) = q_1$, $\delta(q_2, 1) = q_2$, $\delta(q_2, 0) = q_2$ where q_0 is the initial state, q_1 is a terminal (accepting) state and q_2 is a nonterminal (nonaccepting) state. According to these rules, when **A** comes to the state q_1 or q_2, it does not change it, consuming all other symbols from the input word. At the same time, the first two rules allow the automaton **A** to consume any number of symbols from the beginning of the input word. Thus, any path in the tree of processing an input word terminates in q_1 to which 1 is corresponded or in q_2 to which 0 is corresponded. As each path generated by an input word w corresponds to a definite symbol in w, we order all paths according to the order of symbols in the word w.

If we take as the acceptance leaf language (cf. Section 3.3.2) an arbitrary formal language L, then we see that **A** accepts exactly this language.

Theorem 3.3.5 is proved.

As it is explained above, modeling of many computer systems involves processing infinite strings or words. It implies an extension of the concept of a formal language.

Definition 3.3.12. A *formal ω-language L* is any set of infinite words in some alphabet A.

Now there is a developed theory of finite automata working with infinite words (cf., for example, (Vardi and Volper, 1994) and Section 3.3.2).

3.3.4 Finite automata and finite-state machines

Finite automata, Turing machines, neural networks, and cellular automata, when they are considered without output, that is, only in the accepting mode, are special cases of state machines.

Definition 3.3.13. A model of computation consisting of a set of states, an input alphabet, and a transition function that maps input symbols and current states to a next state is called a *state machine*.

It is usually assumed that algorithms are such state machines when they function in the accepting mode. Some include machines that work in the computing mode into the scope of state machines if those machines go from a state to another state in the process of computation. We call such machines extended state machines, having in mind that they while changing their states, also produce some output. Even those systems that work in continuous time changing their states (cf., for example, (Moore, 1996) or (Alur and Dill, 1994)) are such extended state machines.

Definition 3.3.14. A state machine that has a finite set of states is called a *finite-state machine*.

Some think that an infinite-state machine can be conceived but is not practical. As it is explained in Section 3.2, this is not true. Besides, there is an opinion that a finite-state machine is the same as a finite automaton. This is not completely true because a finite-state machine can use (as an external component) some (potentially or actually) infinite memory, while a finite automaton is restricted to its own constituents, which cannot be infinite.

In other sources, finite-state machines that give no output are called finite automata (cf., for example, (Rosen, 1999)). This shows that terminology in the field of finite-state machines is not stable.

Definition 3.3.15. A state machine that has fixed start and final states is called an *adjusted finite-state machine*.

According to a more general understanding, a *state machine* is any device that stores its status or state at a given time and can operate on input to change the state and/or cause an action or give output. Formally, a state machine consists of: (1) a set of possible input events; (2) a set of possible states; (3) a set of possible actions or output events; (4) a function that maps states and input to a new state and output.

Consequently, any automaton is a state machine and any finite automaton is a finite-state machine. When a finite-state machine accepts only symbolic input and gives only symbolic output, it is a finite automaton.

It is necessary to make a distinction between states of the whole machine and states of its elements and parts as partial states of the whole. For example, a state of the tape of a Turing machine T is a partial state of T. A conventional finite automaton does not have parts. So, it does not have partial states.

A computer is basically a finite automaton and each machine instruction is an input that changes one or more states and may cause other actions to take place. Each computer's data register stores a state. The read-only memory from which a boot program is loaded stores a state (the boot program itself is an initial state). The operating system is itself a state and each application that runs begins with some initial state that may change as it begins to handle input.

3.4 Functional programming as know-what: Recursive functions

"Isn't that rather like one of the Rules in Algebra?"
my Lady inquired.

Lewis Carroll, 1832–1898

Recursive functions and finite automata represent two important paradigms in programming. Recursive functions are related to *functional programming*, while finite automata correspond to *procedural programming*. Other paradigms are: *descriptive*, *object-oriented*, and *structured programming*. Procedural, functional and descriptive paradigms are *representational types* of programming systems, while object-oriented and structured programming are *organizational types*. By its properties, functional programming lies between procedural and descriptive or declarative programming.

Functional programming is a style of programming that emphasizes the evaluation of functions, rather than execution of commands (Rabhi and Lapalme, 1999). A functional programming language is a language that supports and encourages functional programming. Mathematical models that correspond to direct effective operation with functions are called recursive functions.

As software becomes more and more complex, it is more and more important to structure it well. When a programming language provides better structuring, it becomes easier to write programs. Well-structured software is also easy to debug. It is usually a collection of connected modules that can be re-used to reduce future programming costs. Conventional programming languages place conceptual limits on the way problems can be modularized. Functional programming languages push in a natural way those limits back. Two features of functional programming languages in particular, higher-order functions and lazy evaluation, can contribute greatly to modularity (Hughes, 1989).

Functions in functional languages have the form of expressions that are built from some collection of basic functions. This technique exactly models structures developed in such approach to computability and algorithm modeling as the theory of recursive functions. This theory emerged from the ideas of Ackermann and was initiated and developed by Gödel, Church, and Kleene (cf. Section 2.2). Usually functions from the set N_0 of all whole numbers into N_0 or from the direct power N_0^k into N_0 are considered and different restrictions are provided in such a way that these functions become in some sense computable. It is not specified who performs computations. However, this approach is oriented mostly at a person and only partially at a device (cf., for example, (Siegelman, 1999)).

The principal technique of this approach consists of three main steps:

1. Choice of some simple functions of whole numbers and assumption that they are computable. They form a basic set for the construction of computable functions. Such functions have, as a rule, very simple rules for calculations. Examples of such functions are functions $0(x) = 0$ for all x and $S(x) = x + 1$.
2. Choice of operations that transform functions and are considered computable or constructive.

3. Formation of the smallest class that is built from the set of chosen functions by means of the chosen operations.

The constructed class is called the class of *computable functions*. Together with implicit rules for calculations of their values, these functions provide a mathematical model of algorithm.

When it is assumed that these operations are finite and are applied a finite number of times for building a new function, we get a class of subrecursive algorithms (subrecursive computable functions). Only such infinite operation as minimization allows one to get recursive algorithms, while limit operations bring us to superrecursive algorithms.

The main operations that are used for building computable functions belong to several classes.

Functional operations:

1. *Sequential composition* or simply, *composition*: For two given functions $f(x)$ and $g(x)$ with one variables x, their sequential composition is determined by the formula $(f \circ g)(x) = f(g(x))$. For $n + 1$ given functions $g_1(z), \ldots, g_n(z)$ each with m variables $z = (z_1, \ldots, z_m)$ and $f(x)$ with n variables $x = (x_1, \ldots, x_n)$, their sequential composition is determined by the formula $(f \circ g)(z) = f(g_1(z), \ldots, g_n(z))$.

2. *Parallel composition* or *definition by cases*: For two given functions $f(x)$ and $g(x)$ with the one variables x, their parallel composition with respect to a functions $h(x)$ and a number a is determined by the formula $(f \vee_{h,a} g)(x) = f(x)$ when $h(x) = a$ and $(f \vee_{h,a} g)(x) = g(x)$ otherwise.

3. *Primitive recursion* or simply, *recursion*: For two given functions $f(x)$ and $g(x)$ with the same vector of variables $x = (x_1, \ldots, x_n)$, their primitive recursion, which gives the function $h(x, y)$, is defined by the following steps: $h(x, 0) = f(x)$ and $h(x, y + 1) = g(x, y, h(x, y))$.

3a. *Bounded recursion*: For three given functions $f(x)$, $g(x)$ and $b(x, y)$ with the same vector of variables $x = (x_1, \ldots, x_n)$, the function $h(x, y)$ is defined as in primitive recursion with the additional condition $h(x, y) \leq b(x, y)$. When this condition is not satisfied, h is not defined. Thus, h is only allowed to grow as fast as another function already in the class that is built with this operation.

4. *Partial projection*: For a given function $f(x_1, \ldots, x_n)$, its partial projection $h(x_1, \ldots, x_m)$ is defined as $h(x_1, \ldots, x_m) = f(u_1, \ldots, u_n)$ where each u_j is either a number or some variable x_i.

Arithmetical operations:

5. *Bounded sum*: For a given function $f(x, y)$, its bounded sum, which gives the function $h(x, y)$, is defined as $h(x, y) = \sum_{z<y} f(x, z)$.

6. *Bounded product*: For a given function $f(x, y)$, its bounded product, which gives the function $h(x, y)$, is defined as $h(x, y) = \prod_{z<y} f(x, z)$.

7. *Cut-off subtraction*: For two given functions $f(x)$ and $g(x)$, their cut-off subtraction, which gives the function $h(x)$, is defined by the following rule: $h(x)$ is

denoted by $f(x) \doteq g(x)$ and is equal to $f(x) - g(x)$ if $f(x) \geq g(x)$ and to 0 if $f(x) < g(x)$.

Order operations:

8. *Minimization*: For a given function $f(x, y)$ with two variables, its minimization $h(x)$ is defined as $h(x) = \min\{y; \ f(x, y) = 0\}$, that is, $h(x)$ is equal to the smallest y such that $f(x, y) = 0$ provided that $f(x, z)$ is defined for all $z \leq y$. If no such y exists, $h(x)$ is undefined.

8a. *Bounded minimization*: For a given function $f(x, y)$, its bounded minimization $h(x, y_{max})$ is defined as $h(x, y_{max}) = \min\{y; \ f(x, y) = 0$ and $y < y_{max}\}$, that is, $h(x, y_{max})$ is equal to the smallest number $y < y_{max}$ such that $f(x, y) = 0$ and $h(x, y_{max}) = y_{max}$ if no such y exists.

Topological operations:

9. *Discrete limit*: For a given function $f(x, y)$ with two variables, its discrete limit is defined by the following formula: $h(x) = \lim_{y \to \infty} f(x, y)$, where the limit $\lim_{y \to \infty} f(x, y)$ is taken in the discrete topology of natural numbers, that is, the limit $\lim_{y \to \infty} f(x, y)$ is defined if there is y_0 such that $f(x, y) = f(x, z)$, provided that $y_0 \leq y \leq z$. If no such y_0 exists, $h(x)$ is undefined.

Usually algorithms work with words or whole numbers, which are traditionally considered in a discrete topology. However, it is possible that the domain of algorithms and their range have nondiscrete topologies.

Let us consider, for instance, some topology τ on the set N_0 of all whole numbers (Kelly, 1957). It is not necessarily discrete topology. For example, there is one-to-one correspondence between N_0 and the set Q of all rational numbers. This correspondence induces on N_0 the natural topology of Q.

10. *Limit*: For a given function $f(x, y)$ with two variables, its limit is defined by the following formula: $h(x) = \lim_{y \to \infty} f(x, y)$, where the limit $\lim_{y \to \infty} f(x, y)$ is taken in the topology τ. If such limit does not exist, $h(x)$ is undefined.

Remark 3.4.1. Operations 1–5 and 6a generate subrecursive algorithms. Minimization extends the class of computable functions to recursive algorithms. Limits take computable functions beyond the class of recursive algorithms. Limit operations allow one to define arbitrary sums and products of functions.

At the same time, only order and topological operations can create a partial function. All the others yield total functions when applied to total functions.

By starting with various basic sets and demanding closure under various properties and operations, researchers have defined various natural classes of subrecursive computable functions. The smallest class of computable functions that have been popular consists of elementary functions, which were introduced by Kalmar (1943).

Definition 3.4.1. An *elementary function* (in the sense of Kalmar) is a function that can be generated from the constant zero function $0(x) = 0$, the successor function $S(x) = x + 1$, addition, cut-off subtraction and projections $P_i(x_1, \ldots, x_n) = x_i$ using any finite number of compositions and the operations of forming bounded sums and bounded products.

In what follows, we call functions from the class E simply elementary.

It is possible to show that usual arithmetical operations belong to the class E. For example, multiplication and exponentiation over N_0 are both in E, since they can be written as bounded sums and products respectively: $xy = \sum_{z<y} x$ and $x^y = \prod_{z<y} x$. Since E is closed under composition, for each m the m-times iterated exponential function $\exp^{[m]}(x)$ is in E, where $\exp^{[m+1]}(x) = 2^{\exp^{[m]}(x)}$ and $\exp^{[0]}(x) = x$.

Although analogue computations are considered in Chapter 4, here we note their relation to elementary functions. One of the approaches to analogue computations (Moore, 1996) is based on a version of Shannon's *general purpose* analog computer (Shannon, 1941). Such analog computer can integrate real functions and differential equations. It is proved in (Campagnolo, Moore, and Costa, 2000) that if the analog computer is allowed to solve only linear differential equations and to treat inequalities in a differentiable way, then it computes exactly the elementary functions.

It is possible to give another characterization of the class E.

Theorem 3.4.1. (Grzegorczyk, 1953). *An elementary function is a function that can be generated from the successor function* $S(x)$ *and the function* x^y *using any finite number of partial projections and limited recursions.*

Elementary functions have strict bounds on their growth and time of computation as the following results show (Cutland, 1980; Rose, 1984; Rogers, 1987).

Theorem 3.4.2. *If* $f \in E$, *then there is a number* m *such that, for all* x, $f(x) \leq \exp^{[m]}(\|x\|)$, *where* $\|x\| = \max\{x_i; x = (x_1, \ldots, x_n)$ *and* $i = 1, \ldots, n\}$, *that is,* f *cannot grow faster than* $\exp^{[m]}(x)$ *for some fixed* m.

Theorem 3.4.3. $f \in E$ *if and only if* f *is computable in elementary time.*

In spite of these restrictions, the class E is very large, containing the most usable computable functions. It includes, for instance, the connectives of propositional calculus, arithmetical operations, and functions for coding and decoding various sequences of natural numbers such as the prime numbers, as well as factorizations, and many useful number-theoretic and metamathematical functions.

The class E is not the smallest class of computable functions. It contains the class SE of elementary functions in the sense of Scolem (1962).

Definition 3.4.2. An *elementary function* in the sense of Scolem is a function that can be generated from the constant zero function $0(x) = 0$, the successor function $S(x) = x + 1$, addition, cut-off subtraction, and projections $P_i(x_1, \ldots, x_n) = x_i$ using any finite number of compositions and the operations of forming bounded sums.

Remark 3.4.1. It is possible instead of $S(x)$ to take the constant one function $1(x) = 1$ for generating elementary functions in the sense of Kalmar and Scolem (cf. Malcev, 1965).

Proposition 3.4.1. (Scolem, 1962). *If $f \in$ SE, then there is a number m such that, for all $x_1, \ldots, x_n$, $f(x) \leq (2 + x_1, \ldots, x_n)^m$.*

Proposition 3.4.2. (Scolem, 1962). **SE** *is a proper subclass of* **E**.

Indeed, **SE** $\subseteq$ **E** and elementary in the sense of Kalmar function 2^x does not belong to **SE** by Proposition 3.4.1.

Remark 3.4.2. Malcev (1965) also introduces functions elementary with respect to a given system of functions.

In spite of being large, **E** is smaller than the majority of classes of computable functions that is studied in the classical theory of recursive functions. A much more extended class comprises primitive recursive functions.

Definition 3.4.3. A *primitive recursive function* is a function that can be generated from the constant zero function $0(x) = 0$, successor function $S(x)$, and projections $P_i(x_1, \ldots, x_n)$ using any finite number of compositions and primitive recursions.

A bridge between elementary functions and primitive recursive functions is formed by the Grzegorczyk hierarchy (Grzegorczyk, 1953).

Let us consider the system of functions $q_0(x, y) = x + y$, $q_1(x) = x^2 + 2$, and $q_{n+1}(0) = 2$, $q_{n+1}(x) = q_n \left(q_{n+1}(x - 1) \right) = q_n^{[x]}(2)$ for all $n = 1, 2, \ldots$.

Definition 3.4.4. The *Grzegorczyk hierarchy*:

 a. E_0 is the class of functions that can be generated from the constant zero function $0(x) = 0$, successor function $S(x)$, and projections $P_i(x_1, \ldots, x_n)$ by applying any number of compositions and bounded recursion.
 b. for $n > 0$, E_n is the class of functions that can be generated from the constant zero function $0(x) = 0$, successor function $S(x)$, projections $P_i(x_1, \ldots, x_n)$, and functions $q_0(x, y), \ldots, q_{n-1}(0)$ by applying any number of compositions and bounded recursion.

Theorem 3.4.4. $E_3 = E$.

Let **PR** be the class of all primitive recursive functions.

Theorem 3.4.5. $\mathbf{PR} = \bigcup_{n=0}^{\infty} E_n$.

The Grzegorczyk hierarchy is also studied in the context of real number computations (Gakwaya, 1996; 1997).

Although the class of primitive recursive functions is much bigger than the class of elementary functions, the former class is much smaller than the class of general recursive functions.

Definition 3.4.5. A *general recursive function* is a function that can be generated from the constant zero function $0(x) = 0$, successor function $S(x) = x + 1$, and projections $P_i(x_1, \ldots, x_n) = x_i$ using any finite number of compositions, primitive recursions, and bounded minimizations.

Lemma 3.4.1. *Any general recursive function is defined for all whole numbers.*

It is possible to give another characterization of general recursive functions with one variable.

Theorem 3.4.6. (Robinson, 1950). *Any general recursive function with one variable can be generated from the constant zero function* $0(x) = 0$, *successor function* $S(x)$, *and the function* $q(x) = x - \left[\sqrt{x}\right]^2$, *using any finite number of compositions and limited recursion.*

Here we also encounter another meaning for the word subrecursive function, which is related to time of computation (Rose, 1984).

Definition 3.4.6. A *subrecursive* (in the sense of Rose) *function* f is a function that can be so computed that the number of steps of computation is a less computationally complex function of the input than the function f itself.

Theorem 3.4.7. *All functions in the Grzegorczyk hierarchy are subrecursive in the sense of Rose.*

The most extended conventional class of computable functions is formed by partial recursive functions.

Definition 3.4.7. A *partial recursive function* is a function that can be generated from the constant zero function $0(x) = 0$, successor function $S(x) = x + 1$, and projections $P_i(x_1, \ldots, x_n) = x_i$ using any finite number of compositions, primitive recursions, and minimizations.

According to the Church–Turing thesis in the form of Church, **computable functions are exactly partial recursive functions**.

Partial recursive functions were initially considered as numerical functions from N to N or from N_0 to N_0. However, there is a natural correspondence between numerical functions and alphabetical functions, which map words into words.

Really, each natural number is represented by some word when it is used. Different alphabets are used for such representations. For instance, the binary number system, which is used by computers, has the alphabet $\{1, 0\}$. The ternary number system, which is the most optimal for arithmetical calculations, has the alphabet $\{0, 1, 2\}$. The decimal number system, which is used by people, has the alphabet $\{0, 1, 2, 3, 4, 5, 6, 7, 8, 9\}$. Computers use the binary number system, while people prefer the decimal number system.

A representation of numbers by words $\mathbf{c} \colon N \to A^*$ allows one to correspond to each numerical function $f \colon N \to N$ the alphabetical function $g \colon A^* \to A^*$. This correspondence is presented by the diagram a) from Figure 3.4 where $g = \mathbf{c} f \mathbf{c}^{-1}$.

On the other hand, it is possible to enumerate by natural numbers all words in any finite alphabet A. For example, taking all words in the alphabet $\{1, 0\}$, we correspond 1 to the empty word ε and define $\mathbf{e}(w)$ for a nonempty word w equal to the number for which $1w$ is its binary representation. Thus, $\mathbf{e}(00) = 100_2 = 4$ or $\mathbf{e}(101) = 1101_2 = 25$.

Figure 3.4. Correspondences between numerical and alphabetical functions.

An enumeration $e: A^* \to N$ allows one to correspond to each alphabetical function $g: A^* \to A^*$ a numerical function $f: N \to N$. This correspondence is presented by the diagram b) from Figure 3.4 where $f = ege^{-1}$.

The established correspondence can be restricted to the class of all computable functions. As a result, we may assume that Turing machines work not only with words but also with natural or whole numbers and consider recursive and partial recursive functions not only for numbers but also for words.

Theorem 3.4.8. (Turing, 1936). *A function f is partial recursive if and only if f is computable by some Turing machine.*

Corollary 3.4.1. *Partial recursive functions are equivalent to (that is, have the same computing power as) Turing machines.*

Definition 3.4.8. A partial recursive function that is everywhere defined is called a *recursive function*.

Theorem 3.4.9. (Kleene, 1936). *A partial recursive function is everywhere defined if and only if it is a general recursive function.*

This result is based on the following property of partial recursive functions.

Theorem 3.4.10. (Kleene, 1936). *For any partial recursive function $f(x)$ there are such general recursive functions $p(x)$ and $g(x, t)$ such that $f(x)$ has the form (the Kleene normal form):*

$$f(x) = p\left(\min\{t;\, g(x, t) = 0\}\right).$$

There is an interesting class of partial recursive functions, which are called prefix functions and are used in the theory of Kolmogorov complexity (cf. Section 5.3).

Definition 3.4.9. A partial recursive function $f(x)$ is called a *prefix function* if for any for any elements x and z from the definability domain of f, $x \subset z$ implies $f(x) = f(z)$.

For two words x and y, $x \subset z$ means that the word x is a beginning of the word z, for example, $101 \subset 1010101111$ or $ab \subset abcbad$.

Although superrecursive algorithms are treated in the next Chapter, here we consider some functions computed by superrecursive algorithms. These functions are obtained from general and partial recursive functions by application of limit operations. Gold and Putnam introduced in 1965 concepts of limit recursive and limit partial recursive functions.

Definition 3.4.10. A *limit partial recursive function* is a function that is the discrete limit of a sequence of partial recursive functions.

Definition 3.4.11. A *limit recursive* function is a function that is the discrete limit of a sequence of recursive functions.

Any partial recursive function is the discrete limit of a sequence of recursive functions, and thus, it is a limit recursive function. However, the class of limit recursive functions is much larger than the class of partial recursive functions. Besides, there are more limit partial recursive functions than limit recursive functions (Gold, 1965; Putnam, 1965). If we use a general limit operation for constructing new functions, we can obtain much more (Burgin, 1992).

In classical recursive function theory, more general than functions objects are defined. They are called functionals and are mappings of the form $2^{N_0} \to N_0$ or $(2^{N_0})^k \times N_0^t \to N_0$. In similar way to functions, it is possible to define elementary, primitive recursive, general recursive, and partial recursive functionals when we use different operations discussed in this section. Recursive and partial recursive functionals, as well as limit recursive and limit partial recursive functions allow one to increase expressive power and efficiency of functional programming.

Now functional programming, which is based on the theory of recursive functions, is rather popular. There is even a special Journal of Functional Programming. There are many functional programming languages. LISP is the most popular of them (Allen, 2001). It is the main language for artificial intelligence. LISP is based on the use of expressions that represent functions. It is possible to compare functional languages with more popular programming languages. Thus, Pascal, C^{++}, or Fortran are classified as procedural or operational or imperative programming languages. These languages are instruction-oriented, the programs consisting of a sequence of instructions. LISP programs, as originally defined, were specified entirely as expressions. Recursion is the central tool for LISP programming. Due to their regular, recursive structure, LISP programs tend to be short and elegant. At the same time, to be able to program effectively in LISP, a different kind of thinking is required, in particular, one must learn to think recursively. This is very different from instruction-oriented thinking required for procedural languages such as Pascal, C^{++}, or Fortran. However, absence of procedures in LISP restricted essentially programming possibilities, making it less flexible. To improve the language, current implementations of LISP have extensions that allow LISP programs to be more instruction-oriented. Thus, programming experience with different languages shows that it is necessary to combine in a universal language different forms of function process representation. To achieve this goal, we need to go from programming languages to more abstract than they programming metalanguages. Vast opportunities for utilization of different programming languages in one program are provided by such programming meta-language as the block-scheme language (Burgin, 1996).

4

Superrecursive Algorithms: Problems of Computability

Shallow ideas can be assimilated;
ideas that require people to reorganize
their picture of the world provoke hostility.
James Gleick, *Chaos*

Recent development of the theory of algorithms allowed researchers to overcome limitations of the Church–Turing thesis, discovering superrecursive algorithms and explicating hypercomputation. New mathematical models for algorithms and computation emerged that are more powerful than models for recursive algorithms such as Turing machines, partial recursive functions, λ-calculus or cellular automata. Higher computing power means that superrecursive algorithms can solve such problems that no recursive algorithms, even such efficient as quantum algorithms, can solve in principle. There are several directions in the area of superrecursive computation. The most important of them (listed in the chronological order) are: inductive computation, computation with real numbers, interactive and concurrent computation, fixed point models, topological computation, neural networks with real number parameters, infinite-time computation, and dynamical system computation. New models of algorithms and computation gave birth to new logics of computation. In addition to greater computational power, some of superrecursive models provide for higher speed and efficiency of computation, decreasing at the same time computational complexity.

In this chapter, we consider the following problems:

◇ What is the general situation with superrecursive algorithms, their origin, and problems of their representation? The general situation shows necessity to have mathematical models for superrecursive algorithms and to develop an efficient theory of superrecursive algorithms (Section 1).

◇ What is the general situation with mathematical models of superrecursive algorithms, their origin, relations to conventional models, and computing power? Different directions in the theory of superrecursive algorithms and hypercomputation are considered (Section 2).

◇ What is and what can do the closest to conventional algorithms superrecursive model, which is called inductive Turing machine? The most general inductive Turing machines have the computing power of Turing machines with oracles, that is, they can compute any function for finite words and decide any formal language; more constructive inductive Turing machines represent functioning of modern and future computers and networks; hierarchies of such machines allow one to build the arithmetical hierarchy and to prove all true sentences in the formal first order arithmetic in contrast to the famous Gödel incompleteness theorem, which states impossibility to obtain such proofs by conventional mathematical methods (Section 3).

◇ How are local and global networks, cluster computers, and the emerging GRID to be modeled? Achieving this goal, a new computational model called grid automaton is developed and studied. Grid automata allow one to synthesize in one model computation and communication, providing for representation and study of the natural synergy of humans and computers (Section 4).

4.1 What superrecursive algorithms are and why we need them

In practice, bravery accumulates gradually
during the process of solving problems
that seem unsolvable.

G. Altshuller, *The Innovation Algorithm*

Most of what we understand about algorithms and their limitations is based on our understanding of Turing machines and other conventional models of algorithms. The famous Church–Turing thesis claims that Turing machines give a full understanding of computer possibilities. However, in spite of this Thesis, conventional models of algorithms, such as Turing machines, do not provide a relevant representation for the notion of algorithm. That is why several extensions of conventional models have been developed. These new models have different levels of constructivity or realizability. The model, or more exactly, the class of models, that does not involve actual infinity or nonconstructive operations is based on the following observation. One of the basic stereotypes for algorithms states that an algorithm has to stop when it gives a result. This is the main restriction hindering the development of computers. When we understand that computation can go on, but we can get what we need, we then go beyond our prejudices and immensely extend computing power, introducing superrecursive algorithms.

Definition 4.1.1. Any class of algorithms that is more powerful than the class of recursive algorithms (such as Turing machines) is called a *class of superrecursive algorithms*.

Superrecursive algorithms define and control hypercomputations.

Definition 4.1.2. *Hypercomputation* is a computational process (including processes of input and output) that cannot be realized by recursive algorithms.

There are three kinds of reasons for the development of superrecursive algorithms and their theory. *Theoretical reasons* emphasize problems that exist with definitions of computability, algorithms, and constructivity in general, as well as provability and truth in mathematics. *Methodological reasons* attract attention of researchers to existence of many processes that are in some or another sense algorithmic, but cannot be represented by recursive algorithms. *Practical reasons* are dealing with problems of adequate and efficient modeling existing computers, networks and other IPS, as well as with finding new ways for their development. Let us consider some of these reasons.

We begin with a practical question whether recursive algorithms provide an adequate model of modern computers. To our surprise, we find that people do not see correctly how computers are functioning. An analysis demonstrates that while recursive algorithms gave a correct theoretical representation for computers at the beginning of "computer era", superrecursive algorithms are more adequate for modern computers.

To understand the situation, let us look at conventional models of algorithm. We can see that in comparison with the informal notion, an extra condition appears in formal definitions of algorithm, that is, after giving a result algorithm stops (cf., for example, (Harel, 2000)). This condition looks rather natural. Indeed, what you have to do more after you have what you wanted. However, if we analyze thoroughly what is going on with real computers, we have to change our mind.

At the beginning, when computers appeared and were utilized for some time, it was necessary to print out data produced by computer to get a result. After printing, computer stopped functioning or began to solve another problem. Now people are working with displays. A computer produces its results on the screen of a monitor. Those results on the screen exist there only if computer functions. If computer stops, then the result on its screen disappears. This is the contrast to the condition on ordinary (recursive) algorithms that demands to stop to give a result. At the same time, it is exactly the case of such superrecursive algorithms as inductive Turing machines (cf. Sections 4.2.2 and 4.3) because their results are obtained, as a rule, without stopping. A possibility to print some results and switch off the computer only shows that recursive algorithms can be modeled by inductive Turing machine.

This feature of modern computers that results exist (on the screen) only when computer continues to work has been developed further in building nonstop computers (Gray, 1985; Flavin, 1991). The aim of such computers is to achieve high reliability and high availability for cost-effective computer systems. Large applications involving database transactions such as automated teller machines (ATM), credit card transactions, and securities transactions demand distributed computer systems that have mean time failures measured in years. In response to these needs, Nonstop Tandem computer systems have been created. Working in this nonstop, superrecursive mode, Tandem computers have achieved mean time to failures measured in tens of years.

Such misunderstanding of modern computers that restricts them to the conditions of the Church–Turing thesis is not unique and similar "blindness" is not new in society. Thus, people thought for thousands of years that the Sun rotated around the Earth and only in the 16th century Copernicus proved different.

It is necessary to remark that in some cases recursive algorithms are sufficient. For example, working with a printer, we reduce superrecursive algorithms to recursive algorithms. It demonstrates that recursive algorithms are relevant for modeling those computers and programs that display their final results by means of a printer. This situation was true for all computers many years ago but it does not correspond to reality now. Users who consider only intermediate results and treat them as final results, in fact, reduce superrecursive algorithms to recursive ones.

To show other shortcomings of the classical model of algorithms, let us consider some examples of contemporary computer utilization. One of the important applications of computers is simulation used for prediction. However, no single computer run or computer output can be considered to be a definitive forecast of what will happen (Karplus, 1992). It is necessary to have many simulations resulting in the form of stacks of computer outputs in order to make more or less valid prediction. Consequently, in the sequence of these simulations, there is no, as a rule, such a moment when the researcher who carries out these simulations can stop computer and say, "Here is the final result." Even when some conclusions are made basing on the output data of simulation, it is possible that after some time the researcher repeats simulation procedure one or several times more. The goal of such repetitions is, as a rule, to obtain more exact or adequate results, to achieve better understanding, or to test some other hypothesis. This situation evidently demonstrates that a conventional algorithm can adequately represent and direct only one run of computer simulation, while the whole process has a very different nature, which demands superrecursive algorithms.

Such big networks as the Internet give another important example of a situation in which conventional algorithms are not adequate. Algorithms embodied in a multiplicity of different programs organize network functioning. It is generally assumed that any computer program is a conventional, that is, recursive algorithm. However, a recursive algorithm has to stop to give a result, but if a network shuts down, then something is wrong and it gives no results. Consequently, recursive algorithms turn out to be too weak for the network representation, modeling and study.

Even more, no computer works without an operating system. Any operating system is a program and any computer program is an algorithm according to the general understanding. But while a recursive algorithm has to stop to give a result, we cannot say that a result of functioning of operating system is obtained when computer stops functioning. On the contrary, when computer is out of service, its operating system does not give an expected result. Moreover, any operating system does not produce a result in a form of some word, while this is an essential condition for any recursive algorithm. Although, from time to time, operating system sends some messages (strings of words) to a user, the result of operating system is reliable functioning of the computer. Stopping computer is only a particular result. Consequently, the real result of the operating system functioning is obtained only when computer does not stop (at least, potentially). Different authors have noticed it. For example, Vardi and Volper (1994) write that dealing with concurrent or nonterminating processes (like those that are supported by operating systems) there is a need to reason about infinite computations. Thus, we come to a conclusion that it is not necessary for algorithm to halt to produce a result.

Recently, Sloman (2002) explained why recursive models of algorithms, such as Turing machines, are irrelevant for artificial intelligence.

These examples and many others vividly demonstrate why a problem of advancing conventional models of algorithm has been so essential for a long period of time. The solution was given by elaborating superrecursive algorithms.

From a **methodological perspective**, there are rules for calculation that are naturally considered algorithms but do not satisfy the termination or halting condition of recursive algorithms (cf. Section 2.5). The Wallis' algorithm for computing π provides a good example. It may be used to calculate successive and exact values of the decimal representation of π, but it cannot specify a final step terminating in the full decimal representation of π. The same is actually true for computations of any irrational real number, either an algebraic one such as $\sqrt{2}$, or a transcendental one such as e.

In general, methods used in numerical analysis are superrecursive algorithms that are only approximated by recursive algorithms. Such constructions have been used in the definition of constructive real numbers (Rice 1951; Mostowski 1957). Numerical methods form a class of superrecursive algorithms distinct from inductive Turing machines because they are working in a domain with a continuous topology. These algorithms are limit Turing machines (Burgin, 1992; 2001b) that work with a domain with a topology, for instance, the topology of the field of real numbers.

A biological population gives another example, where superrecursive algorithms are important as a tool of investigation. Simulation of their functioning essentially involves infinite processes though contemporary methods of modeling in biology and ecology ignore this fact. Consequently, utilization of superrecursive algorithms provides new powerful facilities for simulation of such processes.

The same is true for investigation, evaluation, and simulation of social processes (Burgin 1993) or for social, political, and/or ecological monitoring. Many optimization problems, which are solved with or without an aid of a computer, demand superrecursive representation (Burgin and Shmidskii 1996).

Kelly and Schulte (1997; 1997a) demonstrate necessity of superrecursive algorithms, such as inductive inference, and hypercomputation for modeling cognitive processes and solve problem related to learning and cognition in general.

Another methodological argument for superrecursive algorithms stems from comparison of informal and formal definitions of algorithms. For example, Prasse and Rittgen (1998) write that in computer science, an algorithm is understood as *a procedure which precisely describes in a finite form the transformation of given input data into defined output data where the result is fully determined by the input.* This also covers probabilistic and nondeterministic algorithms as long as the output is still uniquely determined. Then Prasse and Rittgen state that thus, an algorithm is a 'recipe' for computation which satisfies three conditions:

(1) only a finite number of steps (instructions) may be specified;
(2) the first and last instruction can be uniquely identified; the effect of every step is defined, and so is its successor;
(3) every instruction can be carried out in the given context.

However, the informal definition does not imply either of these conditions. As we will see later, it is possible that a result is fully determined by the input, finite description does not mean finite number of steps, the last instruction in the description of algorithm always exist, while the last instruction that is executed may not exist, and definitely, not every instruction even of a Turing machine can be carried out in the given context. For instance, in a deterministic case each context allows, at most, one instruction to be carried out.

Thus, we come to a conclusion that formalizations of informal definitions of algorithm can contain many superfluous conditions. Elimination of such conditions, as a rule, results in new classes of superrecursive algorithms.

Theoretical necessity for extending recursive algorithms to more powerful ones was considered by different authors.

Kalmar (1959) was, may be, the first of those who, aiming at disproving the plausibility of the Church–Turing thesis, pointed at some unnecessary limitations of recursive algorithms. To achieve his goal, Kalmar considers Kleene's instance (Kleene, 1936) of a function defined by the following formula:

$$\psi(x) = \mu_y(\varphi(x, y)) = 0 = \begin{cases} \text{the least natural number } y \text{ for which} \\ \varphi(x, y) = 0 \text{ if there is such a } y, \\ \\ 0 \quad \text{if there is no natural number } y \\ \text{such that } \varphi(x, y) = 0 \end{cases}$$

With an appropriate general recursive function φ of two arguments, this function is not general recursive. By the Church–Turing thesis, φ is not effectively calculable. Kalmar shows that this fact implies consequences that are incompatible with a general intuition.

Indeed, as Kalmar writes, on the one hand, for any natural number p for which a natural number y with $\varphi(p, y) = 0$ exists, an obvious method for the calculation of the least such y, that is, of $\psi(p)$, can be given: calculate in succession the values $\varphi(p, 0)$, $\varphi(p, 1)$, $\varphi(p, 2)$, ... (each of which can be calculated, on account of the general recursivity of φ, in a finite number of steps), until we obtain a natural number q for which we have $\varphi(p, y) = 0$ and take this q. On the other hand, for any natural number p for which we can prove, not in the frame of some fixed postulate system but by means of arbitrary — of course, correct — arguments that no natural number y with $\varphi(p, y) = 0$ exists, we have also a method to calculate the value $\psi(p)$ in a finite number of steps: prove that no natural number y with $\varphi(p, y) = 0$ exists, which requires in any case but a finite number of steps, and gives immediately the value $\psi(p) = 0$. Hence, supposing that ψ is not effectively calculable and applying the *tertium non datur* — which has been utilized already in the definition of the function ψ — we infer the existence of a natural number p for which, on the one hand, *there is no natural number y such that $\varphi(p, y) = 0$*, on the other hand, *this fact cannot be proved by any correct means* — a consequence of Church's thesis which seems very unplausible.

The problem here is with the principle *tertium non datur*, which is essentially nonconstructive in many cases. However, the argument of Kalmar shows that recursive functions (and algorithms) are too restrictive for encompassing the general notion of computability. In turn, this implicitly implies necessity for superrecursive algorithms.

Greenleaf (1995) gives an example of another function, the, so-called, busy beaver function, noncomputability of which looks unnatural. His suggestion to make this function computable leads to inductively constructive data and the idea of computation in the limit.

Recently Yao (2003) has demonstrated that there is fundamental tension between the Extended Church–Turing thesis (cf. Section 2.2) and the existence of numerous seemingly intractable computational problems arising from classical physics.

From a different perspective Minsky in his classic textbook on computer science (1967) considered the same problem. Explaining the termination or halting condition of recursive algorithms, he writes that our concern regarding algorithmic processes is not with the question of whether a process terminates with a correct answer, or even ever halts.

Minsky does not stop, however, with correctly noting that our concept of effectiveness needs to include nonterminating procedures. He further stipulates that the effectiveness of a procedure has to be construed in terms of "whether the next step is always clearly determined in advance." In other words, as Cleland (2001) stresses, the effectiveness of a procedure depends on how well its instructions specify the actions they prescribe. This amounts to an emphasis on the condition that the rules constituting an algorithm are unambiguous (definite), simple to follow (effective), and have simple finite description (are constructive), while excluding from the conditions for algorithm to be terminating. According to Cleland (2001), it is important to keep in mind that Minsky does not view himself as introducing a new proposal for understanding the concept of an algorithm as an effective procedure; he sees himself as explicating the received view among computer scientists. Nevertheless, his remarks lead directly to many models of superrecursive algorithms, such as infinite-time computations, recursion with real numbers, and, especially, limiting recursive functions and inductive Turing machines (cf., Section 4.2). The latter give results in a finite time even without termination their processes.

Thus, when Cleland (2001) writes that from a preanalytic, intuitive standpoint, a procedure is effective, or *is an algorithm*, if correctly followed, it reliably yields a definite outcome, this interpretation justifies inductive Turing machines as algorithms. Although, in many cases these machines give results for such data that no conventional Turing machine is able to process.

It is necessary to remark that the term *algorithm* is also used as a name for any sequence of efficient actions, which may or may not terminate. This essentially extends the concept of algorithm, including into its scope algorithmic schemes such as transrecursive operators studied in (Burgin and Borodyanskii, 1991; 1993; 1994).

4.2   Mathematical models of superrecursive algorithms and why we need them

Everything's got a moral, if only you can find it.

Lewis Carroll, 1832–1898

Although superrecursive algorithms existed and were utilized before the first mathematical models of superrecursive algorithms were created, they were not differentiated from the recursive algorithms and as a result, their utilization was inefficient. In turn, it has resulted in the loss of many possibilities of computers and of the Internet when it was created. Only mathematical models of superrecursive algorithms provide for efficient utilization of modern computers, networks, and their software.

4.2.1   Historical remarks

Truth is stranger than fiction.

A proverb

It is interesting that Turing was the first who went beyond the "Turing" computation that is bounded by the Church–Turing thesis. In his 1938 doctoral dissertation "Systems of logic based on ordinals," Turing introduced the concept of a Turing machine with an "oracle". This, work was subsequently published in 1939. Turing called these enhanced Turing machines "O-*machines*". Some think that "oracles" are logical black boxes for carrying out noncomputable tasks (Cleland, 2001). However, according to the definition, an oracle is a system (a device, black box, person, whatsoever) that contains knowledge about the values of some, computable or noncomputable, function $f(n)$. If a Turing machine T has an oracle for $f(n)$, than T has an operation of supplying the oracle with an arbitrary number n and receiving from the oracle the value $f(n)$. Having an oracle for an noncomputable function $f(n)$, the Turing machine T, as a rule, "computes" a noncomputable function. An oracle for a computable function does not extend boundaries of computability, can be useful for speeding up the computational process. Formal definitions and the legitimacy of Turing machines with an oracle are considered later.

Another approach that went beyond the Turing-Church Thesis was developed by Shannon (1941), who introduced the *differential analyzer*, a device that was able to perform continuous operations with real numbers, and namely, such as operation differentiation. As we will see in Section 4.2.8, even simple operations with real numbers allow an abstract automaton to compute any function on natural numbers.

However, mathematical community did not accept operations with real numbers as tractable because irrational numbers do not have finite numerical representations. Besides, as it is proved in Section 4.2, when algorithms are allowed to work with real numbers, such algorithms can compute any function $f: N \to N$. Consequently, the majority of mathematicians and later scientists have not considered Shannon's differential analyzer as an algorithmic device.

In 1957, Grzegorczyk introduced a number of equivalent definitions of computable real functions. In one of them, for instance, a real number is computable if it can be effectively represented by a rational sequence. A continuous function is computable if it preserves sequential computability and it is effectively uniformly continuous.

Three of Grzegorczyk's constructions have been extended and elaborated independently to superrecursive approaches to computation: the, so-called, "domain approach" (cf. Abramsky and Jung, 1994; Edalat, 1997), "type 2 theory of effectivity" or "type 2 recursion theory" (cf. Ko, 1991; Weihrauch, 2000), and the "polynomial approximation approach" (Pour-El and Richards, 1989; Pour-El, 1999).

In the approach of Pour-El and Richards (1989), classical mathematics is accepted. The point is to see how computable objects and operators look like in ordinary mathematics. As a result, they develop functional scheme, that is, a function is regarded as computable if it can be effectively approximated by rational coefficient polynomials with respect to the norm of a function space, such as a Banach space or a Frechet space.

In 1963, Scarpellini introduced the class M_1 of functions that are built with the help of five operations. The first three are elementary: substitutions, sums and products of functions. At the same time, two other operations are performed with real numbers. They are integrals over finite intervals and taking solutions of Fredholm integral equations of the second kind. The initial set of functions M_0 consists of functions $P(e^{i\alpha 1}, \ldots, e^{i\alpha s}) \, Q^{-1}(e^{i\alpha 1}, \ldots, e^{i\alpha s})$, where P and Q are polynomials with real coefficients, $Q \neq 0$. Starting from the Davis normal form for recursively enumerable predicates, Scarpellini demonstrated that every recursively enumerable predicate is represented by a function in M_1. As in the case with the differential analyzer, functions from M_1 went far beyond partial recursive functions and were not considered as algorithms, even superrecursive.

In 1967 Zadeh introduced fuzzy algorithms. Later Wiedermann (2000) built a mathematical model for fuzzy algorithms, which is called fuzzy Turing machine, and proved that fuzzy computations are more powerful than conventional computations.

In all these constructions suggested by Turing, Shannon, Zadeh and Scarpellini, the model of computation can go beyond the Church–Turing thesis and conventional Turing machines only because they have some very powerful and questionably constructive (or efficient) operations. This invalidates these algorithmic schemes as actual algorithms. Thus, the problem of building more powerful *algorithms* that had been so important to various practical, methodological, and theoretical issues of algorithms and computers remained unsolved. At the same time, it became clear that the conventional halting restriction on algorithms is not reasonable and it is not necessary for an algorithm to stop after getting a result.

So far, so good, but how do we determine a result when the algorithm does not stop functioning?

Mathematicians found an answer to this question. Moreover, a result of non-stopping computation may be defined in different ways. One of them is inductive computation, which is analyzed in Sections 4.2.2 and 4.3. It is the closest kind to the conventional recursive computation because it works with constructive objects and the result is obtained in a finite time.

In addition, they use the most direct way to determine a result when the algorithm does not stop functioning. Namely, a computational process gives as a result some output (word) w if there exists such a number n that after n steps of computation this process gives the same output w at each step. This output w is the result of the functioning of the algorithm.

Two other approaches are the theory of limit computations based on limit Turing machines (Burgin, 1991; 2001) and theory of infinite-time computations, which has four main branches: infinite-time computation realized by infinite-time Turing machines (Hamkins, 2000; Hamkins and Lewis, 2000; Welch, 2000; Davies, 2001), continuous time computation realized by general dynamical systems (da Costa and Doria, 1996; Bournez, 1999; Stewart, 1991), continuous time computation realized by hybrid systems (Gupta et al., 1999), and continuous time computation realized by special computing devices, such as the differential analyzer (Shannon, 1941; Moore, 1990; 1996).

From this perspective, the first genuine superrecursive algorithms were introduced in 1965. Two American mathematicians Mark Gold and Hillary Putnam brought in concepts of limiting recursive, limiting partial recursive functions, and trial-and-error predicates. Their papers were published in the same issue of the Journal of Symbolic Logic, although Gold had written about these ideas before. It is worth mentioning that constructions of Gold and Putnam were rooted in the ideas of nonstandard analysis originated by Abraham Robinson (1966) and inductive definition of sets (Spector, 1959). As a matter of fact, Gold was a student of Robinson.

In 1967, Gold produced a new version of limiting recursion, also called called inductive inference, and applied it to problems of learning. Ideas of Gold and Putnam gave birth to a direction of active research that is called inductive inference (cf., for example, Gasarch and Smith, 1997) and is a fruitful direction in machine learning and artificial intelligence (cf., for example, Angluin, 1992; Luna, 1996).

Schubert (1974), developing ideas of Gold, introduced the concept of iterated limiting recursion and applied it to the program minimization problem. The main constructions of Schubert are k-limiting recursive predicates and functionals. They are defined by repeated application of Gold's limit operator, which is formalized by a discrete limit operation in Section 2.4. Gold regarded limiting function identification (more generally, "black box" identification) as a model of inductive thought. Intuitively, iterated limiting identification might be regarded as higher-order inductive inference performed collectively by an ever-growing community of lower order inductive inference machines.

Limiting recursive, limiting partial recursive functions and methods of inductive inference are superrecursive algorithms and as such can solve such problems that are unsolvable by Turing machines. Although, being in a descriptive and not constructive form, they were not accepted as algorithms for a long time. Even introduction of a device that was able to compute such functions (Freyvald, 1974) did not change the situation. Consequently, this was an implicit period of the development of theory of superrecursive algorithms.

It is interesting that a direct embodiment of the trial-and-error predicates of Putnam (1965) into a model of automaton were done much later than for limiting recursive functions: in 1998 Hintikka and Mutanen built their trial-and-error machines.

Theory of recursive algorithms demonstrates that for the most programs the minimization problem is not recursively solvable and even not inductively solvable, using inductive inference. At the same time, as proved by Schubert, the program minimization problem is limiting recursively solvable for finite input-output lists and 2-limiting recursively solvable for infinite input-output lists, with weak assumptions about the measure of program size. In a way, 2-limiting recursion allows one to get such results that cannot be achieved by simple, first order inductive inference or limiting recursion.

It is necessary to remark that inductive Turing machines of order k, which are considered in Section 4.3, can realize k-limiting recursion, as well as to compute k-limiting recursive predicates and functions.

In 1983 Burgin, independently of inductive inference and limiting recursion introduced inductive Turing machines with structured memory that included all previous models of inductive computation and inference. The main goal was to develop algorithms for computation of some important functions, such as the Kolmogorov complexity (cf. Section 5.3). Only later Burgin discovered that inductive computations realized by inductive Turing machines on the lowest level of inductive hierarchy (cf. Section 4.3) were studied in the theory of inductive inference and limiting recursion.

From the beginning, inductive Turing machines were treated as algorithms. Thus, it was not by chance that their implications for the Church–Turing thesis and the famous Gödel incompleteness theorem were considered (Burgin, 1987) refuting the Thesis and changing understanding of the theorem. This was the beginning of the explicit stage for theory of superrecursive algorithms.

Approximately at the same time, Wolfram (1983; 1984) studied limiting behavior of cellular automata. He found three classes of cellular automaton behavior that were analogous to three classes of behavior found in the solutions to differential equations (continuous dynamical systems). Automata in these classes exhibit behavior analogous to limit points, limit cycles and chaotic attractors. For some differential equations, the solutions obtained with any initial conditions approach a fixed point at large intervals of time. This behavior is analogous to the first class cellular automaton behavior. In a second class of differential equations, the limiting solution at large intervals of time is a cycle in which the parameters vary periodically with time. These equations are analogous to the second class cellular automata. Finally, some differential equations have been found to exhibit complicated, apparently chaotic behavior depending in detail on their initial conditions. With the initial conditions specified by decimals, the solutions to these differential equations depend on progressively higher and higher order digits in the initial conditions. This phenomenon is analogous to the dependence of a particular site value on progressively more distant initial site values in the evolution of a cellular automaton from the third class. Later (1984) the fourth class was discovered, which was probably capable of universal computation and had undecidable properties of its infinite-time behavior.

Although Wolfram did not relate his results to hypercomputation and superrecursive algorithms, these patterns of limiting behavior can be used to define corresponding types of topological computations, some types of which are considered in (Burgin, 2001). For example, the first class of cellular automata, which exhibit behavior analogous to limit points, defines inductive computation.

In addition to this, the concept of computation has been, as a rule, considered in a much broader sense. Related at first only to arithmetical operations with numbers, this concept drastically extended its scope. Researchers can take as computation any process that transforms some information entities. The cause of such changes in understanding the term *computation* is the unimaginable advancement of computers. Being initially invented to speed-up arithmetical operations, computers expanded their function to an unbelievable degree. Naturally, computation has been perceived as actions that computers have been doing or have been able to do. Computers solve such a diversity of problems that almost any kind of information processing may be considered as computation. Now computation and communication, that is, information exchange, become combined in a common process. This is considered as an outstanding achievement of modern technology, although backward analysis shows that any computation, even the simplest, includes communication, while any communication includes computation, for example coding and decoding messages.

Algebraic or abstract categories form a universal tool in mathematics. So, it is not by chance that categories were used for building models of computations and automata for a long time beginning from sixtieth (Riguet, 1965). However, active research in this area was postponed until seventieth when the development of the theory of automata and computation in categories became very intensive. Many constructions from the traditional theory of automata and computation were adapted to categories and many properties of these constructions were obtained (cf. Adamék, 1974; 1975; Arbib and Manes, 1974; 1975; 1975a; Trnková, 1974; Budach and Hoehnke, 1975). For some time, categories as the context and constructive tools for the development of models for computation and different kind of automata have become very popular (cf., for example, (Rydeheard, 1988), (Rosolini, 1987) or (Adámek, J. and Trnková, 1990)).

Another approach in the theory of algorithms, automata and computation is based on general system theory and dynamical systems (Mesarovic and Takahara, 1975). Dynamical systems theory views a process as a sequence of state transformations that are controlled by equations of motion in the state space. In this framework, any automaton is a dynamical system and the main idea of the dynamical system approach to computation is to represent computational process by some dynamical system, which is usually more general than traditional models of algorithms. For example, the idea of quantum computation (Feynman, 1982; 1986) is to use quantum dynamical systems as such models and to realize these theoretical models by means of physical quantum systems.

Field computation, which was suggested by MacLennan (1990; 1999; 2001), is close to dynamical system computation. In this model, data are represented by fields, a mathematical model of which is a continuous function over a bounded domain, and information processing is performed by field transformations.

In addition, several theories of computation over abstract structures have been developed. See, for example, (Friedman, 1971; Moschovakis, 1974; Skordev, 1974; 1976; 1982; 1992; Tiuryn, 1979; Ivanov, 1986; Engeler, 1993). These general approaches both exploit and explore the logical properties of procedures. But, as it is written in (Blum et al., 1998), when applied to specific structures such as the real numbers, they do not yield the concrete mathematical results.

Tucker and Zukker (1988; 1992; 2002) suggested and developed another approach, building the theory of computable (partial recursive) functions and relations over many-sorted algebras. They obtained many mathematical models of algorithms and computation, including computations by finite-dimensional and general machines of Blum, Shub, and Smale (1989), as well as Type 2 Turing machines (Weihrauch, 2000) as special instances of computations over many-sorted topological partial algebras.

Definition 4.2.1. A many-sorted algebra $\mathbf{U}$ with a system of operations Σ and a collection of sorts I is a set U (the support of $\mathbf{U}$) which is a union $\cup \{A_i; i \in I\}$ of sets A_i and each operation is a mapping having the form $f: A_{i1} \times A_{i2} \times \ldots \times A_{ik} \to A_i$.

From the algebraic perspective, a deterministic finite automaton A is a many-sorted algebra, which has the support $\{\Sigma, Q, \Omega\}$, two binary operations $\delta: \Sigma \times Q \to Q$, and $\sigma: \Sigma \times Q \to \Omega$, and several unary operations $\sigma_0, \sigma_1, \ldots, \sigma_k$ on the set $Q: \sigma_0 = q_0, \sigma_1 = q_1, \ldots, \sigma_k = q_k$ with $q_1, \ldots, q_k \in F$. We will denote this algebra by $\mathrm{Al}(A)$.

In addition, another many-sorted algebra is corresponded to a deterministic finite automaton A. It has the support $\{\Sigma^*, Q, \Omega^*\}$, two binary operations $\delta^*: \Sigma^* \times Q \to Q$, and $\sigma^*: \Sigma^* \times Q \to \Omega^*$, and several unary operations $\sigma_0, \sigma_1, \ldots, \sigma_k$ on the set $Q: \sigma_0 = q_0, \sigma_1 = q_1, \ldots, \sigma_k = q_k$ with $q_1, \ldots, q_k \in F$. We will denote this algebra by $\mathrm{EAl}(A)$.

Other examples of many-sorted algebras are linear spaces, linear algebras, modules, and polygons (that is, sets on which monoids act), polyadic or Halmos algebras (Halmos, 1962), nonhomogeneous polyadic algebras (Leblanc, 1962), relational algebras, and state machines.

Many-sorted algebras were studied by several authors under different names. To mention only some of them, it is necessary to name algebras with a scheme of operators introduced by Higgins (1963; 1973), heterogeneous algebra in the sense of (Birkhoff and Lipson, 1970) and (Mathienssen, 1978), multibase universal algebras (Glushkov et al., 1974; Shaposhnikov, 1999; Karpunin and Shaposhnikov, 2000), and many-sorted algebras studied by Plotkin (1991). The term "heterogeneous algebras" is used more often than other related terms. Heterogeneous (multibase or many-sorted) algebras represent the next level of the development of algebra. Namely, in ordinary (or homogeneous) universal algebras operations are defined on a set, while in heterogeneous algebras operations are defined on a named set (Burgin, 1990). This makes possible to develop more adequate models for many processes and systems. For example, many-sorted algebras are extensively used for mathematical modeling information processing by computers. (Tucker and Zukker, 1988; 1992; 2002;

Gurevich, 1995). In addition, relational algebras are extensively used for modeling relational databases (Plotkin, 1991).

It is necessary to remark that transition from a set to a named set (Burgin, 1990; 1997) as a basic structure (a carrier for other structures) for the development of different fields in mathematics is peculiar not only for algebra but for other fields. Thus, in many cases fibers which are special cases of topological named sets, replace topological spaces in topology. Multivalued and multisorted logics are becoming more and more popular in logic. Manifolds are used instead of Euclidean spaces in mathematical analysis. Modern combinatorics is built not on sets but on multisets (Knuth, 1981), which are special cases of named sets (Burgin, 1990).

Such broad understanding of computation resulted in that many abstract processes described only theoretically has been called computations. Many new directions appeared. The most important of those that go beyond the Church–Turing thesis (listed in the chronological order) are: analogue, or continuous time, computation, fuzzy computation, inductive computation, computation with real numbers, interactive and concurrent computation, topological computation, and neural networks with real number parameters. The main advantage of inductive Turing machines is that they work with finite objects and obtain the results in a finite period of time. Other models either work with infinite objects (for example, Turing machines, neural networks, and topological algorithms working with real numbers) or need infinite time to produce results which are beyond the Church–Turing thesis (for example, infinite-time Turing machines and, persistent Turing machines).

Although by the discovery of superrecursive algorithms the Church–Turing thesis was refuted as an absolute and universal principle, it is reasonable to search in what conditions the Thesis is valid. In the same way, scientists look for conditions of validity for natural laws. Such validation of the Thesis has to go into three directions: test it for actual computers, verify it for theoretical computing schemes, and examine its consistency for axiomatic theories. For example, this Thesis may be proved in some axiomatic contexts and disproved in others. A relevant context for such studies of the Thesis might be provided by some theory of formal computations like the axiomatic theory of algorithms (Burgin, 1985) or theory of computations on abstract structures (Moschovakis, 1974). For example, choosing appropriate axioms, it is possible to prove the Church–Turing thesis in the theory of transrecursive operators (Burgin and Borodyanskii, 1991). One of these axioms states that the result of a computation is obtained after a finite sequence of steps and we know when it happens. Without this axiom, we come to the class of all inductive Turing machines with recursive memory. In some sense, these machines are such superrecursive algorithms that are the closest to the recursive algorithms. More exactly, it is possible to say that inductive Turing machines are the most powerful among those superrecursive algorithms, which lie one step from conventional models of algorithm, and are the most realistic among the most powerful superrecursive algorithms.

Thus, superrecursive algorithms appeared when the theory of recursive algorithms was actively developing and producing new and new models of algorithms. As a result, superrecursive algorithms were built by modification of recursive algorithms. Researchers used three principal ways to build superrecursive algorithms enhancing some computational scheme of recursive algorithms:

1. *Modification of hardware* or addition of new components to machines (abstract automata). In such a way, Turing machines with oracles, Turing machines with advice, inductive Turing machines with structured memory, and interactive Turing machines have been constructed.
2. *Modification of software* or computational rules. In such a way, Turing machines and neural networks operating with real numbers, Turing machines with integration operation, Turing machines with advice, inductive Turing machines with structured memory, and interactive Turing machines have been constructed.
3. *Modification of metarules for input and/or output* (by extending output possibilities, limiting recursive and limiting partial recursive functions, inductive Turing machines, and limit Turing machines have been constructed).

There are three main ways of extending capabilities of algorithms without changing the computational scheme:

1. *Extending the computational process* of machines (abstract automata). In such a way, infinite-time Turing machines and recursion on ordinal numbers have been built.
2. *Extending the input and/or output process* of machines (abstract automata). In such a way, finite automata, Turing machines and other recursive automata working with infinite words have been developed.
3. *Inclusion of interaction* in the process of machine functioning. In such a way, persistent Turing machines and other Turing machines that interact with the environment (Goldin and Wegner, 1988; Van Leeuwen and Wiedermann, 2001) have been constructed.

Table 4.1 reflects different directions of the theory of superrecursive algorithms related to three types of mathematical models: algorithmic, semialgorithmic, and abstract.

4.2.2   Limiting recursive functions and inductive computation

> *"Now, here, you see,*
> *it takes all the running you can do,*
> *to keep to the same place."*
> Lewis Carroll, 1832–1898

It is possible to separate three forms of inductive algorithms:

◇ Limiting recursive functions;
◇ Inductive centralized computations;
◇ Inductive net computations.

Directions in the superrecursive theory of computability

A. Algorithmic Models
1. **Limiting recursive functions** (Gold, 1965; Putnam, 1965)
2. **Inductive inference** (Gold, 1967; Blum, and Blum, 1975)
3. **Inductive centralized computations** (Freuvald, 1974; Burgin, 1983; Burgin, 1987; Burgin, 1999)
4. **Inductive net computations** (Garson and Franklin, 1989; Garson, Franklin, Bagget, Boyd, and Dickerson, 1992)
5. **Topological computations** (Burgin, 1983; 1992; Pour-El and Richards, 1989)
6. **Fuzzy computations** (Zadeh, 1969; Wiedermann, 2000)
7. **Interactive and concurrent computations** (Milner, 1973; 1980; Milne, 1985; Hoare, 1985; Wegner, 1995)

B. Semialgorithmic Models
1. **Computations with an oracle** (Turing, 1939)
2. **Infinite precision net computations** (Pollack, 1987; Hartley and Szu, 1987; Siegelman and Sontag, 1991)
3. **Recursion theory on real numbers** (Abramson, 1971; Blum, Shub, and Smale, 1989; Moore, 1996)
4. **Analogue computations** (Shannon, 1941; Scarpellini, 1963; Moore, 1996)
5. **Infinite time Turing machines** (Hamkins and Lewis, 2000; Welch, 2000)
6. **Trans-recursive operators** (Burgin and Borodyanskii, 1992)
7. **Dynamical system and field computations** (Mesarovic and Takahara, 1975; MacLennan, 1990; da Costa and Doria, 1996; Bournez, 1999)
8. **Fixed point models** (Scott, 1971; Manna and Vuillemin, 1972; Abramsky and Jung, 1994; Edalat, 1997; Edalat and Sünderhauf, 1998)

C. Abstract Models
1. **α-recursion or recursion on ordinals** (Takeuti, 1960; Machover, 1961; Levy, 1963)
2. **Generalized computability** (Moschovakis, 1969; Skordev, 1976; Tucker and Zukker, 1988; 2002)
3. **Recursion in higher types** (Kleene, 1959; Kleene, 1963)
4. **Induction in abstract structures** (Moschovakis, 1974)
5. **Computations in categories** (Riguet, 1965; Adamék, 1974; Arbib and Manes, 1974; Trnková, 1974)

Table 4.1. Directions and areas in the superrecursive theory of computability.

In the beginning, limiting recursive functions (Gold, 1965) and their version, trial-and-error predicates (Putnam, 1965) were introduced. We provide the following definitions as the first forms of inductive algorithms.

Definition 4.2.2. A partial function $f(x)$ is called *limiting recursive* if there is a total recursive function $g(x, n)$ such that

$$f(x) = \lim_{n \to \infty} g(x, n) \tag{1}$$

Definition 4.2.3. A partial function $f(x)$ is called *limiting primitive (partial) recursive* if $g(x, n)$ in (1) is primitive (partial) recursive.

Definition 4.2.4. A predicate $P(x_1, x_2, x_3, \dots, x_n)$ is called *trial-and-error predicate* if there is a general recursive function $g(x_1, x_2, x_3, \dots, x_n, y)$ such that

$$P(x_1, x_2, x_3, \dots, x_n) = \lim_{y \to \infty} g(x_1, x_2, x_3, \dots, x_n, y).$$

As in the theory of recursive functions and algorithms, a class of algorithms (functions) defines sets that are decidable (enumerable) with respect to this class.

Definition 4.2.5. A set X is called *limiting recursive (limiting recursively enumerable)* if its characteristic function $\chi_S(x)$ (its partial characteristic function $C_S(x)$) is limiting recursive.

Definition 4.2.6. A set X is called *limiting primitive (partial) recursive (limiting primitive (partial) recursively enumerable)* if its characteristic function $\chi_S(x)$ (its partial characteristic function $C_S(x)$) is limiting primitive (partial) recursive.

Following results that relate the limiting recursion to the arithmetical hierarchy levels (cf. Section 4.3.4) were proved by Gold (1965).

Theorem 4.2.1. *The following statements are equivalent:*

(1) S *is limiting recursive (LRS).*
(2) S *is limiting primitive recursive (LPRS.*
(3) S *is limiting partial recursive (LGRS).*
(4) S *belongs to the level* $\Sigma_2 \cap \Pi_2$ *of the arithmetical hierarchy.*

Theorem 4.2.2. *The following statements are equivalent:*

(1) S *is limiting recursively enumerable (LRES).*
(2) S *is limiting primitive recursively enumerable (LPRES).*
(3) S *belongs to the level* Σ_2 *of the arithmetical hierarchy.*

Theorem 4.2.3. *The class of limiting partial recursively enumerable sets (LPRES) is contained in* Σ_3, *contains* $\Sigma_2 \cup \Pi_2$, *and is not closed under complementation.*

The exact location of the class of limiting partial recursively enumerable sets in the arithmetical hierarchy was not defined by Gold.

Taking the classes **LRS, LPRS, LGRS, LRES, LPRES**, and **LGRES** of all limiting recursive, limiting primitive recursive, limiting partial recursive, limiting recursively enumerable, limiting primitive recursively enumerable, and limiting partial recursively enumerable sets, correspondingly, we have the following inclusions according to the properties of recursive, primitive recursive, and partial recursive functions (cf. Section 3.4):

$$\textbf{LPRS} \subseteq \textbf{LRS} \subseteq \textbf{LGRS} \text{ and } \textbf{LPRES} \subseteq \textbf{LRES} \subseteq \textbf{LGRES}.$$

However, some of these inclusions are not strict. Namely, we have

LPRS = **LRS** = **LGRS** (cf. Theorem 4.2.1) and **LPRES** = **LRES** (cf. Theorem 4.2.2).

Ausiello and Protasi (1990) and Ausiello, Protasi, and Angelaccio (1991) studied limiting polynomial approximation of complexity classes, obtaining a representation of polynomial space and subexponential time complexity classes as limiting polynomially decidable sets. The development of this direction brought researchers to the following concept (cf., for example, (Apsïtis et al., 1999)).

Definition 4.2.7. An *inductive inference machine* (IIM) M is a generating procedure that requests inputs from time to time. It produces some words as its partial output also from time to time. These words produced by the machine after receiving each portion of data are called *conjectures*. The final result of the inductive inference machine M is the word w to which the computational process of M converges.

Definition 4.2.8. If $u(n)$ denotes the conjecture produced by an inductive inference machine M after receiving the portion of input data with the number n, then the computational process of the inductive inference machine M *stabilizes* (or *converges*) to a word w if there exists a number $n_0 \in N$ such that $u(n)$ equals w for any $n > n_0$.

Here procedure means some algorithmic scheme such as Turing machine that is described by effective operations, but for which it is not specified how the result is obtained. This means that an inductive inference machine is a superrecursive algorithm.

In (Apsïtis et al., 1999), inductive inference machines are used for algorithmic generation of recursive real valued functions, or more exactly, algorithms that compute such functions.

Definition 4.2.8. An *inductive inference machine for real functions* (IIMrf) is a procedure that requests inputs from time to time and produces, as conjectures, algorithms that compute recursive real-valued functions from time to time.

In a similar way, general Turing machines are defined to model programs that never halt and to study Turing machines that do not need halt instructions (Schmidhuber, 2000). The definition is given for both finite and infinite words and is based on a concept of convergence.

If x is an infinite word or string, then $l(x) = \infty$. For any word w, $(w)_n$ denotes the prefix (initial part) of w that has length equal to n.

Definition 4.2.9. A *general Turing machine* has the structure of a conventional Turing machine with one output tape and at least two work tapes (sufficient to compute everything traditionally regarded as computable). After each step (with number n), some word w_n is written in the output tape. The output *stabilizes* and *converges* towards the finite or infinite word x if for each n satisfying $0 \le n \le l(x)$ there is a natural number t_n such that for all $t \ge t_n$, we have $(w_t)_n = (x)_n$ and $l(w_t) \le l(x)$.

All these approaches and constructions are synthesized in the concept of inductive Turing machine. At first, we consider simple inductive Turing machines. They realize *inductive computation of the first level*. The goal is to demonstrate one of their principal distinctions from Turing machines. Much more powerful inductive Turing machines with structured memory are treated in Section 4.3.

The simplest realistic inductive Turing machine has the same structure as a conventional Turing machine with three tapes and three heads: input, working, and output (cf. Figure 4.1).

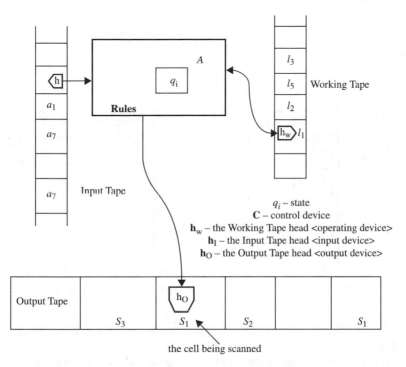

Figure 4.1. The structure of a simple inductive Turing machine.

This structure is much closer to the architecture of modern computer than the structure of a Turing machine with one tape. In a generalized form, the architecture of modern computer is presented in Figure 4.2.

Both inductive and ordinary Turing machines make similar steps of computations. Their differences lie in how they determine their outputs. We know (cf. Section 2.3) that a conventional Turing machine produces a result only when it halts. We assume that this result is a word written on the output tape. A simple inductive Turing machine also produces words as its results. In some cases, it stops at its final state and gives a result like a conventional Turing machine. The difference begins when the machine does not stop. An inductive Turing machine can give a result without

THE STRUCTURE OF A COMPUTER

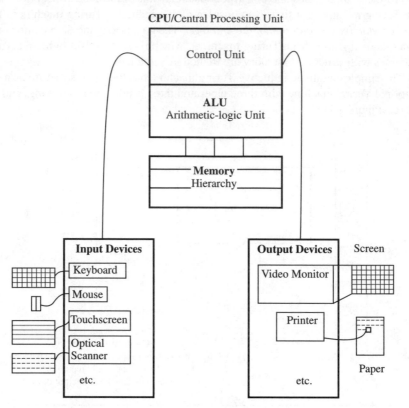

Figure 4.2. The generalized architecture of a modern computer.

stopping. To show this, we consider the output tape and assume that the result has to be written on it.

It is possible that in the sequence of computations, the word that is written on the output tape after some step is not changing although the machine continues to work. Then the last reached (unchanging) word, is taken as the result of this computation. Thus, an inductive Turing machine does not halt but it still produces a definite result after a finite number of computing operations. It explains the name "inductive." In induction we also proceed step by step checking if some statement P is true for an unlimited sequence of cases. When it is found that P is true for each case whatever number of cases is considered, we conclude that P is true for all cases.

While working without halting, an inductive Turing machine can occasionally change its output. However, people are not put off by machine that occasionally change outputs. They can be satisfied that the result just printed is good enough, even if another (possibly better) result may arrive in the future. And if you continue computing, it will eventually come. Another example is a program that outputs successively better approximations to a number. Once a few digits of accuracy are attained, the user can use the output generated even if the machine is not "done". All these properties essentially extend the possibilities and indicate uses of inductive Turing machines, as well as of related limit Turing machines (cf. Section 4.2.5).

Theorem 4.2.4. *For any Turing machine T, there is an inductive Turing machine M such that M computes the same function as T, that is, M and T are functionally equivalent.*

Indeed, if we have a Turing machine T, it is possible to assume that it has one tape and one head (Hopcroft et al., 2001). We use this tape in M and add to the memory of M a new output tape L_M. We use all states of T as the states of the control device A of M, adding to them duplicates of all final states of T, which become final states of M. We also use all rules of T as the rules of M, changing them so as to achieve the following functioning of M. For simplicity, we assume that T works with the alphabet $\{1, 0\}$.

At the beginning, M writes 0 in the first cell of the tape L_M. Then it begins to imitate T. When T makes a step, M makes the same step and then changes the symbol in the first cell of the tape L_M. If it was 0, it becomes 1. If it was 1, it becomes 0. When T stops in a state that is not final, then M also stops in the same state and consequently, both T an M give no result.

When T stops in a final state, then M rewrites the word from the tape of T into L_M and also stops in a final state. The construction of M implies that for any input both T an M give the same result.

Theorem 4.2.4 is proved.

Theorem 4.2.4 demonstrates that Turing machine is, in some sense, a particular case of inductive Turing machine.

To show that inductive Turing machines are more powerful than ordinary Turing machines, we need to find a problem that no ordinary Turing machine can solve and to explain how some inductive Turing machine solves this problem. To do this, let us take the problem, which was found one the first to be unsolvable and now is one of the most popular in the theory of algorithms. This is the halting problem for an arbitrary Turing machine with a given input. Turing proved that no Turing machine can solve this problem for all Turing machines (cf. Section 2.5).

However, there is an inductive Turing machine **M** that solves this problem. This machine **M** contains a universal Turing machine **U** as a subroutine. Given a word u and description D(**T**) of a Turing machine **T**, machine **M** uses machine **U** to simulate **T** with the input u. While **U** simulates **T**, machine **M** produces 0 on the output tape. If machine **U** stops, and this means that **T** halts being applied to u, machine **M** produces 1 on the output tape. According to the definition, the result of **M** is equal to 1 when **T** halts and the result of **M** is equal to 0 when **T** never halts. In such a way, **M** solves the halting problem.

So, even the simplest inductive Turing machines can be more powerful than conventional Turing machines. At the same time, the development of their structure allowed inductive Turing machines to achieve much higher computing power than the simplest inductive Turing machines described above. This is in contrast to such a property of a conventional Turing machine that by changing its structure, we cannot get greater computing power.

Now after we learned a little bit about limiting recursion and inductive Turing machines, it is possible to give examples of superrecursive algorithms that are right now actually used on our computers.

Example 4.2.1. Program performance checker **PCH**.

Let P be some program. Then **PCH** works in the following manner:

1. After each step of computer functioning, **PCH** checks whether the program P performs some instruction at this step.
2. **PCH** produces (prints, displays, and so on) 0 if the program P did not perform any instruction at that step, then **PCH** goes to its stage 1; otherwise **PCH** starts producing (printing, displaying, and so on) 1.

In such a way, **PCH** checks if the program P has been used. As it is possible that the program is never used, **PCH** is a superrecursive algorithm, which is defined for all inputs admissible by a software that includes the program P.

Example 4.2.2. Current user interference checker **UNCH**.

Let P and Q be some programs that always produce a result. Then **UNCH** works in the following manner:

1. **UNCH** checks whether there is an instruction from the user A;
2. If there is such an instruction, **UNCH** starts the program P and gets its result. Then **UNCH** displays the output of the program P and goes to step 1;
3. If there no instruction from A, **UNCH** starts the program Q and gets its result. Then **UNCH** display the output of the program Q and goes to step 1.

As any user can give only a finite number of instructions, **UNCH** is a superrecursive algorithm, which always displays the result of the program Q.

Example 4.2.3. Complete user interference checker **UCCH**.

Let P and Q be some programs that always produce a result. Then **UCCH** works in the following manner:

1. **UCCH** checks whether there has been an instruction to the computer from the user A;
2. If there is such an instruction, **UNCH** starts the program P and gets its result. Then **UNCH** displays the output of the program P and continues to do this without stopping.
3. If there no instruction from A, **UNCH** starts the program Q and gets its result. Then **UNCH** display the output of the program Q and goes to step 1.

As any user either gives some instruction to the computer or not, **UCCH** is a superrecursive algorithm.

We can see that these algorithms solve definite problems, are useful in many situations, especially, in those related to security issues, may be programmed for a contemporary computer, and then performed by this computer.

To conclude with the first level of inductive algorithms, we describe relations between different models of such algorithms.

Theorem 4.2.5. *The following classes of algorithms have the same computing power for finite words, that is, they can compute the same classes of functions on sets of finite words:*

(a) limiting partial recursive functions;

(b) general Turing machines;

(c) simple inductive Turing machines;

(d) inductive Turing machines of the first order (cf. Section 4.3);

(e) trial-and-error machines (Hintikka and Mutanen, 1998);

(f) inductive inference machines with a recursive generating procedure.

An extensive philosophical analysis of relations between recursive and inductive computations of the first level in the form of inductive inference is given by Kelly and Schulte (1997).

4.2.3 Infinite-time computation and supertasks

> *"One sees certain objections to it," she said.*
> *"But how did you work with the Metropolitan trains?*
> *None of them go infinitely fast, I believe."*
>
> Lewis Carroll, 1832–1898

Various philosophers and physicists have investigated such concept as supertask (cf., for example, Thomson, 1954 or Benacerraf, 1970).

Definition 4.2.10. A *supertask* is a task that demands infinite time for its execution.

Physicists analyzing supertasks have constructed theoretical models in which supertasks can apparently be carried out (Pitowsky, 1990; Xia, 1992; Hogarth, 1992; 1994; Earman, 1995; Earman and Norton, 1996). For example, Erman (1995) analyzes possibility of infinite-time computations. He writes that if the strongest form of cosmic censorship (that is the requirement of global hyperbolicity) fails, it is often possible for a point to contain in its past a future-directed half curve of infinite proper length. This suggests the possibility that a task that takes infinite time (for example, checking the truth of the Goldbach conjecture for all possible integers) could be carried out by some observer, and the result be observed by another within a finite amount of his lifetime. The ramifications of this property and the physicality of such processes are discussed, the conclusion being that it is probably not physically possible to take advantage of the global structure to carry out such supertasks.

Thus, no realistic solution has been suggested. However, like real numbers can be a good model for huge quantities of rational numbers (cf. Section 4.2.4), supertasks their solutions might give a reasonable approximation to modern superfast computers whose speed seems to be increasing without bound. From this perspective, a problem of developing a computational theory of supertasks that involves infinite-time computations and ideal infinitely fast computers becomes rather sensible. Following Thomson (1954), it is possible to consider a supertask as an infinite sequence of tasks.

A theoretical foundation on which to treat this problem is provided by the model for infinite-time computation suggested by Kidder and Hamkins in 1989 (cf. (Hamkins and Lewis, 2000)). The goal of the theory of infinite-time Turing machines is not so much finding what is physically possible to compute in a supertask as what is mathematically possible. Though the physicists may explain how it is possible or impossible to carry out a supertask in a finite amount of time, Hamkins and Lewis (2000), being focused on the algorithm, regard computations with infinitely many steps and extend the conventional Turing machine concept into transfinite ordinal time.

Thus, it is assumed that such machines perform infinitely many steps of computation, and can go on to more computation after that. The number of performed steps may be equal to the smallest infinite ordinal number ω, which corresponds to the set of all natural numbers, or any other ordinal number.

For simplicity, we consider only such infinite-time Turing machines that work with the alphabet $\{1, 0\}$. As a device, an infinite-time Turing machine has the usual Turing machine hardware, including the same uniform infinite tape that consists of separate cells and a head. This head moves mechanically back and forth reading and writing 0's and 1's on a tape according to a finite algorithm (set of rules) p. As in the case of inductive Turing machines, an infinite-time Turing machine has three separate tapes, one for input, one (or more) for scratch work, and one for output. What is new is the definition of the machine behavior at limit ordinal moments of time.

The infinite-time machine starts, like some forms of Turing machines, with the head resting in anticipation in the first cell, while the control device is at the beginning in a special state called the *start* state. The input is written on the input tape, and the scratch tape and the output tape contain only empty cells. At each step of computation, the head reads the cell values, reflects on its state, consults the program p of the machine about what should be done in an encountered situation and then carries out the instructions: it writes 0 or 1 in any cell from any of the tapes and moves left or right, while the control device switches to a new state accordingly. Thus, an infinite-time Turing machine performs each operation just like a conventional Turing machine. This procedure determines the machine configuration at stage $\alpha + 1$, given the configuration at stage α, for any ordinal number α.

A limit of these computations is taken to identify the configuration of the machine at stage ω and, more generally, at limit ordinal stages in the supertask computation. To set up such a limit ordinal configuration, the head is plucked from wherever it might have been racing towards, and placed in the first cell, while the control device is placed in a special distinguished *limit* state. Taking a limit of the cell values on the tape is done cell by cell according to the following rule: if the values appearing in a cell have stabilized, that is, if they are either eventually 0 or eventually 1 before the limit stage, then the cell preserves this value at the limit stage. Otherwise, in the case that the cell values have alternated from 0 to 1 and back again unboundedly often, the limit cell value is determined as 1. This is equivalent to making the limit cell value the *lim sup* of the cell values before the limit. Such a process completely describes the machine configuration at any limit ordinal stage β, and the machine can go on computing to $\beta + 1$, $\beta + 2$, and so on, eventually taking another limit at $\beta + \omega$ and so on through the ordinals. If at any stage the machine finds itself in the special *halt* state, then computation ceases, and whatever is written on the output tape becomes the official output. Otherwise, the infinite-time machine will compute endlessly as the ordinals fall one after another through the transfinite hourglass.

In this way, for every infinite-time Turing machine program p determines a partial function. If the machine with program p takes an input x, begins functioning, and, eventually, halts, there will be some output, which is denoted by $\text{Out}_p(x)$. The domain of Out_p is simply the collection of x which lead to a halting computation. The natural input for these machines is an *infinite* binary string, or sequence, x that belongs to the set 2^ω of all infinite sequences. Thus, the infinite-time computable functions are partial functions on Cantor space. It is possible to consider the elements of Cantor space 2^ω as real numbers, and think of the computable functions as functions on the real numbers. In particular, the set 2^ω is denoted by R, although the set of all real numbers is obtained by factorization of the set. For example, the sequences 1 and $0.99999\ldots$ denote the same real number. Adding extra input tapes, allows one to have functions of more than one argument.

Definition 4.2.11. A partial function $f(x)$ is *infinite-time computable* when there is an infinite-time Turing machine that computes f.

Definition 4.2.12. A set of real numbers A is *infinite-time decidable* when the characteristic function of A is infinite-time computable.

Definition 4.2.13. A set of real numbers A is *infinite-time semidecidable* when the function which gives the affirmative values, the function with the domain A and constant value 1, is infinite-time computable.

Thus, a set is semidecidable exactly when it is the domain of a computable function, since it is a simple matter to modify a program to change the output to the constant 1. It is possible to stratify the computable sets according to how long the computations goes. A set X is α-decidable for an ordinal number α. when the characteristic function of this set is computable by a machine that on any input takes fewer than α steps. Thus, restricting to the case of finite input and time, a function $f : 2^\omega \rightarrow 2^\omega$ is ω-computable exactly when it is computable in the Turing machine sense.

It is proved (Hamkins and Lewis, 2000) that for infinite-time computations, it is possible to consider only countable ordinals.

Theorem 4.2.6. *Every halting infinite-time computation demands only countable number of steps.*

This result correlates with the Skolem-Lowenheim theorem that states existence of a countable model for formal theories with countable languages (cf. Shoenfield, 1967).

The computability theory of the infinite-time Turing machines leads to a notion of computation on the real numbers, concepts of decidability and semidecidability for sets of real numbers, as well as individual real numbers, two kinds of jump-operator, and a notion of relative computability using oracles. This gives a rich degree structure on both the collection of real numbers and the collection of sets of real numbers.

In comparison with the theory of infinite-time Turing machines, Davies (2001) treats constructions of infinite-time machines as some idealizations. As we know, mathematicians build their systems without any appeal to reality. However, in many cases, these systems have appeared very useful for modeling real systems and processes. It is possible to apply similar arguments to computer science, developing some of its mathematical parts without attempts to find direct links to computers.

Consequently, as infinity now has exact mathematical models, we can use these models to build models for infinite computations. When an algorithm performs infinitely many steps, informally the computations begins with first step and goes to infinity. The most appropriate mathematical theory for such ideal processes and procedure is the theory of ordinal numbers, which was originated by Cantor (1895). In addition, it is reasonable to assume that there is an enumeration of the computation steps. It means that the steps correspond to the sequence of natural numbers $1, 2, 3, \ldots$. Thus, ω steps are made. At the same time, as Cantor (1895) found and contemporary set theory states, there are many kinds of infinities that may be used for step counting and ω is only the smallest one. Thus, we come to algorithms that are able to make α steps where α is some infinite ordinal number. This kind of algorithms, or more precisely, algorithmic schemes were studied in the theory of recursion under the name of α-recursion (Takeuti, 1960; Machover, 1961; Levy, 1963).

As we never find infinity in the physical world, α-recursion and recursion on higher types (Kleene, 1959; 1963; Platek, 1966; Harrington, 1973) is a far reaching idealization of recursion on natural numbers, which gives a relevant model for many types of computation. However, many properties of ordinal numbers are similar to properties of some non-Diophantine arithmetics (Burgin, 1997a). Non-Diophantine arithmetics reflect properties of physical, social, and economical systems. So, it seems interesting to develop a recursion theory in which numbers from such arithmetics are taken for counting steps of computation instead of natural numbers, which are elements of the conventional, or Diophantine, arithmetic.

4.2.4 Computations with real numbers and analogue computation

You never know what you can do till you try.

A proverb

There are three main approaches to computations with infinite precision real numbers: the theory of computing machines over the field of real numbers and other integral domains (Blum, Shub, and Smale, 1989; Blum, Cucker, Shub, and Smale, 1998; Boldi and Vigna, 1998; Weihrauch, 2000), the theory of real recursive functions (Moore, 1990; 1996; Campagnolo, Moore, and Costa, 2000), and the theory of infinite precision net computations (Pollack, 1987; Hartley and Szu, 1987; Siegelman and Sontag, 1994; 1995; Siegelman, 1995; 1999).

We begin with the first direction. It comprises two complementary approaches: in one approach a conventional Turing machine is taken and extended by allowing input and output tape to contain (infinite) representations of real numbers (Weihrauch, 2000), while in the other approach, real numbers are taken as basic atomic entities, on which exact computations and tests are permitted (Blum, Shub, and Smale, 1989).

The second approach is developed even in a more general situation. Let R be a field or an ordered integral domain, that is, a commutative ring without zero divisors with unit and R^n be the module of n-tuples $(x_1, x_2, x_3, \ldots, x_n)$ of elements x_i of R. When R is the space $\boldsymbol{R}$ of all real numbers, R^n is an n-dimensional vector space over $\boldsymbol{R}$.

Definition 4.2.14. A *finite-dimensional machine M over R* has the structure of a finite directed connected graph or flow-chart with which several spaces and mappings are associated. The graph of M has four types of nodes: *input, computation, branch,* and *output*. The unique input node has no incoming edges and only one outgoing edge. All other nodes have possibly several incoming edges. Computation nodes have only one outgoing edge, branch nodes exactly two, **Yes** and **No**, and output nodes none. In the case of a flow-chart, nodes are boxes, which may contain operations of four types: input, computation, branch, and output.

In addition, the machine M has three associated spaces: the *input space I_M, state space S_M*, and *output space O_M* of the form R^n, R^m, and R^l, respectively, where n, m, and l are positive *integers*. Associated with each node of M are maps of these spaces and next node assignments. These assignments correspond one next node to each input and computation node, two next nodes to each branch node, and absence of the next nodes to all output nodes.

1. Associated with the *input node* is a linear map $I: I_{M'} \to S_M$ and unique next node a_q.
2. Each *computation node q* has an associated *computation map*, a polynomial (or rational) map $g_q: S_M \to S_M$, and unique next node a_q. If R is a field, g_q, can be a rational map.
3. Each *branch node r* has an associated *branching function*, a nonzero polynomial function $h_r: S_M \to R$. The next node along the **Yes** outgoing edge, a_q^+, is associated with the condition $h_r(z) \geq 0$ and the next node along the **No** outgoing

edge, a_q^-, with $h_r(z) < 0$. In some cases, a_q^+ is associated with the condition $h_r(z) = 0$ and a_q^- with $h_r(z) \neq 0$.

4. Finally, each *output node* p has an associated linear map $O_p : S_M \to O_M$ and no next node.

The computational model of a finite-dimensional machine was first introduced by Blum, Shub, and Smale (1989). In some sense, it is a real number version of a random access machine (Shepherdson and Sturgis, 1963).

A polynomial (or rational) map $g : R^m \to R^m$ is given by m polynomials (or rational functions) $g_j : R^m \to R$, $j = 1, \ldots, m$. If g is a rational map associated with a computation node (in the case R is a field), we assume that each function g_j is given by a fixed pair of polynomials (p_j, q_j), where $g_j(x) = (p_j(x))/(q_j(x))$.

To a finite-dimensional machine M, one can attach a function from a subset of the input space to the output space. This function Φ_M is called the *input-output map*.

A computation of M is a sequence of mappings that are associated with the sequential nodes of M beginning with the input node. A computation halts if the last mapping is associated with the output node.

However, the construction of a finite-dimensional machine M over R has its limitations. That is why, to be able to build universal machines over R, a more general definition is introduced by Blum et al. (1997). In contrast to a finite-dimensional machine M, which works with finite-dimensional vectors of a fixed dimension, a general machine H over R can process vectors of arbitrary length.

Definition 4.2.15. A *general machine H over R* has the structure of a finite directed connected graph or flow-chart with which several spaces and mappings are associated. The graph of H has five types of nodes: *input, computation, branch, shift,* and *output*. The associated spaces are: the *input space* I_M, *state space* S_M, and *output space* O_M of the form R^∞, R_∞, and R^∞, respectively, where R^∞ is the disjoint union of all spaces R^m, $m = 1, 2, 3, \ldots$ and R_∞ consists of all elements of the form $x = (\ldots, x_{-2}, x_{-1}, x_0, x_1, x_2, \ldots)$, $x_i \in R$, $i \in Z$. Shift operations map R_∞ onto itself. There are two shift operations: one, the right shift, changes all indices i in the element x to $i + 1$, while another, the left shift, changes all indices i in the element x to $i - 1$, with $i \in Z$.

This allows the machine H to work with vectors of finite but unbounded length. Coordinates for such vectors are taken from a ring or field.

This approach is a particular case of the computable functions theory and relations over many-sorted algebras built by Tucker and Zukker (1988; 1992; 2002).

Meer (1992; 1993) considers such flow-chart machines in which linear or trigonometric functions are used for computation instead of polynomial and rational mappings.

Another construction of machines that work with real numbers is chosen by Wehrauch (2000), who introduces Type 2 Turing machine, from which he develops for them a Type 2 Theory of Effectivity (TTE). The structure of a Type 2 Turing machine looks very similar to the structure of an ordinary Turing machine.

Definition 4.2.16. A *deterministic Type 2 Turing machine* consists of the following parts:

1. The control device.
2. One or several heads.
3. A finite number of read-only one-way input tapes (possibly none), each containing at start an infinite word (string) in the alphabet $\{\bar{1}, 0, 1, \cdot\}$; these words represent real numbers from R in a signed binary form.
4. A finite number of write-only one-way output tapes (possibly none), on which the machine is supposed to write representations of elements of R;
5. One or several working tapes, which are initially empty.

This structure is very similar to the structure of an ordinary Turing machine. The only but essential difference with a standard Turing machine is the possibility of filling completely the input and output tapes.

Every Type 2 machine is a nonstopping machine that gives elements of R as its output. It computes a function, which is defined for all inputs which are binary representations of real numbers and for which all cells of the output tapes are eventually written.

Freund (1983) extended the notion of a Type 2 Turing machine by introducing *weak Type 2 Turing machine*. This machine is defined by making the output tape two-way and demanding that in every computation the content of an output tape cell is changed a finite number of times, so every finite prefix of the output tape eventually stabilizes. It is demonstrated (Freund, 1983) that weak Type 2 Turing machines are more powerful than standard Type 2 Turing machines.

In contrast to Type 2 Turing machines, finite-dimensional and general machines that work with real numbers, Moore (1996) develops computational theory for real numbers as an extension of the concept of recursive function considered in Section 2.4. His main idea is that integration is the closest continuous analog to primitive recursion and it is possible to build functions using this operation.

In analogy with the conventional recursive functions on natural numbers N (cf. Section 2.4), it is possible to define recursion for real numbers. At first, we define *R-recursive* functions of the form $f : R^n \to R$.

Operations that are used for building R-recursive functions are:

1. *Composition*: $h(x) = f(g(x))$
2. *Differential recursion* or simply *integration* for a function $h(x, y)$ with two arguments $x \in R^n$ and $y \in R$ is defined by the following rules:

$$h(x, 0) = f(x), \quad \partial_y h(x, y) = g(x, y, h(x, y)).$$

In other words, let $h = f$ at $y = 0$, and then let the derivative of h with respect to y depend on $h(y)$, y, and x. Then

$$h(x, y) = f(x) + \int_0^y g(x, u, h(x, u)) \, du$$

or, in a short form, $h = f + \int_g$.

3. *Minimization* or *μ-Recursion*: For a given function $f(x, y)$ with two variables, its minimization $h(x)$ is defined as $h(x) = \mu, yf(x, y) = \inf\{y; f(x, y) = 0\}$.

Here the operation *infinum* chooses the number y with smallest absolute value when such number is unique, and (by convention) the negative one if there are two or more numbers y satisfying the condition and having the same absolute value. If no such y exists, the value $h(x)$ is undefined.

Remark 4.2.1. Here composition is the same as in the case of recursive functions. Minimization is an extension of minimization for recursive functions (cf. Section 2.4), while integration defines a transition from local description of a mapping (in a form of derivatives) to the global unified description (in a form of a general law).

Definition 4.2.17. An **R-recursive function** $f: R^n \to R$ is a function that can be generated from the constant zero function $0(x) = 0$ and the constant one function $1(x) = 1$ using any finite number of compositions, differential recursions, and minimizations.

Vector-valued R-recursive functions $f: R^n \to R^m$ of one argument and $h(x, y)$ with two arguments $x \in R^n$ and $y \in R^k$ are determined by defining their components.

However, there is a problem with differential recursion. First, a solution to the differential equation need not be unique. For example, the boundary problem $f(0) = 0$, $df(x)/dx = 2f(x)$ is solved by $f(x) = ax^2$ for any real number a. Second, the function obtained by integration can diverge, such as $g(0) = 0$, $dg/dx = g^2 + 1$, for which $g = \tan x$ is only defined on the interval $(-\pi/2, \pi/2)$. To eliminate these inconsistencies, Moore (1996) demands that the function $h(x, y)$ is only defined where a finite and unique solution that includes the point $h(x, 0) = f(x)$ exists.

It is possible to develop theory of real number computation based on some modifications of Turing machines. The great flexibility of Turing machines allows us to define those machines that work with real numbers.

Let us take some class P of mappings of R into R or R^m into R^m.

Example 4.2.4. P consists of all polynomials with one variable.

Example 4.2.5. P consists of all polynomials with one variable that have degree not larger than n.

Example 4.2.6. P consists of all rational functions with one variable.

Example 4.2.7. P consists of arithmetical operations and all trigonometric functions.

Example 4.2.8. P consists of all arithmetical operations and integrals of the form $\int_0^y g(x, u, h(x, u)) \, du$.

Definition 4.2.18. A *real-number Turing machine* T over P is a Turing machine that in addition to its regular components has: 1) one or more tapes that can store real numbers, and 2) rules that have the form

$$q_h a_i \rightarrow q_k g(aj)D$$

where a_i is a denotation for an arbitrary real number (real vector) from a class $B \subseteq R$ ($B \subseteq R^m$) and $g(a_j)$ is a denotation of some mapping that belongs to P and D is a direction for moving the corresponding head of T.

Thus, a_i may be a real number, real variable, or real variable with a given range. Type 2 Turing machines are special cases of real-number Turing machines.

Utilization of different operations for data transformation makes real-number Turing machines closer to real computers. Besides, when we take high level programming languages, we see that their instructions may demand many operations in a computer. Those instructions may be even subprograms, that is, programs of any reasonable length. As a result, corresponding steps of computation become arbitrarily complex. This contradicts to the definitions of an algorithm that demand *simplicity* (at least, relative) of separate instructions. Most mathematical models of algorithm, such as Turing machines, finite automata or generative grammars, embody this demand in their structure. They are useful for theoretical research, but makes these models less relevant to computers and networks. Real number Turing machines allow one to overcome this shortcoming and provide reasonable flexibility for choosing operations for the set P to suit better specific problems related to computers and networks.

It is necessary to remark that all considered models of algorithms for real numbers can be naturally extended to algorithms models for complex numbers. As a result, we have an extended theory of real and complex number computation, which works with numbers with infinite precision.

However, from the physical perspective, infinite precision computations seem unrealistic. For example, any real measurement does not provide precise results but only approximate values. It is possible to speculate that in future we will be able to build infinite precision measuring devices, but this hypothesis contradicts some physical laws, in particular, the uncertainty principle.

So, we come to a conclusion that, in general, procedures that operate with infinite precision real numbers are only algorithmic schemes or ideal algorithms. However, such models can be useful and in some cases even necessary for relevant reflection, adequate modeling, and efficient study of real computations.

Blum et al. (1997) give their arguments for introducing algorithms that work with real numbers. They write:

"Recursive algorithms have given a firm foundation to computer science as a subject in its own right. Use of Turing machines yields a unifying concept of the algorithm well formalized. Thus this subject has been able to develop a complexity theory that permits discussion of lower bounds of all algorithms without ambiguity.

Scientific computation is the domain of computation that is based mainly on the equations of physics. For example, from the equations of fluid mechanics, scientific computation helps to better design of airplanes, or assists in weather prediction. The main tool here is numerical analysis.

The situation in numerical analysis is quite the opposite. Algorithms are primarily a means to solve practical problems. There is not even a formal definition of algorithm in the subject."

Thus, algorithms that work with real numbers give tentative models for numerical analysis. The reason is that the modern digital computer operates with a finite set of rational numbers, but they fill up a bounded set of real numbers (for example, between -1000 and 1000) sufficiently densely that viewing the computer as manipulating real numbers is a reasonable idealization, at least, in a number of contexts.

In addition, such model better describes graphical output of computer such as Mandelbrot or Julia sets with their apparently fractal boundaries. For a wide variety of scientific computations, the continuous mathematics that a machine is simulating is a sufficient vehicle for analyzing the operation of the machine itself.

Of course a great many issues such as round-off error must be dealt with. It is possible, as we discuss below, to treat effectively some of these issues if instead of classical mathematics with its completely exact objects and operations, such new fields as interval analysis or neoclassical analysis are utilized.

Here are some other examples when algorithms that work with real numbers might be useful.

There are situations when computers work with such real numbers such as $\sqrt{2}$, $\sqrt{3}$, $\sqrt{5}$, π, e, and some other irrational numbers. If computers multiply $\sqrt{2}$ by $\sqrt{2}$, they get 4 as the exact answer, or adding π to π, they also get exact answer 2π. So, in this operation, computers and even many calculators achieve infinite precision.

In other situations a fixed boundary for approximation of real numbers in computer operations does not exist. Thus, in place of imprecise approximations, theoretical computation models can better work with ideal real numbers having infinite precision. Real numbers become adequate models of their rational approximations.

If such processes of useful real number operations exist, we need to know their regularities and thus, model and study them. Real number algorithms provide an adequate technique for these studies.

The same is true for analogue computations. According to physical laws, they cannot be completely precise. When we measure something, we can achieve only finite precision. The problem is only for what processes, analogue or digital, this precision is better and/or the results are computed faster. The development of digital technology shows its advantages. However, theoretical studies of analogue computations can prompt new direction in the development and utilization of IPS.

At the same time, it is possible to develop models of real number computation, including possible imprecision. To solve this problem, we consider various existing and conceivable approaches to this problem. One way is to take approximate values of real numbers and perform approximate operations with them. Thus, Boldi and Vigna (1998) introduce a version of the finite-dimensional machine over $\boldsymbol{R}$, called a δ-uniform finite-dimensional machine over $\boldsymbol{R}$. In this machine, exact tests are not allowed, and the test for equality with 0 is replaced with a test for membership in an arbitrary ball around 0. The condition of δ-uniformity reduces the full power of the finite-dimensional machine nearly to the level of Turing machines.

Approximate computations studied by Boldi and Vigna (1998) involve new types of real functions: pointwise δ-approximable, uniform δ-approximable, and computable δ-approximable. These functions are closely related to Type 2 computable functions of Wehrauch (2000) and to weakly Type 2 computable functions of Freund (1983).

Theorem 4.2.7. (Meyssonnier et al., 2001). *The class of all Type 2 computable functions coincides with the class of all computably δ-approximable functions.*

Theorem 4.2.8. (Meyssonnier et al., 2001). *The class of all weakly Type 2 computable functions coincides with the class of all pointwise δ-approximable functions.*

Another way of developing imprecise computations with real numbers is to utilize those theories that implant imprecision in the classical real number arithmetic and calculus to reflect inexactness existing in the real world. One of these approaches is interval analysis (Moore, 1966; Alefeld and Herberger, 1983). In it computations and analytical operations are performed with number intervals instead of just numbers. These intervals reflect imprecision of measurement and inaccuracy of data processing.

The second approach utilizes constructions and technique of neoclassical analysis (Burgin, 1995; 2001e). Neoclassical analysis is aimed at making the power of classical calculus applicable to inexactness and imprecision of computation and measurement. In it, ordinary structures of analysis, that is, functions, sequences, series, and operators, are studied by means of fuzzy concepts: fuzzy limits, fuzzy continuity, and fuzzy derivatives. For example, continuous functions studied in the classical analysis become a part of the set of fuzzy continuous functions studied in neoclassical analysis.

In addition to better represent computational structures, neoclassical analysis makes it possible to extend and, in some cases, even to complete many basic results of the classical calculus. This provides us with deeper insight and a better understanding of the classical theory.

Real number computation is closely related to continuous-time computation. As Campagnolo, Moore, and Costa (2000) write, "while a few efforts have been made in the direction of studying computation by continuous-time dynamical systems (Moore, 1990; 1996, Orponen, 1997; 1997a; Siegelmann and Fishman, 1998, Bournez, 1999), no particular set of definitions has become widely accepted, and the various models do not seem to be equivalent to each other. Thus, analogue computation has not yet experienced the unification that digital computation did through Turing's work in 1936."

4.2.5   Topological computations

The proof of the pudding is in the eating.

A proverb

All constructions of models for topological computations are based on some *topology* on the set of all outputs (output words) of a given class of algorithms (cf. Burgin, 1992; 2001b). Here we do not go into detail and give only a definition of a limit Turing machine and some examples of its applications.

A *limit Turing machine T* has exactly the same structure as inductive Turing machine, either simple or with a structured memory (cf. Section 3.3). Although machines of both types work exactly in the same way, they differ in how the result of computation is defined.

Definition 4.2.19. (Burgin, 1992). If $a_1, a_2, a_3, \ldots$ is the sequence of words that are written in succession in the output tape when a *limit Turing machine T* is working after taking an input x, then a limit a of this sequence is a result of computations of the machine T with the input x.

When the limit Turing machine T is deterministic and the topology on the set of all outputs (output words) is Hausdorff (cf. Kelly, 1957), then the result of computation is unique. Otherwise, the machine T represents a multivalued algorithm (Burgin, 1984) with multiple results.

Inductive Turing machines are topological Turing machines that have discrete topology on the set of all outputs.

There are other topological algorithms related to a non-discrete topology. For instance, we consider algorithms that work with rational numbers. We know that it is possible to code each rational number by a finite word. However, the natural topology in this set is not discrete. For another instance, we take algorithms to work with p-adic numbers (cf. Van der Waerden, 1971), which also have a nondiscrete topology.

Different properties of limit Turing machines are considered by Burgin (1992; 2001b).

Discontinuous topologies (Burgin, 2001f) and fuzzy limits (Burgin, 2001e) allow us to extend topological algorithms, defining fuzzy limit Turing machines.

A fuzzy limit Turing machine T has exactly the same structure as limit Turing machine, either simple or with a structured memory (cf. Section 4.3). Although machines of both types work exactly in the same way, they differ in how the result of computation is defined.

Definition 4.2.20. If $a_1, a_2, a_3, \ldots$ is the sequence of words that are written in succession in the output tape when a *fuzzy limit Turing machine T* takes an input x, then a fuzzy limit a of this sequence is a result of computations by the machine T with the input x.

While limit Turing machines are only useful idealizations of computers as computers cannot go to the limit value and compute with absolute exactness, fuzzy limit Turing machines describe many kinds of numerical computations of modern computers as they demand only relative exactness (Burgin and Westman, 2000).

Other models of topological computations are developed in a series of papers (Hotz, Schieffer, and Vierke, 1995; Chadzelek and Hotz, 1997; 1999; Chadzelek, 1998). The most general model is called mathematical machine.

A *mathematical machine* is a structure $M = (A, K, K_a, K_e, K_z, \text{tr}, \text{in}, \text{out})$ where A is the alphabet of M, which is not confined to a finite set; K is the set of configurations or states of M; $K_a, K_e, K \subseteq K$ are the initial, final, and target configurations; $\text{tr} : K \to K$, $\text{in} : A^* \to K_a$, and $\text{out} : K_a \to A^*$ are the transition, input, and output functions of M with tr equal to the identity function on K_e.

A sequence $b = (k_i)_{i=0}^{\infty}$ of states is called a computation applied to k_i if $k_{i+1} = \text{tr}\,(k_i)$ for all $I = 0, 1, \ldots, n, \ldots$. A computation b is called *finite* if there is a number n such that b becomes stationary for all i larger than n. The result of this computation is equal to $\text{out}\,(k_{n+1})$. We call this result the output configuration.

It is possible to define a finite computation relative to an output, or some other type, of machine configurations. Namely, a computation b is called finite relative to an output configuration if there is a number n such that $\text{out}\,(k_i)$ becomes stationary for all i larger than n.

According to this definition, an inductive Turing machine gives a result if and only if its computation is finite relative to the output configuration.

However, when A^* is a metric space, the authors define results not only for finite computations but also for infinite convergent computations. Namely, if a subsequence $a = (h_i)_{i=0}^{\infty}$ of all target configurations in b is infinite, then the result of this computation is equal to $lim_{i \to \infty} \text{out}\,(h_i)$. Computations that have such results are called *infinite convergent computations*. Limit Turing machines (Burgin, 1991) are particular cases of mathematical machines with infinite convergent computations when target configurations are those configurations of a limit Turing machine at which they give a new result into the output register.

Taking mathematical machine as the basic construction and some ring R, which is later reduced to only two cases of the fields Q of rational and $\mathbf{R}$ of real numbers, the authors define some kind of register machines, called R-machines. These machines are equivalent to the model of Blum et al. (1998) in the case of finite computability.

One more model introduced in (Chadzelek and Hotz, 1999) is δ–Q-machine, which takes in real numbers as inputs and use infinite converging computations on more and more precise rational roundings of their inputs. It is demonstrated that infinite converging computations with rational numbers can simulate finite computations with real numbers. In addition, such problems as computability of solutions of ordinary differential equations (ODE) and decidability of stability problems for dynamical systems defined by ODEs.

It is possible to show that constructions and methods of Chadzelek and Hotz (1999) can be extended for computations with hypernumbers (Burgin, 2002d) by means of corresponding $\mathbf{H}$-machines where $\mathbf{H}$ is a set of hypernumbers.

Computational model of Boldi and Vigna (1998) is also based on topological constructions of pointwise δ-approximation, uniform δ-approximation, and computable δ-approximation.

A logical approach to a kind of topological computation, called convergent infinite computation, is developed by Li et al. (2001). Infinite computations in the sense of Li et al. (2001) correspond to non-terminating or very long time running processes. The main characteristics of convergent infinite computations are that they constantly access some huge sets of external data during the run time, and the infinite sequences of running states, which they go through, are convergent to some certain limits as the time goes to the infinity. Computations with finite words can be expressed by first order theories. This allows the authors to express infinite computations by sequences of first order theories and their limits in the sense of (Li, 1992). Logical varieties (Burgin, 1995a) provide a natural context to consider theories and their limits.

4.2.6 Computations with an oracle

> *It is the province of knowledge to speak and*
> *it is the province of wisdom to listen.*
>
> Oliver Wendell Holmes, 1809–1894

As we know, the first computation model that was more powerful than Turing machines was Turing machine with an oracle introduced by Turing in 1939. Let us consider computational power of this model. To do this, we need exact definitions.

Definition 4.2.21. An *function oracle* is a system (a device, black box, person, whatsoever) that contains knowledge about the values of some function $f(n)$.

Definition 4.2.22. A *Turing machine T with an oracle* for a function $f(n)$ is a conventional Turing machine with additional operation of access to the oracle. During this access, the machine T supplies the oracle with an arbitrary number n and receives from the oracle the value $f(n)$.

It does not mean that having an oracle for a noncomputable function $f(n)$, the Turing machine T always "computes" a noncomputable function. For example, the instructions of a Turing machine with an oracle perform the operation of changing any input word w into the output word ww. However, an oracle for a noncomputable function $f(n)$ allows Turing machine to "compute" a variety of noncomputable functions, which are called (cf. Rogers, 1987) functions computable relative to $f(n)$. In such a way, theory of relative computations is built.

Usually, oracles are considered not for functions but for sets (cf. Rogers, 1987).

Definition 4.2.23. A *set oracle* is a system (a device, black box, person, whatsoever) that contains knowledge about the membership in some set X.

Definition 4.2.24. A *Turing machine T with an oracle* for a set X is a conventional Turing machine with additional operation of access to the oracle. During this access, the machine T supplies the oracle with a description of an arbitrary element x (it may be a word, a number etc.) and receives from the oracle the value 1 when x belongs to X and 0 when x does not belong to X.

In general, the legitimacy of unrestricted Turing machines with an oracle as physical devices is rather questionable. Some think (cf., for example, Cleland, 2001) that it would be a mistake to conclude that oracles are of only theoretical interest because we could never confirm that a proposed physical candidate produced something genuinely noncomputable by conventional Turing machines.

It is possible to realize an oracle by using different constructions. One of them is called advice-taking Turing machine (Balcazar, Diaz, and Gabarro, 1988; Schöning, 1988).

Definition 4.2.25. An *advice function* $f(x)$ is some function, the values of which depend only on the length of x and are written on a special tape of a Turing machine.

Hence, advice functions provide external information to the machines, just as do oracles, but the information provided by an oracle may depend on the actual input, whereas the information provided by an advice function does not. Indeed, only the length of the input matters. Consequently, advice-taking Turing machines form a subclass of Turing machines with oracles.

Advice-taking Turing machines are important in complexity theory because definitions and results are often based on special Turing machines that can determine the result of an oracle "for free", that is, in constant time.

Definition 4.2.26. An *advice-taking Turing machine* is a Turing machine enhanced with the possibility to access the tape with the advice in constant time and read from it the value of its advice $f(x)$ also in constant time.

The fact that the value of the advice $f(x)$ can be determined in constant time (while $f(x)$ can be an intractable or even undecidable function) essentially increases the power and efficiency of an advice-taking Turing machine in comparison with a regular Turing machine. For example, an advice-taking Turing machine can calculate in polynomial time many functions that a regular Turing machine cannot (including some intractable ones).

Computations performed by Turing machines with oracles and Turing machines with advice are nonuniform because a different advice string may be defined for every different length of input.

Analyzing the situation with oracles for automata, we come to conclusion that in general, Turing machines and other automata with unbounded oracles are only algorithmic schemes or ideal algorithms. However, such models can be useful and in some cases are even necessary for adequate and efficient reflection of real situations. Here are some examples.

Example 4.2.9. One case is when an oracle is used to model more powerful computer. Few people need supercomputers. They work with ordinary PC or other similar devices. However, sometimes they might need much more computational power. So, their computer T accesses a supercomputer M that solves some problem for the user of T. As result, M plays the role of an oracle for T. Consequently, Turing machine with oracle is a good model for this case.

Example 4.2.10. Another case is when the computer T has access to some database. Then an oracle can model that database. In addition, different types of oracles can model: experimental devices that give to the computer T results of their experiments; utilization in calculations data from some physical system; or testing some hypothesis elaborated by a researcher by providing to T some hypothetical data.

Example 4.2.11. Oracles are useful in the theory of algorithms and computation, where there is a developed theory of relative algorithms and computation (cf., for example, Rogers, 1987). All this shows that Turing machines and other automata with oracles can be very useful if they are used adequately.

Example 4.2.12. Oracle can also simulate a situation in which computer receives data from some source about which it is unknown whether its output may be defined algorithmically or not.

4.2.7 Interactive computation and concurrency

> *There are two sides to every problem.*
>
> A proverb

Interaction takes the theory of automata, algorithms, and computation to the next level of information processing systems. Namely, we consider their processes as components of other automata, algorithms, and computational processes. The purpose of an interactive system is usually not to compute some final result but to react to or interact with the environment in which the system is placed and with other systems. An interactive automaton or algorithm also maintains a well-defined action-reaction behavior. Interactive systems are usually operating on unbounded strings of symbols. Consequently, it is natural to model them by using automata with infinite strings, for which inputs are not specified from the beginning and may depend on intermediate outputs and external sources.

This shows that there is a misconception that algorithms do not interact with their environment. For example, Wegner (1997) writes that interactive tasks, like driving home from work or reserving a seat on an airline, cannot be realized through algorithms.

This is not correct. Even accepting finite automata, which represent one of the simplest forms of algorithms, interact with their environment accepting or rejecting some input from this environment. Automata, as a rule, do not supply themselves with input. A more realistic model of a general finite automaton does not only accept data from the environment, but also gives some output into this environment. So, we have a two-way interaction.

Turing machines are more powerful and flexible in comparison with finite automata. So, there are Turing machines that in contrast to computers do not give output (cf., for example, (Hopcroft et al., 2001). Some theoreticians even study Turing machines that start computation on an empty tape, generate one by one all numbers from N, and compute all values of some function $f: N \to N$. However, these models are nonrealistic and can confuse some readers, bringing them to a conclusion that algorithm models have nothing to do with real computers and programs. Thus, if we do not forget about reality, the models we use always have input and output (cf. Section 1.2 and Section 2.2.3).

Another misconception is that some think that an algorithm receives the whole input before it starts functioning. This is a possible interpretation, but only one of different options for an algorithm. However, this restriction became so prevalent that many attach it to the very definition of algorithm. This is not correct because there are other acceptable interpretations of the input. For example, the standard definition of finite automaton (cf. Section 2.3) assumes that letters of the input word are given to the automaton one by one. The automaton reacts separately to each letter. It is difficult even to imagine that the automaton provides these input letters by itself. So, it is natural to suppose that these letters come from the environment individually and in groups.

Also, the word that is written in the input tape of a Turing machine may be written in parts. Completely fixed input is good, as a rule, only for computing functions. However, most algorithms are constructed for other purposes. Algorithms regulate traffic, optimize production, play chess and checkers, translate texts from one language to another and do many other useful things besides function computation.

When some people restrict algorithms to some of their applications, it does not mean that another application changes algorithm into something different. To do so is similar to saying that a wheel used in a car or truck for moving is a wheel, but a steering wheel in the same car or in a plane or a wheel in a clock is not a wheel. We may use additional words like "a sprocket," but this is still a wheel with some additions. In a similar way, it is more relevant to the social practice to call a variety of definite models of algorithms by the name "Turing machine." So, even an inductive Turing machine is, in some sense, an unconventional kind of Turing machines.

In theory, it is possible to interpret the process of automata functioning as construction of a mapping, but this is only *a posteriori* formalization. So, conventional finite automata and algorithms equivalent to Turing machines can be interpreted and used as interactive devices in a multitude of situations.

However, what is really true is that these interactive features of classical models of algorithms have not been studied explicitly and those researchers who do it now contribute to the general development of the theory of algorithms. Moreover, new models that explicitly include interaction are necessary to reflect better interactive peculiarities of contemporary computation and functioning of automatic devices in general.

Several such models have been suggested. Wegner (1997) defines interaction machines as Turing machines with input and output. This notion is formalized in (Goldin and Wegner, 1988) by persistent Turing machine.

Definition 4.2.27. A *persistent Turing machine* is a Turing machine with input and output tapes (cf. Figure 4.1) that receives from time to time some input (from the environment), but does not begin to process the next input until it finishes to work with the previous input coming to a final state.

Thus, in contrast to conventional Turing machines, a persistent Turing machine does not halt coming to a final state, but instead it begins to process a new portion of information. This is the *interactive persistent mode* of functioning of a Turing machine. It is possible to define and find the advantages of the interactive persistent mode of functioning for an inductive Turing machine in comparison with those of a persistent Turing machine.

Remark 4.2.2. Researchers considered separate input and output tapes not only for persistent Turing machine but also for different kinds of abstract automata (cf., for example, (Burgin, 1983; 1988; Stockmeyer, 1987)).

Other interactive modes of functioning include:

Interactive functioning with priorities when a task with a higher priority interrupts execution of the task with a lower priority, occupying all computational resources.

Interactive functioning with control points at which priorities of tasks are checked and changes in task execution are made. It is possible to have control points in time. For instance, each second a check is done and necessary changes are performed. It is possible to have control points in the process structure. For instance, at each branching operator a check is done and necessary changes are performed.

Interactive functioning with a queue when a task with a higher priority interrupts execution of a task with a lower priority and shifts the latter to the queue.

Interactive functioning with a background mode when a task with a higher priority is executed in the foreground mode, while tasks with lower priorities are executed in the background mode.

There are different models of interactive automata. Van Leeuwen and Wiedermann (2000) suggest a simple model of interactive computing that consists of a computing component C and environment E interacting with one another by using single streams of input and output signals. This pair satisfies the condition that C is guaranteed to give some meaningful output within finite time any moment of time after receiving a meaningful input from E and vice versa. It is assumed that C is a program with unbounded memory in which the memory contents is building up over time and never erased (unless the component explicitly does so).

Van Leeuwen and Wiedermann (2000a) describe another model of interactive computing as a *global Turing machine* (or *Internet machine*). It is finite. However, it consists of time-varying set of communicating site machines that can compute ad infinitum. It reflects the situation when a global computer may be evolving over time without limit. In this context, *site machines* are defined, by augmenting Turing machines with a communication facility.

Grid automaton (cf. Section 4.4) is a formal model for such interactive systems. Internet machines of van Leeuwen and Wiedermann are grid automata in which all nodes are Turing machines.

Another kind of interactive computational model computation on the Web is suggested by Abiteboul and Vianu (2000). It is called a Web machine.

A *Web machine* has the structure of a Turing machine with three linear tapes: input, output, and working tapes. The first two are infinite only in one direction, while the third is infinite in both directions. Initially, the input tape contains an infinite word that represents an encoding of the Web instance, while the working and output tapes are empty. The input tape head is positioned at the first cell. The moves are standard for a Turing machine, except that the output tape head can only move to the right (so nothing can be erased once it is written on the output tape). Thus, a Web machine works much like a Turing machine, but takes as input an infinite string and may produce an infinite answer. The Web is represented as an infinite database over the fixed relational schema. All infinite structures in this model are countable.

Based on the Web machine, Abiteboul and Vianu (2000) define the notions of computability and eventual computability of queries. The latter notion arises from the fact that infinite answers to queries are allowed. A query is *computable* if its answer is always finite and computable by a halting Web machine. A query is *eventually computable* if there is a Web machine, possibly nonterminating, which eventually outputs each object in the answer to the query.

The Web machine captures a very general form of computation on the Web. However, two particular modes of computation on the Web are prevalent in practice: browsing and searching. To represent these modes, Abiteboul and Vianu (2000) define two more machine models: a browser machine, which models browsing and browse/search machine, which models browsing and searching combined, allowing searching in the style of search engines.

Similar interactive computational model for querying the Web is suggested by Spielmann, Tyszkiewicz, and Van den Bussche, (2002). It is called a Web automaton. Formally, a Web automaton is a variant of a register automaton equipped with an additional communication component.

An interesting approach to understanding interaction is suggested by Prasse and Rittgen (1998). They write that interaction machines are not algorithms according to the conventional computability theory, which is concerned with computations of functions, but, instead, are components of interactive systems. These systems (protocols), and not the interaction machines themselves, define a computation model with respect to certain interaction behaviors and accepting rules.

Consequently, we come to a conclusion that it is correct to say that conventional models of algorithms focus only on the processor of information processing systems (cf. Figure 1.1), while a complete model must include input and output components with a description of their functioning. As result, we obtain an interactive model of algorithms and computation that give more complete algorithmic representation of various processes (Wegner, 1998). Grid automata (see Section 4.4) give the most general and comprehensive model for interactive computation.

4.2.8 Computational power of superrecursive algorithms and algorithmic schemes

> *What is now proved was once only imagined.*
>
> William Blake, 1757–1827

Let X be some alphabet and X^* be the set of all words in X.

Theorem 4.2.9. *For any function $f: X^* \to X^*$ ($f: N \to N$), there exists an advice-taking Turing machine M that computes f.*

Proof. To construct the advice-taking Turing machine T, we use two conventional Turing machines, $T_{X,1}$ and $T_{1,X}$, and an advice-taking Turing machine T_1. The machine $T_{X,1}$ enumerates words from the set X^*, establishing one-to-one correspondence between X^* and N and representing each number in the alphabet with one symbol 1. Let $g_{X,1}: X^* \to N$ be the function that is computed by the machine $T_{X,1}$. Then it has the inverse function $g_{X,1}^{-1}: N \to X^*$ and we take $h = (g_{X,1})f(g_{X,1}^{-1})$: $N \to N$ as the advice function of T_1. It is possible to do so because the function h satisfies the condition that all words of the same length have the same advice value. We also take a machine $T_{1,X}$ that computes the function $g_{X,1}^{-1}$.

Now we explain how the machine T works. When the input of T is equal to x, then x is transmitted to $T_{X,1}$, which transforms x into a word in the alphabet $\{1\}$. The result n of this transformation goes to the machine T_1, which assigns to n the value $h(n)$ using the oracle. Then this value is transmitted to $T_{1,X}$, which transforms $h(n)$ into a word z in the alphabet X. The output z of $T_{1,X}$ is taken as the output of the whole T. By definition of corresponding machines, we have

$$z = g_{X,1}^{-1}(h(n)) = g_{X,1}^{-1}\big(h(g_{X,1}(x))\big) = g_{X,1}^{-1}\big((g_{X,1})f(g_{X,1}^{-1})(g_{X,1}(x))\big) = f(x).$$

So, the value computed by T is equal to $f(x)$. As x is taken arbitrarily, the machine T computes the function f. The theorem is proved. $\square$

Corollary 4.2.1. *For any set $Z \subseteq X^*$, there is an advice-taking Turing machine that accepts Z.*

Corollary 4.2.2. *For any set $Z \subseteq X^*$, there is an advice-taking Turing machine that decides Z.*

As advice-taking Turing machines form a subclass of Turing machines with oracles, we have the following result.

Theorem 4.2.10. *For any function $f: X^* \to X^*$ ($f: N \to N$), there exists a Turing machine M with an oracle that computes f.*

Corollary 4.2.3. *For any set $Z \subseteq X^*$, there is a Turing machine with an oracle that accepts Z.*

Corollary 4.2.4. *For any set $Z \subseteq X^*$, there is a Turing machine with an oracle that decides Z.*

Theorem 4.2.11. *For any function $f: X^* \to X^*$ ($f: N \to N$), there exists a system of two randomly interacting Turing machines that computes f.*

Let us consider a class **K** of algorithms or abstract automata that work with real numbers. In addition, we assume that **K** is closed under subprogram compositions of its algorithms, that is, any algorithm or automaton from **K** may be included into another algorithm or automaton from **K** as a subprogram of the program or as a part of the automaton. We assume that the class **K** contains algorithms or automata that multiply real numbers by 10, subtract natural numbers from real numbers, count numbers of digits, and compare real numbers with integers.

Lemma 4.2.1. *Any function $f: N \to N$ ($f: X^* \to X^*$) is computable in **K**.*

Proof. At first, we codify an arbitrary function $f: N \to N$ as a real number a_f. It has the following structure: $f(1)$ zero's, then 1; then $f(2)$ zero's, then 1; then $f(3)$ zero's, then 1 and so on. For example, if $f(n)$ is a constant function identically equal to 3, then $a_f = 0.0001\,0001\,0001 \ldots$ is a periodic real number. We call a_f the number of $f(n)$.

To prove Lemma 4.2.1, we consider two cases in which an arbitrary real number can be:

(1) an arbitrary real number can be given as an input;
(2) an arbitrary real number can be implemented as an active parameter of the hardware.

An example of the latter is given by neuron networks that include neurons with real weights. Another example is an oracle of a Turing machine that has a real number as its parameter.

In the first case, the class **K** contains: an automaton or algorithm M that is called a multiplier and can accept and multiply any real number in the decimal representation by 10; an automaton or algorithm S that is called a subtractor and can subtract 1 from any real number; an automaton or algorithm C that is called a comparator and compares real numbers with natural numbers; an automaton or algorithm D that is called an adder, stores its results, and adds 1 to a given or previously stored natural number.

To build an automaton or algorithm A that computes the function $f(n)$, we combine the following into one scheme (cf. Figure 4.3): one multiplier M, one subtractor S, one comparator C, and two adders D_1 and D_2.

There are two options: (a) input of A contains two or more real numbers; (b) input of A consists of a single real number. At first, we consider the first situation, that is, the input of the automaton or algorithm A consists of two or more real numbers. To find the value $f(n)$ by means of A, we take a_f as the first input and n as the second input of A.

The number a_f goes to the automaton M, which multiplies a_f by 10. The result u_1 goes to the automaton C, which checks whether $u_1 > 1$ or not. When $u_1 > 1$, C sends 1 to D_1, which adds 1 to its initial stored value, which is equal to 0, and stores the result. Then the automaton C checks whether the result of D_1 is equal to

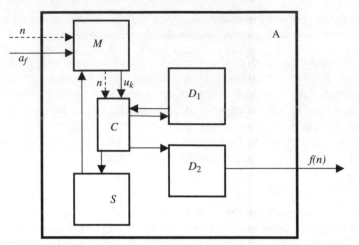

Figure 4.3. The structure of the machine A.

n or not. If it is equal, then the automaton A gives 0 as its final output and stops. The definition of a_f implies that $f(1) = 0$ and A gives this result. If the result of D_1 is not equal to n, then C sends u_1 to S and 0 to M and to D_2. At the same time, the automaton S subtracts 1 from the number u_1 and M multiplies the difference by 10, getting a number u_2; while the automaton D_2 makes its own stored value equal to 0.

When $u_1 < 1$, C sends 0 to M and 1 to D_2. The automaton D_2 adds 1 to the initial value, which is equal to 0, and stores the new result. The automaton M multiplies u_1 by 10, getting u_2.

In the case when $u_1 > 1$, the result u_2 goes to the automaton C, which checks whether $u_2 > 1$ or not. When $u_2 > 1$, C sends 1 to D_1, which adds 1 to its stored value and stores this result. Then the automaton C checks whether the result of D_1 is equal to n or not. If it is equal to n, then C sends the special symbol of output to D_2 and the automaton A gives as its output the stored value of D_2, which is equal to 0 at this moment, and stops. The definition of a_f implies that $f(2) = 0$ and A gives this result. If the result of D_1 is not equal to n, then C sends u_2 to M and 1 to D_2. The automaton M multiplies u_2 by 10, getting u_3. The automaton D_2 adds 1 to its initial stored value, which is equal to 0, and stores the new result.

In the case when $u_1 < 1$, the result u_2 goes to the automaton C, which checks whether $u_2 > 1$ or not. When $u_2 > 1$, C sends u_2 to M and 1 to D_2. At the same time, the automaton D_2 adds 1 to its stored value and stores this result; while the automaton M multiplies u_2 by 10, getting u_3 and sending it to C for comparison with 1. This cycle is repeated until C finds that $u_k > 1$ for some k.

When we have $u_k > 1$, C sends 1 to D_1, which adds 1 to its stored value and stores the result. Then the automaton C checks whether the result of D_1 is equal to n or not. If it is equal, then C sends the special symbol of output to D_2 and A gives as its output the stored value of D_2 and stops. The definition of a_f implies that $f(1) = k - 1$ and the automaton A gives this result. If the result of D_1 is not equal to n, then C sends u_k to S and 0 to M and to D_2. At the same time, the automaton S subtracts 1 from number u_k and M multiplies the difference by 10, getting u_{k+1}; while the automaton D_2 makes its stored value equal to 0.

If the stored value of D_1 is not equal to n, we continue this process until the stored value of D_1 becomes equal to n. Then the automaton A gives the result of D_2 as its final output and stops. The definition of a_f implies that A computes the value $f(n)$. By induction, we prove this for all numbers n, that is, that A computes the value $f(n)$.

The situation when input of the automaton A consists of a single number is reduced to the previous case by sending n and a_f one after another and storing the first input.

Now let us consider the case when an arbitrary real number can be implemented as an active parameter of the hardware of the automata from the class **K**. Then to get an automaton A that computes the function $f(n)$, we take the number a_f as a parameter of hardware that makes it possible to use this parameter in computations and build an automaton presented in Figure 4.3. We may assume that a_f is a parameter of the multiplier M. Then we give n as the input to A and repeat the whole procedure described above. The same arguments show that that A computes the value $f(n)$.

Lemma 4.2.1 is proved.   □

Lemma 4.2.1 implies corresponding results for several models of real number computations.

Theorem 4.2.12. *For any function $f : N \to N$ ($f : X^* \to X^*$), there exists a finite-dimensional machine M over the set R of all real numbers that computes f.*

Corollary 4.2.5. *For any set $Z \subseteq N$ ($Z \subseteq X^*$), there exists a finite-dimensional machine over R that accepts Z.*

Corollary 4.2.6. *For any set $Z \subseteq N$ ($Z \subseteq X^*$), there exists a finite-dimensional machine over R that decides Z.*

Theorem 4.2.13. *For any function $f : N \to N$ ($f : X^* \to X^*$), there exists a real-number Turing machine M that computes f.*

Corollary 4.2.7. *For any set $Z \subseteq N$ ($Z \subseteq X^*$), there exists a real-number Turing machine that accepts Z.*

Corollary 4.2.8. *For any set $Z \subseteq N$ ($Z \subseteq X^*$), there exists a real-number Turing machine that decides Z.*

It is possible to obtain similar results for Type 2 Turing machines.

Theorem 4.2.14. *For any function $f : N \to N$ ($f : X^* \to X^*$), there exists a real number neural network that computes f.*

Corollary 4.2.9. *For any set $Z \subseteq N$ ($Z \subseteq X^*$), there exists a real number neural network that accepts Z.*

Corollary 4.2.10. *For any set* $Z \subseteq N$ $(Z \subseteq X^*)$*, there exists a real number neural network that decides* Z*.*

Theorem 4.2.15. *For any function* $f : N \to N$*, there exists a real number recursive function in the sense of Moore (1996) that is equal to* f*.*

Corollary 4.2.11. *For any set* $Z \subseteq N$*, there exists a real-number Turing machine that accepts* Z*.*

Corollary 4.2.12. *For any set* $Z \subseteq N$*, there exists a real-number Turing machine that decides* Z*.*

These results show that models working with real numbers are omnipotent for computing functions on finite words, that is, they can compute any such function. In some cases when we want to extend the scope of computability, this feature is good. In other cases, it does not allow one to discern computable functions from noncomputable and decidable sets from undecidable in the realm of finite words.

However, for infinite words and for objects that are represented by infinite words (such as real and complex numbers) these models provide essential opportunities to study algorithmic and computational properties of such domains. Existence of different algorithmic and computational models that work with inherently infinite objects reflects the diversity of situations when these models may be useful.

4.3 Emerging computation, inductive Turing machines, and their computational power

> *"Oh, you're sure to do that," said the Cat,*
> *"if you only walk long enough."*
>
> Lewis Carroll, 1832–1898

Recursive, inductive, and infinite-time computational processes represent three forms of computations:

◇ Recursive computations are *accomplished processes* as they terminate giving the result.
◇ Inductive computations are *emerging processes* as they produce the result without stopping, that is, the final result emerges through a sequence of intermediate results.
◇ Infinite-time computations are *potential processes* as it is possible to have the result they produce only after an infinite number of steps.

Our main interest here is in emerging processes. In Section 4.2.2, we got acquainted with the lowest level of inductive computations: limiting recursion, inductive inference, general Turing machines, and simple inductive Turing machines. The most powerful model for inductive computations is an advanced inductive Turing machine, as covered in this section.

There are different types and kinds of advanced inductive Turing machines: with a structured memory, structured program or rules (or control device), and structured head (operating device) (Burgin, 2001a). The main emphasis here is made on inductive Turing machines with a structured memory.

4.3.1 The structure of inductive Turing machine with a structured memory

– but there's one great advantage in it,
that one's memory works both ways.

Lewis Carroll, 1832–1898

As we already know, the structure of inductive Turing machine in its simplest form is similar to the structure of Turing machine. However, such simple forms of inductive Turing machines do not allow us to compute much more than conventional Turing machines. That is why we develop here an advanced form that is called inductive Turing machine with a structured memory. In what follows, the term "inductive Turing machine" always means "inductive Turing machine with a structured memory". Simple inductive Turing machines are considered as inductive Turing machine with a rather primitive structured memory, consisting of three linear tapes.

Similar to Turing machine, any kind of inductive Turing machine is an abstract automaton, which is similar in many aspects to a modern computer. We first describe the structure of an inductive Turing machine with a structured memory and then we explain how it functions.

The structure of any inductive Turing machine, as an abstract automaton, consists of three components: *hardware*, *software*, and *infware*. We begin with the infware, that is, with a description and specification of information processed by an inductive Turing machine. Computer infware consists of information, or more exactly, data processed by the computer. An inductive Turing machine M is an abstract automaton, which works with symbolic information in a form of words of formal languages. Consequently, formal languages with which M works constitute its infware. Usually, these languages are divided into three categories: input, output, and working language(s). In contrast to the languages of everyday life (such as English, German or French), inductive Turing machines use formal languages.

A *formal language* **L** consists of three parts: the alphabet A of **L**, which is a finite set of symbols; the set A^* of all words in A, which are finite strings of symbols; and the subset L of the set A^*. Elements from L are called the words of the language **L**. The set L is often represented by generating rules R_G, that is, the rules that build words from L, or by selection rules R_S, that is, the rules that separate words that belong to L from all other words in A^*.

The language **L** of an inductive Turing machine consists of three parts $\mathbf{L} = (\mathbf{L_I}, \mathbf{L_W}, \mathbf{L_O})$ where $\mathbf{L_I}$ is the *input* language, $\mathbf{L_W}$ is the *working* language, and $\mathbf{L_O}$ is the *output* language of M. Each of them has the following structure $\mathbf{L_X} = (A_X, R_X, L_X)$ where A_X is the *alphabet*, R_X is the set of *generating rules*, and L_X is the *set of all words* of the language $\mathbf{L_X}$ where X is one of the symbols I, O or W.

Usually the generating rules for formal languages as a whole consist of one operation, which is called concatenation and combines two words into one. For example, if x and y are words, then xy is the concatenation of x and y. Taking the alphabet $A_X = \{1, 0\}$ with two words $x = 1001$ and $y = 001$ in this alphabet, we have 1001001 as the result of concatenation. The set A^* of all finite strings in the alphabet A is also a formal language; it includes the empty word ε that contains no symbols. Because a formal language is an arbitrary subset of A^*, it is possible to consider the languages of an inductive Turing machine M as one language $\mathbf{L}(M)$, which consists of three parts: $\mathbf{L_I}$, $\mathbf{L_W}$, and $\mathbf{L_O}$.

Now let us look at the *hardware* or *device* $\mathbf{D}$ of the inductive Turing machine M with a structured memory. What is hardware of a computer? It consists of all devices (the processor, system of memory, display, keyboard, etc.) that constitute the computer. In a similar way, the inductive Turing machine M has three abstract devices: a *control device* A, which is a finite automaton and controls performance of the machine M; a *processor* or *operating device* H, which corresponds to one or several *heads* of a conventional Turing machine; and the *memory* E, which corresponds to the *tape* or tapes of a conventional Turing machine. These devices are presented in Figure 4.1, which gives the structure of the simplest kind of an inductive Turing machine, in which the memory consists of three linear tapes, and the operating device consists of three heads, each of which is the same as the head of a Turing machine and works with the corresponding tape.

The control device A has the *state structure* or *configuration* $S = (q_0, Q, F)$ where Q is *the set of states* or *the state space* of A and of M, q_0 is an element from Q that is called the *start* or *initial state*, and F is a subset of Q that is called the set of *final* (in some cases, *accepting*) states of M. It is possible to consider a system Q_0 of start symbols from Q, but this does not change the computing power of an inductive Turing machine. The automaton A regulates the state of the whole machine M, the processing of information by H, and the storage of information in the memory E.

The memory E is divided into different but, as a rule, uniform cells. It is structured by a system of relations that provide connections or ties between cells. Each cell can contain a symbol from an alphabet of the languages of the inductive Turing machine M or it can be empty. Formally, $E = (P, W, K)$ where P is the set of all cells from E, W is the set of connection types, and $K \subseteq P \times P$ is the binary relation on P that provides connections between cells. In such a way, K structures the memory E. Each of the sets P and K is also structured. The set P is enumerated, that is, a one-to-one mapping ν of P into the set N of all natural numbers is given. A type is assigned to each connection from K by the mapping $\tau \colon K \to W$.

In a general case, cells from the set P also may be of different types. This stratification is represented by the mapping $\iota \colon P \to V$ where V is the set of cell types. Different types of cells may be used for storing different kinds of symbols. For example, binary cells, which have type B, store bits of information represented by symbols 1 and 0. Byte cells, which have type BT, store bytes of information represented by strings of eight binary digits. Symbol cells, which have type SB, store symbols of the alphabet(s) of the machine M. Conventional cells in Turing machines have this type. Natural number cells, which have type NN, are used in random access machines (Aho et al., 1976). Cells in the memory of quantum computers, which have type QB, store q-bits or quantum bits. When different kind of devices are combined into one, this new device has several types of memory cells. This is just the case of grid automata, which are considered in Section 4.4. In addition, different types of cells facilitate modeling of the brain with its sophisticated structure of neurons by inductive Turing machines.

Likewise, the set of cells P is divided into three disjoint parts P_I, P_W, and P_O, where P_I consists of the *input* registers, P_W is the *working* memory, and P_O consists of the *output* registers of M. Correspondingly, K is divided into three parts K_I, K_W, and K_O which define connections between the cells from P_I, P_W, and P_O. Usually, input registers are used only for reading, while output registers are used only for writing. For simplicity, we consider P_I as one register and P_O as one register, which are, as a rule, one-dimensional tapes. Besides, it is possible to consider only such inductive Turing machines that have the read-only input register or tape because such machines are functionally equivalent to the general case of inductive Turing machines.

At the same time, to model a modern computer with its advanced hierarchical memory, the set P has to be subdivided into more than three components.

Each cell from a linear two-sided tape has two neighbors left and right. The first cell in a one-sided tape has only one neighbor. The structure of a linear tape, which is standard for Turing machines, is realized by the relation **Lin** with connections of two types: R and L. Each cell with the number i is connected to the cell with the number $i + 1$ with the connection R, and each cell with the number $i + 1$ is connected by the connection L to the cell with the number i ($i = 1, 2, 3, \ldots$). To get a two-sided linear tape, we re-enumerate the corresponding part of P by integer numbers and use the similar connections. To get a two-dimensional tape, we can use ties of four types $W = \{R, L, U \text{ and } D\}$. Enumeration of cells by natural numbers is transformed to labeling the cells by pairs of integer numbers. Then each cell (i, j) is connected to four of its neighbors: to $(i + 1, j)$ by the connection R, to $(i - 1, j)$ by the connection L, to $(i, j + 1)$ by the connection U, and to $(i, j - 1)$ by the connection D.

It is possible to realize an arbitrary structured memory of an inductive Turing machine, using only one linear one-sided tape L. To do this, the cells of L are enumerated in the natural order from the first one to infinity. Then L is decomposed into three parts according to the parts P_I, P_W, and P_O of the structured memory. After this nonlinear connections between cells are installed according to the relation K and the mapping $\tau : K \rightarrow W$. When an inductive Turing machine with this memory works, the head (processor) is not moving to the right or to the left cell from a given cell, but uses the installed nonlinear connections.

Such realization of the structured memory allows us to consider an inductive Turing machine with a structured memory as an inductive Turing machine with conventional tapes in which additional connections are established. This approach has many advantages. One of advantages is that inductive Turing machines with a structured memory can be treated as multitape automata that have additional structure on their tapes. Then it is conceivable to study different ways to construct this structure.

In addition, this representation of memory allows us to consider any configuration in the structured memory E as a word written on this unstructured tape.

In a similar way, it is feasible to build, study and utilize Turing machines with a structured memory. They have almost the same hardware (they do not necessarily need the output tape, but always have final states) and the same software as inductive Turing machines with a structured memory. But in contrast to inductive Turing machines, Turing machines have to stop to produce a computation result.

Some authors consider a random-access deterministic Turing machine (Tourlakis, 2000). It adds random access machines (RAM) to the family of Turing machines and makes possible to move the head from any cell to any other cell. A relevant structured memory allows a Turing machine to perform even more complex access activities.

If we look at other devices of the inductive Turing machine M, we can see that the processor H performs information processing in M. However, in comparison to computers, this operational device performs very simple operations. When H consists of one unit, it can change a symbol in the cell that is observed by H, and go from this cell to another using a connection from K. This is exactly what the head of a Turing machine does.

It is possible that the processor H consists of several processing units similar to heads of a multihead Turing machine. This allows in a natural way one to model various real and abstract computing systems by inductive Turing machines. Examples of such systems are: multiprocessor computers; Turing machines with several tapes; networks, grids and clusters of computers; cellular automata; neural networks; and systolic arrays. However, such representation of information processing systems is not always efficient. This is why, other models of IPS have been constructed and utilized.

Connections between the control device A and the processor H may be organized in different way:

1) The processor H may be rigidly connected to A. In this case, the memory E or its part moves when it is necessary to observe the next cell (compare to Figure 2.2). This is similar to the way we use a floppy disk or CD.
2) The connection between A and H may be flexible, allowing H or its parts to move from one cell to another under the control of A (compare to Figure 2.3). This structure is virtually realized when data from the RAM of a computer are transferred to registers of arithmetic units in the same computer.
3) The processor H or its parts function autonomously from A, only sending to A information about the content of cells (compare to Figure 2.4). In this case, H or its parts contain those instructions from the software to operate H or its parts. This operation mode models the intelligent agent approach to computation. There an agent moves to the location of data and performs its operation at this new site.

As we know, different programs constitute computer software. Programs tell the system what to do (and what to not do). The *software* R of the inductive Turing machine M is also a program in the form of simple rules. The traditional representation assumes that the processor H functions as one unit. The rules for functioning have the following form:

$$q_h a_i \rightarrow a_j q_k, \tag{1}$$

$$q_h a_i \rightarrow C q_k \tag{2}$$

It is also possible to use only rules of one form:

$$q_h a_i \rightarrow a_j q_k c \tag{3}$$

Here q_h and q_k are states of A; and a_i and a_j are symbols of the alphabet of M, while c is a type of connection from K.

Each rule directs the inductive Turing machine M to perform one step of computation. For instance, the rule (1) means that if the state of the control device A of M is q_h and the processor H observes in the cell the symbol a_i in the cell, then the state of A becomes q_k and the processor H writes the symbol a_j in the cell where it is situated. The processor H then moves to the next cell using a connection of the type c.

Like Turing machines, inductive Turing machines can be both deterministic and nondeterministic. For a *deterministic inductive Turing machine*, there exists at most one connection of any type from any cell. In a *nondeterministic inductive Turing machine*, several connections of the same type may originate in the same cell, connecting it with (different) other cells. If there is no connection of this type going from the cell to be observed by H, then H stays in the same cell. There may be connections of a cell with itself. Then H also stays in the same cell. It is possible that H observes an empty cell. To represent this situation, we use the symbol Λ. Thus, it is possible that some elements a_i and/or a_j are equal to Λ in the rules from R. Such rules describe situations when H observes an empty cell and/or when H simply erases the symbol from some cell, writing nothing in it.

The rules allow an inductive Turing machine to rewrite a symbol in a cell and to make a move in one step. Other representations of rules treat such operations as separate steps. Rules of inductive Turing machine define the transition function of M and describe changes in A, H, and E. Consequently, they also determine the transition functions of A, H, and E.

When the processor H consists of several processing units or heads, we have several functioning modes:

Uniform synchronized processing (processor units function synchronously): At each step of M each unit performs one operation; they all are controlled by the same system of rules.

Uniform concurrent processing (processor units function concurrently): Units perform operations independently of one another, but all of them are controlled by the same system of rules.

Specialized synchronized processing: Each processor unit has its own system of rules, but all of them function synchronously, that is, at each step of M each unit performs one operation.

Specialized concurrent processing: Each processor unit has its own system of rules and they perform operations independently of one another.

In what follows, we consider for simplicity only the case when processor H consists of one unit and M always starts functioning in the same state. Thus, the functioning of the inductive Turing machine M begins when the control device A is in the start state q_0, the working and output memories are empty, and the processor H observes such a cell in the input register P_I that this cell contains some symbol and has the least number of all nonempty cells in the input register. It is possible that nothing is written in the input register P_I. In this case, H observes an arbitrary cell. When H observes an empty cell, we denote the content of this cell by the symbol Λ.

A general step of the machine M has the following form. At the beginning of any step, the processor H observes some cell with a symbol a_i in some cell (for an empty cell the symbol is Λ) and the control device A is in some state q_h.

Then the control device A and/or the processor H choose from the system R of rules the rule r to make with the left part equal to $q_h a_i$ and then perform the prescribed operation. If that rule in R with a corresponding left part, the machine M stops functioning. If there are several rules with the same left part does not exist, M works as a nondeterministic Turing machine (cf., for example, (Hopcroft, Motwani, and Ullman, 2001)) performing a range of possible operations. When A comes a final state from F, the machine M also stops functioning. In all other cases, it continues operation without stopping.

For an abstract automaton, as well as for a computer, we consider two important things. Specifically, not only how it functions, but also how it obtains its results. In contrast to Turing machines, inductive Turing machines obtain results even in the case when their operation is not terminated. This results in an increase in the performance of systems of algorithms.

The result of computation performed by the inductive Turing machine M is the word written in the output register P_O of M. We have two options for this: when M halts, after its control device A is in some final state from F, or when M never stops but at some step the content of the output register P_O becomes fixed and does not change although the machine M continues to function. In all other cases, M gives no result.

Theorem 4.3.1. *Any (inductive) Turing machine T with a recursive memory can be simulated by a (inductive) Turing machine D with one conventional tape, that is, the machine D computes the same function as T, imitating all moves of T.*

Proof. This is similar to that for the equivalence of different classes of Turing machines. It is done by using the standard procedure in which D writes on a special tape consequent instantaneous descriptions of the machine T (cf., for example, Hopcroft et al., 2001).   □

Some mathematicians and computer scientists object that, in contrast to Turing machines, inductive Turing machines does not always inform a user that a result was obtained. This is the cost that we have to pay for its essentially higher computational power. However, mathematicians and computer scientists encountered similar situation with Turing machines. Having the class of all Turing machines or any other class of recursive algorithms, one never knows whether the given machine will produce the necessary result or not. In contrast to this, the condition that an algorithm always gives a result is often demanded (cf. Section 2.1). Trying to limit ourselves to recursive algorithms that always give a result brings us to the following situations: either we have a sufficiently powerful class but one cannot distinct algorithms from this class from others or we can build all such algorithms but they have insufficient computational and decision power. Thus, we have to make a choice: either to use more powerful algorithms or to know more about algorithms that are used. From this perspective, inductive Turing machine is the next step in the ever-going trade off between knowledge and power.

4.3.2 Inductive Turing machine versus Turing machine

> *Forewarned is forearmed.*
>
> A proverb

We have seen that inductive Turing machines give results either with or without halting. The question is whether it is really necessary for an inductive Turing machine sometimes to stop to give a result. We can see that this not the case.

Lemma 4.3.1. *For any inductive Turing machine M, there is an inductive Turing machine G such that G never stops and computes the same function as M, that is, M and G are functionally equivalent.*

Proof. The machine G has the alphabet and the same set of states as M.
To eliminate termination in finite states, we do the following transformations:

(1) We make the system of terminating states of G empty, and then
(2) for each symbol q from F of the final states of M and each symbol a_i from the alphabets of M (including the symbol ε), we add the rule $qa_i \rightarrow a_iq$ to the system $\boldsymbol{R}_M$ of rules of the machine M.

In such a way, we obtain a new inductive Turing machine V. By the definition of the functioning of an inductive Turing machine, if M obtains some result w by terminating in a final state, then V obtains the same result without termination. As a result of the change, the machine V stops only when there is no rule in $\boldsymbol{R}_M$ for continuation. This means that for some pairs (q_k, a_i) there are no rules in $\boldsymbol{R}_M$ with such left part. To eliminate this termination, for each such pair, we do the following transformations:

(1) we define connections X of all cells to some cell in the output register $\boldsymbol{P}_O$,
(2) for each symbol q_k from Q, we add the symbol p_k to the set Q of all states of A,
(3) we add two new symbols b and c to the alphabet of the machine M,
(4) for each such pair (q_k, a_i) that there are no rules in $\boldsymbol{R}_M$ with such left part, we add the rules $q_k a_i \rightarrow Xp_k$, $a_i p_k \rightarrow bp_k$, $bp_k \rightarrow cp_k$, and $cp_k \rightarrow bp_k$ to the system $\boldsymbol{R}_V$ of rules for the machine V; and each time one of the symbols is written in the working tape, the same symbol is written in the output tape.

In such a way, we obtain a new inductive Turing machine G. As a result of this change, the machine G never stops and gives exactly the same results as M.

Lemma 4.3.1 is proved.   □

All mathematical models of algorithms, which existed before appearance of superrecursive algorithms, demanded to stop functioning to give a result. As we see, this is different for inductive Turing machines. They can continue to work and give a result without stopping. The new definition of the result of a computation is based on the intrinsic property of emergence. To explain this, let us consider the conventional approach to emergent computation, description of which is given by Forrest (1991). She writes that the result of emergent computations are difficult or even impossible to predict. The whole process is performed by a collection of agents, each following explicit instructions. Interactions among agents form implicit global patterns that are not predetermined by the algorithm of this computation and there is no one address where one can read out the result of computation.

However, when we analyze this description, we encounter essential problems. Indeed, if we assume that computation is emergent when it displays some pattern that is not predetermined by the algorithm of this computation, then only nondeterministic computational models can produce emergent computation. Thus, almost all models that are included in the scope of emergent computation (connectionist models, classifier systems, cellular automata, deterministic biological models, and artificial-life models) are deterministic and do not satisfy this condition. For example, the result of any complex network of deterministic artificial neurons with deterministic interactions is predetermined by the structure of the neurons and network, including rules for interaction, and by the input. At the same time, nondeterministic Turing machines satisfy the condition from the description of emergent computation and thus, their computations have to be considered as emergent. Moreover, as it is demonstrated by Machlin and Stout (1991) even simple Turing machines, which are usually contrasted to neural networks, can have very complex behavior, which satisfies the above definition of emergent computation.

To eliminate this discrepancy, we introduce a slightly different definition of emergent computation.

Definition 4.3.1. A computation is emergent when it is a complex process and to predict the pattern (result) that is generated by this computation is as complex as to perform the whole computation itself.

Taking inductive Turing machines, we see that although in some cases the results are predictable, in general the result emerges only in the corresponding computational process. There are situations when the result is already obtained but the inductive Turing machine cannot stop functioning. As a consequence, the process of emergence becomes in some sense infinite, although the result is always computed in a finite time.

Inductive Turing machines present a transition from terminating computation to intrinsically emerging computation, making a leap from the "being" mode of recursive algorithms to the "becoming" mode of superrecursive algorithms. All kinds of computations considered in (*Emergent Computation*, 1991) become really emergent only when the machine does not need to stop.

Inductive Turing machine introduces a new type of infinity into mathematics. In contrast to contemporary mathematics and philosophy, philosophers of the ancient Greece considered three types of infinity: *actual infinity*, which already exists as an infinity; *potential infinity*, which can or may exist as an infinity; and *becoming* or *emerging infinity*, which is in a process of transition from finite to infinite. After the Golden Age of the Greek intellectual achievements, only two concepts of infinity (actual and potential) survived and were formalized in mathematics. The third type (emerging infinity) disappeared even from philosophy, to say nothing about mathematics. Synergetics brought back interest of scientists to *becoming systems* (cf., for example, (Prigogine, 1980)). Super-recursive algorithms give a mathematical model for the concept of becoming infinity.

Any advice function, which is described in Section 4.2, can be codified by connections in the structured memory and then decoded. Thus, a relevant structured memory can realize any kind of an oracle or advice. Namely, the following result is true.

Theorem 4.3.2. *For any Turing machine T with an advice, there exists a Turing machine M with a structured memory that computes the same function as T.*

The same is true for machines with arbitrary oracles.

Theorem 4.3.3. *For any subset X of N and any Turing machine T with the oracle Or(X), there is an inductive Turing machine M with a structured memory such that M computes the same function as T, that is, M and T are functionally equivalent.*

Corollary 4.3.1. *For any Turing machine T with an oracle (with an advice), there exists an inductive Turing machine M with a structured memory that computes the same function as T.*

Corollary 4.3.2. *For any function $f: X^* \to X^*$ ($f: N \to N$), there exists a Turing machine T with a structured memory that computes f.*

Theorems 4.3.2 and 4.2.5 imply the following result, which was originally demonstrated by A.N. Kolmogorov in 1983.

Theorem 4.3.4. *For any function $f: X^* \to X^*$ ($f: N \to N$), there exists an inductive Turing machine M with a structured memory that computes f.*

Such high computing power of inductive Turing machines with an arbitrary structured memory makes it reasonable to introduce inductive Turing machines with reasonable constructive restrictions on the memory. This is done in the following sections of this Chapter. The choice of restrictions for the machine memory makes these machines much closer to conventional algorithms than arbitrary inductive Turing machines with a structured memory.

4.3.3 Arithmetical hierarchy and definability of sets

> *The different branches of Arithmetic —*
> *Ambition, Distraction, Uglification, and Derision.*
>
> Lewis Carroll, 1832–1898

To measure decision and computing power of inductive Turing machines, we use the arithmetical hierarchy. To use this hierarchy, we need its properties. According to the conventional construction, the arithmetical hierarchy consists of sets of natural numbers that are defined as relations of definite form. The hierarchy is usually developed with the universal $\forall$ and existential $\exists$ quantifiers restricted to the natural numbers because this hierarchy was built to study the arithmetic of natural numbers. However, the arithmetical hierarchy can be defined using other types of quantifiers.

Elements of the arithmetical hierarchy are relations of natural numbers.

Definition 4.3.2. An $(m + n)$-ary relation $R(x_1, \ldots, x_n, z_1, \ldots, z_m)$ or a relation with n variables on the set N of all natural numbers is any set elements of which are vectors or strings that have the form $(a_1, \ldots, a_n, b_1, \ldots, b_m)$ where all $a_1, \ldots, a_n$, $b_1, \ldots, b_m$ are natural numbers.

The set of all recursive relations is taken as the base for building the arithmetical hierarchy.

Definition 4.3.3. (Rogers, 1987). A relation $R(x_1, \ldots, x_n)$ is called *recursive* if it possesses a recursive characteristic function, that is to say, $R(x_1, \ldots, x_n)$ is recursive if and only if there exists a recursive function f such that $f(x_1, \ldots, x_n) = 1$ if the string $(x_1, \ldots, x_n)$ belongs to $R(x_1, \ldots, x_n)$ and $f(x_1, \ldots, x_n) = 0$ if the string $(x_1, \ldots, x_n)$ does not belong to $R(x_1, \ldots, x_n)$.

For example, relations $=$, $<$, and $>$ are recursive on the set N of all natural numbers. However, they are not recursive on the set of all real numbers.

Levels in the arithmetical hierarchy are labeled as Σ_n if they consist of all relations $\exists x_1 \, \forall x_2 \, \exists x_3 \, \ldots \, forall x_{n-2} \, \exists x_{n-1} \, \forall x_n \, R(x_1, \ldots, x_n, z_1, \ldots, z_m)$ limited to $n - 1$ pairs of alternating quantifiers starting with $\exists$ and recursive $R(x_1, \ldots, x_n, z_1, \ldots, z_m)$. Similarly the class of all relations $\forall x_1 \, \exists x_2 \, \forall x_3 \, \ldots \, \exists x_{n-1} \, \forall x_n \, R(x_1, \ldots, x_n, z_1, \ldots, z_m)$ that start with $\forall$ and have $n - 1$ alternations of quantifiers is labeled as Π_n and recursive $R(x_1, \ldots, x_n, z_1, \ldots, z_m)$.

Only alternating pairs of quantifiers are counted because two quantifiers of the same type occurring together are equivalent to a single quantifier. Indeed, there is a one-to-one mapping $r: N^2 \to N$ from all pairs of natural numbers onto all natural numbers (Rogers, 1987; §5.3). Such mapping r may be given by the formula $r(n, m) = 2^n(2m + 1) - 1$. Using this mapping r, one can convert a representation with two quantifiers $\forall x \, \forall y \, R(x, y, z_1, \ldots, z_m)$ to the representation with one quantifier $\forall z \, R(z = r(x, y), z_1, \ldots, z_m)$. Both expressions represent the same relation, while the second one has less universal quantifiers $\forall$. In such a way, we can eliminate all repetitions of the quantifier $\forall$. The same technique applies to two or more consecutive existential quantifiers $\exists$.

As the result, the classes Σ_0 and Π_0 are defined as having no quantifiers, consist of all recursive relations and thus, are equivalent. The classes Σ_1 and Π_1 are defined as having a single quantifier: relations from Σ_1 have the form $\exists x\ R(x, y)$ and relations from Π_1 have the form $\forall x\ R(x, y)$ where $R(x, y)$ is a recursive relation and y is an arbitrary vector of natural numbers. The classes Σ_2 and Π_2 are defined as having two quantifiers: relations from Σ_2 have the form $\exists x\ \forall y\ R(x, y, z)$ and relations from Π_2 have the form $\forall x\ \exists y\ R(x, y, z)$ where $R(x, y, z)$ is a recursive relation and z is an arbitrary vector of natural numbers. By the definition, we have inclusions $\Pi_m \subseteq \Sigma_n \cap \Pi_n$ and $\Sigma_m \subseteq \Sigma_n \cap \Pi_n$ any $n > 1$ and any $m < n$.

Theorem 4.3.5. (The Hierarchy Theorem.) (Kleene, 1955). *For any number $n > 1$, there is a relation R that belongs to Σ_n and does not belong to Π_n and thus, belongs neither to Π_m nor Σ_m for all $m < n$. There is also a relation Q that belongs to Π_n and does not belong to Σ_n and thus, belongs neither to Π_m nor Σ_m for all $m < n$.*

Mathematical logic studies formal theories, the majority of which are taken from mathematics, although there are results concerning formalizations of some physical theories. Arithmetic is one of the basic mathematical theories. A formalization of arithmetic by means of logic is called Peano arithmetic or elementary arithmetic **Ar**. Formulas of elementary arithmetic are built from the following symbols: $+, \times, 0, 1,$ $2, 3, \ldots, =$, variable symbols (such as x, y, z), sentential connectives ($\rightarrow$ (implies), $\wedge$ (and), $\vee$ (or), and $\rceil$ (not)) and two quantifier symbols $\exists$ and $\forall$.

For instance, we have formulas $\exists x\ R(x, z), \exists x\ P(x), \exists x\ \forall z\ R(x, z), \forall x\ \exists y$ $\forall z\ Q(x, y, z), \exists x_1\ \forall x_2\ \exists x_3 \ldots \forall x_{n-2}\ \exists x_{n-1}\ \forall x_n\ T(x_1, \ldots, x_n, z)$, and $\forall x_1\ \exists x_2$ $\forall x_3 \ldots \exists x_{n-1}\ \forall x_n\ T(x_1, \ldots, x_n, z)$ where $R(x, z) = (x < z)$, $P(x) = (x = 5)$, $Q(x, y, z) = (x < y \wedge y < z)$, and $T(x_1, \ldots, x_n, z) = (x1 + -x_2 + x_3 + \ldots + x_n = z)$.

Definition 4.3.4. Formulas in which all variables are related to quantifiers are called *closed*. Other formulas are called *open*.

For instance, formulas $\exists x\ P(x), \exists x\ \forall z\ R(x, z)$, and $\forall x\ \exists y\ \forall z\ Q(x, y, z)$ are closed, while formulas $\exists x\ R(x, z), \exists x_1\ \forall x_2\ \exists x_3 \ldots \forall x_{n-2}\ \exists x_{n-1}\ \forall x_n\ T(x_1, \ldots, x_n, z)$, and $\forall x_1\ \exists x_2\ \forall x_3 \ldots \exists x_{n-1}\ \forall x_n\ T(x_1, \ldots, x_n, z)$ are open.

Open formulas define sets and relations. For instance, the formula $\exists x\ (x < z \wedge x,$ $z \in N)$, defines the set of all natural numbers for which a smaller number exists, that is, the set $2, 3, \ldots, n, \ldots$.

Definition 4.3.5. A set X is definable in a formal theory **T** if there is a formula in **T** that defines X.

Each formal theory has its own symbols and uses logical symbols. For example, arithmetic has symbols of order relations $\leq, \geq, <$ and $>$, as well as symbols of arithmetical operations $+, -, \times$ or $\div$, etc. Formulas of a given theory **T** use only symbols of this theory and logical symbols.

It is necessary to remark that there are different logical theories that use different logical symbols. For example, in addition to conventional logical symbols, modal logics (cf. Feys, 1965) use additional symbols $\square$ (necessity) and $\diamond$ (possibility), while

Arithmetical Hierarchy

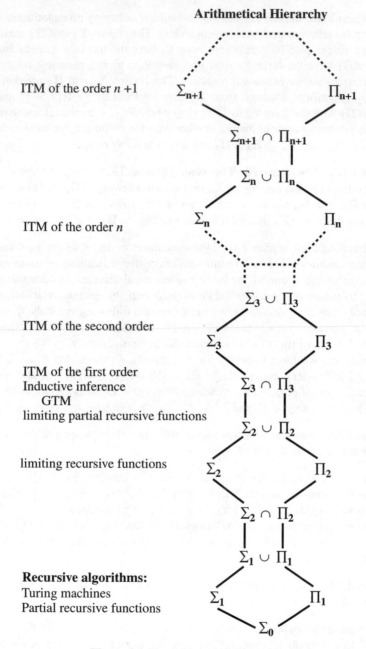

ITM of the order $n+1$ Σ_{n+1} Π_{n+1}

$\Sigma_{n+1} \cap \Pi_{n+1}$

$\Sigma_n \cup \Pi_n$

ITM of the order n Σ_n Π_n

$\Sigma_3 \cup \Pi_3$

ITM of the second order

Σ_3 Π_3

ITM of the first order
Inductive inference
    GTM
limiting partial recursive functions $\Sigma_3 \cap \Pi_3$

$\Sigma_2 \cup \Pi_2$

limiting recursive functions Σ_2 Π_2

$\Sigma_2 \cap \Pi_2$

$\Sigma_1 \cup \Pi_1$

Recursive algorithms:
Turing machines
Partial recursive functions Σ_1 Π_1

Σ_0

Figure 4.4. The Arithmetical Hierarchy.

the intuitionistic **B**-logic of Gödel (1933) contains additional symbol **B**. Existence of different logics results in a possibility to build different formalizations of a given mathematical theory.

The following result explains why the considered hierarchy of sets is called the arithmetical hierarchy.

Theorem 4.3.6. (Arithmetical Representation Theorem.) (Rogers, 1987). *For any relation R, R is in the arithmetical hierarchy if and only if R is arithmetical, that is, definable in the elementary arithmetic.*

It is possible to take a more general base for the arithmetical hierarchy construction. We fix some set X and take as the base all relations that are recursive relative to X. By the definition (Rogers, 1987), a relation R on the set of natural numbers is recursive relative to X if given information about membership in X, the characteristic function of R becomes computable by a Turing machine. Such Turing machine is called a Turing machine with the oracle X (cf. Section 4.2) and the corresponding construction in which recursive relative to X sets are taken as the base is called the arithmetical hierarchy in X. The ordinary, or absolute arithmetical hierarchy, which is considered above, is a particular case when X is the empty set.

Theorem 4.3.7. (Relative Arithmetical Representation Theorem. (Rogers, 1987). *For any set $X \subseteq N$ and any relation R, R is in the arithmetical hierarchy in X if and only if R is arithmetical in X, that is, definable in elementary arithmetic augmented by X.*

It is necessary to remark that the arithmetical hierarchy is interesting and important in its own way. First, it is closely related to an important mathematical field such as arithmetic. Second, it displays an explicit model of exact definitions in mathematics. There are two main ways in mathematics to define formal structures: descriptive and constructive. According to the first approach, a mathematical object is defined by some formula (expression of a mathematical theory). It may be a logical formula (for example, $\exists x \, \forall y \, (y \in N \Rightarrow x > y)$) or a formula of calculus (for example, $x = lim_{i \to \infty} a_i$) or of any other mathematical theory. According to the second approach, a mathematical object is defined by an algorithm that allows one to construct this object. Relations between the arithmetical hierarchy and systems of algorithms reflect the place of these constructions in mathematics.

What we are concerned about is the first aspect of the arithmetical hierarchy importance. Namely, it is generally accepted that: 1) arithmetic of natural numbers is one of the main intellectual means of human civilization; 2) formal arithmetic is the basic mathematical theory; and 3) arithmetical hierarchy contains all sets that are definable in this theory by conventional logical means. Thus, it is important to understand how sets are described in arithmetic and to compare their descriptions to the means of their construction and validation. For example, the famous Gödel undecidability theorem (Gödel, 1931) states that it is impossible to validate truth for all true statements about arithmetical objects. This is a consequence of the fact (Turing, 1936) that it is impossible build all sets from the arithmetical hierarchy by Turing machines.

It is possible to interpret formulas that define sets in the arithmetical hierarchy as questions about algorithms or programs. Let $T_A(x)$ be the time (number of steps of A) of computation of an algorithm or automaton A with the input x. Then the set of all programs or algorithms that terminate in time $T_A(x) = n$ with the input x is recursive, that is, it belongs to Σ_0. Thus, the question "*Will a computer program halt in fixed time?*" can be answered in Σ_0. The question "*Will a computer program ever halt?*" is more complex. The set of all programs or algorithms that ever terminate, that is, $T_A(x)$ is finite, with the input x is recursively enumerable, that is, it belongs to S_1. So, we can to answer the question "*Will a computer program never halt?*" in Π_1. In a similar way, the question "*Will a computer program have at most a finite number of outputs?*" can be answered in Σ_2, the question "*Will a computer program have an infinite number of outputs?*" is answered in Π_2 and so on.

4.3.4   Hierarchies of inductive Turing machines

All in good time.

A proverb

As we have seen in Section 4.3.2, inductive Turing machines with an arbitrary structured memory have unrestricted computing power. However, the problem is how to build such structure of the memory that solves all problems and computes all functions. It is impossible (at least, now). So, we need some reasonable conditions on the memory to be able to realize it. For example, when all connection between the cells in the memory and naming relations for these cells are built by some Turing machine, it natural to treat such memory as constructible. We extend this approach building the *inductive hierarchy* of inductive Turing machines.

Definition 4.3.6. The memory E is called *recursive* if the relation $K \subseteq P \times P$ that provides connections between cells and all mappings $\nu \colon P \to N$, $\tau \colon K \to W$, and $\iota \colon P \to V$ are recursive.

Here recursive means, as before, that there are some Turing machines that decides all naming mappings and relations in the structured memory.

As the lowest level of a hierarchy of inductive Turing machines, we take machines with recursive memory. They are called *inductive Turing machines of the first order*. Their memory is constructed by Turing machines.

When we have inductive Turing machines of the first order, it is possible to use them for constructing memory for other inductive Turing machines. In such a way, we define 1-inductive memory. Inductive Turing machines with such memory have the second order. If we continue this process, we come to the following constructions.

Definition 4.3.7. The memory E is called *n-inductive* if the relation $K \subseteq P \times P$ that provides connections between cells and all mappings $\nu \colon P \to N$, $\tau \colon K \to W$, and $\iota \colon P \to V$ are defined by some inductive Turing machines of order n.

This means that there are inductive Turing machines of order n that build and decide all naming mappings and relations in the structured memory.

Definition 4.3.8. An inductive Turing machine M has order n when it has the memory E that is $(n-1)$-inductive.

In such a way, using inductive Turing machines of the first order, we build inductive Turing machines of the second order. Then using inductive Turing machines of the second order, we build inductive Turing machines of the third order and so on. Note that if an inductive Turing machine has order n, then it has order m for all $m > n$. As a result of this process, we obtain the infinite *inductive hierarchy* **IM** *of inductive Turing machines with the inductively defined memory*:

$$\mathbf{ITM}_1 \subseteq \mathbf{ITM}_2 \subseteq \ldots \subseteq \mathbf{ITM}_{n-1} \subseteq \mathbf{ITM}_n \subseteq \mathbf{ITM}_{n+1} \subseteq \ldots$$

where $\mathbf{ITM}_n$ is the class of all inductive Turing machines of order n.

We call it the inductive hierarchy because each level is built from a previous one by an inductive process. Any finite part of this hierarchy is called a *recursive hierarchy* of inductive Turing machines because each step in the process of building this hierarchy is recursive.

Thus, only finite parts are recursive, while the whole hierarchy is inductive like natural numbers or arithmetical hierarchy.

In the next section, we show that all inclusions in the inductive hierarchy are proper.

By the definition, we have the following result.

Lemma 4.3.2. *If we change the memory E of an inductive Turing machine M that has order n by means of a Turing machine, then the new machine has the same order.*

For simplicity in what follows, we consider a substantially uniform memory E, in which all cells are the same. In other words, P is a multiset (Knuth, 1997) with indistinguishable cells.

In addition, we suppose that: 1) any cell c from the set P is represented by its number $\nu(c)$; 2) the partition $P = P_I \cup P_W \cup P_O$ is always recursive, that is, given a cell i, it is possible to say to what part it belongs; 3) all parts K_I, K_W, and K_O of K are defined separately.

Turing machines with oracles, which are considered in Section 4.2, are related to relative arithmetical hierarchies. In a similar way, we introduce an inductive Turing machine with an oracle. Given a set X, the oracle $\mathrm{Or}(X)$ answers the question whether an arbitrary element a belongs to X or not.

If X is a subset of N, we add to a given inductive Turing machine the oracle $\mathrm{Or}(X)$ and receive in such a way an inductive Turing machine with the oracle $\mathrm{Or}(X)$. Technically, we can realize such an oracle by adding to the memory E of an inductive Turing machine M a new tape $\mathbf{L}_X$ in which all numbers from N are written. In addition to this tape, we build special connections of the new type $\mathbf{t}_X$. They connect each element from X with a cell in which 1 is written and each element that do not belong to X with a cell in which 0 is written. When M performs its computation and has to ask the oracle about some number n, whether it belongs to X or not, the processor H of M goes to the cell from the tape $\mathbf{L}_X$ that contains n. After this the head performs a transition along the connection of the type t_X, coming in such a way to a cell containing either 1 or 0. In such a way, H finds the answer to the question about the number n in this cell, to which the head comes after this transition.

Let $IM(X)$ denotes the class of all machines that are obtained from the machines from the hierarchy **IM**, by augmenting them with the oracle $Or(X)$. This gives us a relative hierarchy of inductive Turing machines with oracles:

$$\mathbf{ITM}_1(X) \subseteq \mathbf{ITM}_2(X) \subseteq \ldots \subseteq \mathbf{ITM}_{n-1}(X) \subseteq \mathbf{ITM}_n(X) \subseteq \mathbf{ITM}_{n+1}(X) \subseteq \ldots$$

In contrast to the absolute hierarchy **IM**, it is possible that in the relative hierarchy $IM(X)$ not all inclusions are proper. For example, if X is a universal set for the level n, then all levels $\mathbf{ITM}_i(X)$ with $i < n$ coincide.

When the structured memory E of an inductive Turing machine T is constructed by some algorithm (another inductive Turing machine) T_E, then it is natural to consider how this algorithm T_E functions and interacts with the functioning of T. It is possible to assume that T_E determines functioning of the memory E. Taking into account modes of the structured memory functioning, we obtain two memory types:

Fixed or *static* memory, in which all connections are fixed in the hardware and are not changing at all, and *active* or *dynamic* memory, in which connections are built by a convenient automaton A_E that realizes algorithm T_E.

Active memory of the inductive Turing machine T can function in three modes:

1. *Preprocessing* when all connections are established before the machine T begins its computation.

Preprocessing of information, similar to the preprocessing in the memory of an inductive Turing machine, is often used in biological systems, making them more efficient. For example, we see this in organization of human vision. It is known that information from visual receptors (rods and cones) does not go directly to the brain for processing. At first, this information is preprocessed by three kinds of neurons (bipolar cells, horizontal cells, and amacrine cells) in the retina. Then it is preprocessed by ganglion cells, which transmit the resulting information to the brain (Gray, 1994).

In a similar way, data preprocessing is used in computation. By the definition, data preprocessing is any type of computational process that is performed on raw data to prepare these data for another processing procedure. Data preprocessing software performs data selection, cleaning, integration, and merging, format conversion and filtering, verifying data quality and other operations before higher level analysis. For example, in data mining practice, data preprocessing transforms the data into a format that will be more easily and effectively processed for the purpose of the user, for example in a neural network. There are a number of different tools and methods used for preprocessing. They include: sampling, which selects a representative subset from a large population of data; transformation, which manipulates raw data to produce a single input; filtering, which removes noise from data; normalization, which organizes data for more efficient access; and feature extraction, which pulls out specified data that are significant in some particular context.

2. The *synchronous mode* when connections are built on a request which is sent by the computational process through the control device of the machine T.

The synchronous mode of memory functioning reflects definite aspects of information processing by the brain, as well as the work contemporary computers and networks, in which connections to the necessary elements of the computer memory or of the network are established on a request.

3. The *concurrent mode* when the automaton A_E and the mainframe BT of the machine T work concurrently. When some demanded by BT connection is not established and BT needs it for continuation, T stops functioning and waits until the necessary connection is constructed by A_E, or halts forever if A_E does not build the connection. In the latter case, T does not give a result of the computation.

The concurrent mode of memory functioning reflects the principles of the modern search engines, in which the index form such an active structured memory of the engine and this memory works in the concurrent mode. Theoretical investigation of properties of this functioning mode might help to create much better search engines.

Having inductive Turing machines, we extend the concepts of recursive computability, acceptability, and decidability.

Definition 4.3.9. A function f is *inductively (n-inductively) computable* if there is an inductive Turing machine (of order n) that computes this function.

Definition 4.3.10. A set X is *inductively (n-inductively) acceptable* if there is an inductive Turing machine (of order n) that accepts this set.

Definition 4.3.11. A set X is *inductively (n-inductively) decidable* if there is an inductive Turing machine (of order n) that decides this set, or in other words, when the characteristic function of X is inductively (*n-inductively*) computable.

Definition 4.3.12. A set X is *inductively (n-inductively) semidecidable* if there is an inductive Turing machine (of order n) that semidecides this set, or in other words, when the function which gives the affirmative values, the function with the domain X and constant value 1, is inductively (*n-inductively*) computable.

4.3.5 Spaces of inductive Turing machines and their computational power

Truth lies at the bottom of a well.

A proverb

Let us get some properties of inductive Turing machines. We need these properties to find computing and deciding power of different classes of inductive Turing machines.

Lemma 4.3.3. *Given m inductive Turing machines $M_1, \ldots, M_m$ of order n, it is possible to build an inductive Turing machine of order n such that it has an infinite sequence of independent tapes (structured memories) with enumerated cells and models all machines $M_1, \ldots, M_m$.*

Proof. Let us take at first two inductive Turing machines M_1 and M_2 of order n and build an inductive Turing machine M of order n that simulates common functioning of M_1 and M_2. For simplicity, we suppose that each of the machines M_1 and M_2 has a one-sided tape L with sequentially enumerated cells as its memory. At first, we show that it is possible to restructure L into two tapes L_1 and L_2 with sequentially enumerated cells. Indeed, we relate all cells with odd numbers to L_1 and all cells with even numbers to L_2. The set Q of all states is divided into two groups $Q_1 = \{q_i;$ $i = 1, \ldots, k\}$ and $Q_2 = \{p_j; \, j = 1, \ldots, m\}$. When the head is in the first tape, the state of the control device of M belongs to the first set, that is, it is some q_i. When the head is in the second tape, the state of the control device of M belongs to the second set, that is, it is some p_j. Then we add to each set Q_k duplicates of all states: $Q_1 = \{q_i;$ $i = 1, \ldots, k; \, q'_i; \, i = 1, \ldots, k\}$ and $Q_2 = \{p_j; \, j = 1, \ldots, m; \, p'_j; \, j = 1, \ldots, m\}$. In addition, if the set of rules $\boldsymbol{R}_1$ of the machine Q_1 has a rule $q_h a_i \rightarrow Cq_k$, in which C is either R (right) or L (left), then we change this rule for the following rules $q_h a_i \rightarrow Cq'_h$ and $q'_h a_j \rightarrow Cq_k$ for all a_j from the alphabet of the machine Q_2. In a similar way, we change the rules $\boldsymbol{R}_2$ for the machine Q_2. If the set of rules $\boldsymbol{R}_2$ for the machine Q_2 has a rule $p_h a_i \rightarrow Cp_k$, in which C is either R or L, then we change this rule for the following rules $p_h a_i \rightarrow Cp'_h$ and $p'_h a_j \rightarrow Cp_k$ for all a_j from the alphabet of the machine Q_1. These changes provide for the following feature of M: performing the rules of Q_1, the processors works only with cells that have odd numbers, while performing the rules of Q_2, the processors works only with cells that have even numbers. As a result, the new inductive Turing machine M models both machines M_1 and M_2 and has by Lemma 4.3.2 the same order as these machines because the new partition of the memory L is defined by a finite automaton and thus, is recursive.

If we have more than two inductive Turing machines $M_1, \ldots, M_m$, then we use for the first machine M_1 the cells from L that have odd numbers. For the second machine M_2, we use the cells from L that have numbers of the form $2(2n + 1)$. For the third machine M_3, we use the cells from L that have numbers of the form $4(2n + 1)$ and so on. Each time we repeat the procedure that we have done with the initial tape L. Then we change the rules so that the new machine models each M_i, utilizing the corresponding tape L_i. As a result, the new inductive Turing machine M models all machines $M_1, \ldots, M_m$ and has by Lemma 4.3.2 the same order as these machines because the new partition of the memory L is defined by a finite automaton and thus, is recursive.

To conclude the proof, it is sufficient to remark that in the described way we have formed infinitely many tapes for a new machine, which can model any amount of other inductive Turing machines due to the fact that each tape may be structured independently into an $(n - 1)$-inductive memory.

Lemma 4.3.3 is proved.   □

Lemma 4.3.4. *For any inductive Turing machine M of order n, there is an inductive Turing machine T of order n that perform all operations of M and in addition, accepts and performs instructions of the form "go to the cell with number i."*

Proof. For an inductive Turing machine M with an arbitrary structured memory E, the solution is simple. We add to the system of connections of M, a new set of connections (i, j), which connect any two cells i and j from the set P. Each pair (i, j) is labeled by the type $C(j)$. In this setting, the rule

$$q_l a_i \to C(j)q_h \tag{4}$$

means that the processor H has to go to the cell that is connected to the given cell by the connection $C(j)$. This is exactly the cell with number j. So, the rule (4) is equivalent to the instruction "go to the cell with number i."

The situation is different when the structure memory E is locally bounded (that is, the number of connections that go from any cell is bounded by some number k) or locally finite (that is, the number of connections that go from any cell is finite). In these cases, we cannot tie to each cell infinitely many connections as before. Thus, we utilize Lemma 4.3.3 and organize in E several additional tapes. One of them L_1 is used for storage of the address i for the transition of H. Another tape L_2 is used as a counter for the number of steps that the processor H makes in the memory. To perform the instruction of the form "go to the cell with number i," where i is stored in L_1, the processor H goes, at first, to cell with number 1. Then H starts moving from the cell 1 to the cell 2 to the cell 3 and so on. With each step, the counter of steps adds 1 to the number stored in L_2 and compares the result with the number i stored in L_1. All operations are recursive and do not change the order of the inductive Turing machine. When both numbers become equal, the processor H is situated in the cell with number i.

Lemma 4.3.4 is proved. $\square$

Random access machines or RAM form an important class of mathematical models of algorithms (Aho, Hopcroft, and Ullman, 1976). Lemma 4.3.4 implies the following property of Turing machines and inductive Turing machines, demonstrating how they are related to RAM.

Corollary 4.3.3. *It is possible to completely model any RAM (with an oracle $Or(X)$) by a Turing machine (with an oracle $Or(X)$) and thus, by an inductive Turing machine (with an oracle $Or(X)$).*

The following result shows that, in contrast to Turing machines, application of an inductive Turing machine to organization of the memory another inductive Turing machine causes increase in computing power of machines. In some sense, the first inductive Turing machine performs preprocessing of information for the second inductive Turing machine. This is a nonlinear composition of inductive Turing machines, which extends the computability space from one order to another many times. The result of this process is a hierarchy of inductive Turing machines isomorphic as a computability space to the arithmetical hierarchy.

Lemmas 4.3.3 and 4.3.4 are used to prove this result.

Theorem 4.3.8. (Superrecursive Representation) *a) For any arithmetical relation Y, there exists an inductive Turing machine M such that it computes the characteristic function of Y.*

b) If Y belongs to the level n of the arithmetical hierarchy, that is, $Y \in \Sigma_n \cup \Pi_n$, then there is an inductive Turing machine M of order n such that it decides Y.

Proof. As we are going to compare results of computations of inductive Turing machines with the arithmetic hierarchy, we assume that inductive Turing machines and Turing machines work with natural numbers. Initially they work with strings of symbols, but it is possible to codify natural numbers by strings consisting of symbols 1 and 0. We may use, for example, the conventional binary coding. In it, the string 101 means or denotes the decimal number 5 and the string 1011 means or denotes the decimal number 11.

Arithmetical hierarchy consists of relations with arbitrary number of variables. However, we can recursively reduce any decision problem for a relation $Q(x_1, \ldots, x_n)$ with n variables to a decision problem for a relation with one variable.

Indeed, there is a recursive one-to-one mapping $r_n: N^n \to N$ from all n-vectors of natural numbers onto the set N of all natural numbers (Rogers, 1987; §5.3). It is possible to build this mapping r_n by iteration of the one-to-one mapping $r: N^2 \to N$ that is described in Section 4.3.3. Then a vector $(x_1, \ldots, x_n)$ belongs to a relation $Q(x_1, \ldots, x_n)$ if and only if the number $r_n(x_1, \ldots, x_n)$ belongs to the set $r_nQ = \{z \in N; \exists(x_1, \ldots, x_n) \in Q(x_1, \ldots, x_n) (r_n(x_1, \ldots, x_n) = z)\}$. As r_n is a recursive mapping, a Turing machine can decide whether $(x_1, \ldots, x_n) \in Q(x_1, \ldots, x_n)$ if and only if (may be another) Turing machine can decide whether $r_n(x_1, \ldots, x_n) \in r_nQ$. The same is true for inductive Turing machines of any fixed order. Namely, an inductive Turing machine can decide whether $(x_1, \ldots, x_n) \in Q(x_1, \ldots, x_n)$ is true or not if and only if (may be another) inductive Turing machine of the same order can decide whether $r_n(x_1, \ldots, x_n) \in r_nQ$ is true or not.

Thus, it is sufficient to prove the theorem only for relations with one free variable, which have the form $\exists x_1 \forall x_2 \exists x_3 \ldots \forall x_{n-2} \exists x_{n-1} \forall x_n R(x_1, \ldots, x_n, z)$ or $\forall x_1 \exists x_2 \forall x_3 \ldots \exists x_{n-1} \forall x_n R(x_1, \ldots, x_n, z)$. Actually, relations with one free variable are some sets of natural numbers.

For the proof, we use induction on the level n of the arithmetical hierarchy, that is, we show that for arbitrary natural number n and any relation that belongs either to Σ_n or to Π_n, there is an inductive Turing machine that computes the characteristic function of this relation.

1. As the base for induction, we take sets Σ_1 and Π_1. Let us consider an arbitrary relation $Q(z)$ that belongs to Σ_1. It means that $Q(z) = \exists x R(x, z)$ where $R(x, z)$ is a recursive relation. Thus, there is a Turing machine T_R such that given a pair (a, b) of natural numbers, it produces 1 when $(a, b) \in R(x, z)$ and produces 0 when $(a, b) \notin R(x, z)$.

To build an inductive Turing machine M_Q that computes the characteristic function of the relation $Q(z)$, we use a finite automaton T_N with feedback that generates consecutively all natural numbers. Receiving the first symbol 0, T_N generates 1 and sends it to itself as input. Receiving 1, T_N generates 2 and sends it to itself as input. In such a way, it generates all natural numbers.

The inductive Turing machine M_Q contains as its parts a pairing automaton AP, a Turing machine T_R^1, which is an isomorphic copy of the Turing machine T_R, and a finite automaton T_N^1, which is an isomorphic copy of the finite automaton T_N. Isomorphic machines perform the same operations and give the same results. They completely model one another and are functionally equivalent. A pairing automaton AP has two inputs and one output. Given natural numbers a for one input and b for another, AP combines them into the pair (a, b). All parts of the machine M_Q are realized as subroutines. A technique for realization of subroutines that are presented by Turing machines in another Turing machine is described in (Hopcroft, Motwani, and Ullman, 2001), (Davis and Weyuker, 1983), and (Ebbinghaus et al., 1970). The same technique works for inductive Turing machines because the structure of inductive Turing machine is similar to, but at the same time, richer than the structure of Turing machine.

The structure of the machine M_Q is presented in the Figure 4.5.

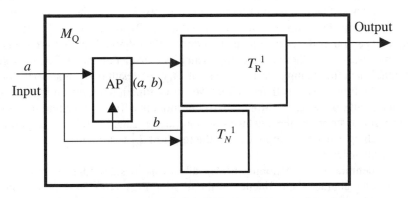

Figure 4.5. The structure of the inductive Turing machine M_Q.

Given a number a as an input, the machine M_Q works in the following manner. The element a is sent to the automaton AP and 0 is sent to T_N^1. The automaton T_N^1 produces number 1, which is sent to AP. The automaton AP combines a and 1 into the pair $(1, a)$ and sends it to the machine T_R^1. The machine T_R^1 checks whether $(1, a) \in$ R(x, z). When $(1, a) \in$ R(x, z), T_R^1 gives 1 as its result, and when $(1, a) \notin$ R(x, z), T_R^1 gives 0 as its result. The result of T_R^1 goes to the output tape of M_Q. If this result is 1, then either M_Q stops functioning, informing that it has the final result, or, when we want to have an inductive Turing machine that always works without stopping, the machine M_Q continues to function without stopping, but writes nothing else on its output tape.

If the result of T_R^1 is 0, then T_N^1 begins the next cycle of the machine M_Q. The automaton T_N^1 produces number 2, which is sent to AP. The automaton AP combines a and 2 into the pair $(2, a)$ and sends it to the machine T_R^1. The machine T_R^1 checks whether $(2, a) \in$ R(x, z). When $(2, a) \in$ R(x, z), T_R^1 gives 1 as its result, and when $(a, b) \notin$ R(x, z), T_R^1 gives 0 as its result. The result of T_R^1 goes to the output tape of M_Q. If this result is 1, then either M_Q stops, informing that it has the final result, or M_Q continues to function without stopping, but writes nothing else on its output tape. If the result of T_R^1 is 0, then T_N^1 begins the next cycle of the machine M_Q.

This process continues and in all possible cases the machine M_Q gives some result. By the definition of the functioning of the machine M_Q, the result that it produces for a is 1 when there is some number x for which the pair (x, a) belongs to $R(x, z)$ and the result is 0 when such number x does not exist. In other words, as a is any natural number, the machine M_Q computes the characteristic function of the relation $Q(z) = \exists x\ R(x, z)$. The relation $Q(z)$ is arbitrary in the class Σ_1, and M_Q is an inductive Turing machine of the first order because AP and T_N^1 are finite automata, while T_R^1 is a conventional Turing machine that decides the relation $R(x, z)$. Consequently, we have demonstrated that any relation from the class Σ_1 is decidable by some inductive Turing machine of the first order.

The same is true for the class Π_1. Indeed, if a relation $P(z)$ from this class is given, we know from the definition that $P(z) = \forall x\ R(x, z)$ where $R(x, z)$ is a recursive relation. This allows us to build an inductive Turing machine M_P of the first order that computes the characteristic function of the relation $P(z)$. This machine has the same structure as the machine M_Q, which is considered above. The only slight difference is in functioning. As in the case of M_Q, the result of T_R^1 goes to the output tape of M_P. If this result is 0, then either M_P stops functioning, informing that it has the final result, or, when we want to have an inductive Turing machine that always works without stopping, the machine M_P continues to function without stopping, but writes nothing else on its output tape. If the result of T_R^1 is 1, then T_N^1 begins the next cycle of the machine M_P.

This definition of functioning of M_P allows us to show that the machine M_P computes the characteristic function of the relation $P(z) = \forall x\ R(x, z)$. The relation $P(z)$ is arbitrary in the class Π_1 and M_P is an inductive Turing machine of the first order. Consequently, we have demonstrated that any relation from the class Π_1 is decidable by some inductive Turing machine of the first order.

This completes the first step of our induction.

2. Now we have to make a general inductive step, building inductive Turing machines for classes Σ_n and Π_n. At the beginning, we assume that for any relation from the class Σ_{n-1} or the class Π_{n-1}, there is an inductive Turing machine of order $n-1$ that builds the characteristic function of this relation. Let $Q(z)$ be an arbitrary relation that belongs to the class Σ_n. It means that $Q(z) = \exists x_1\ \forall x_2\ \exists x_3 \dots \forall x_{n-2}\ \exists x_{n-1}\ \forall x_n\ R(x_1, \dots, x_n, z)$, where $R(x_1, \dots, x_n, z)$, is a recursive relation. Then, by the definition, the relation $K(x_1, z) = \forall x_2\ \exists x_3 \dots \forall x_{n-2}\ \exists x_{n-1}\ \forall x_n\ R(x_1, \dots, x_n, z)$ belongs to the class Π_{n-1}.

To make an inductive step, we need for an arbitrary relation $Q(z)$ that belongs to Σ_n such an inductive Turing machine M_Q that computes the characteristic function of $Q(z)$ and an arbitrary relation $P(z)$ that belongs to Π_n such an inductive Turing machine M_P that computes the characteristic function of $P(z)$. We begin with the relation $Q(z)$ from Σ_n. By the definition, $Q(z) = \exists x_1\ K(x_1, z)$ where the relation $K(x_1, z)$ belongs to the class Π_{n-1}. By the inductive assumption, there is an inductive Turing machine M_K of order $n-1$ such that given a pair (a, b) of natural numbers, M_K produces 1 when $(a, b) \in K(x_1, z)$ and produces 0 when $(a, b) \notin K(x_1, z)$. This machine M_K is used to build ties in the inner memory of the machine M_Q.

By Lemma 4.3.3, it is possible to assume that M_Q has infinitely many working tapes $L_1, L_2, \ldots, L_t, \ldots$. We use the tape L_1 for keeping 1 in the first cell and 0 in the second cell. The inductive Turing machine M_K computes the following relation for memory connections: the cell with a number m in the tape L_2 is connected to the first cell of the tape L_1 if $m = r(a, b)$ and $(a, b) \in K(x_1, z)$ and connected to the second cell of the tape L_1 if $m = r(a, b)$ and $(a, b) \in K(x_1, z)$. Here $r: N^2 \to N$ is a one-to-one recursive mapping from all pairs of natural numbers onto all natural numbers (cf. Section 4.3.3). We label all these connections of cells from L_2 to cells from L_1 by the letter C. Other tapes $L_3, L_4, \ldots, L_t, \ldots$ are used for counters and organization of access of the head of M_Q to any cell by its address in its tape.

In addition to tapes, the inductive Turing machine M_Q contains, as its parts, the pairing automaton AP, a Turing machine T_r, which given a pair (a, b) converts it into the number $r(a, b)$, and the finite automaton T_N^1, which is an isomorphic copy of the finite automaton T_N. All parts of the machine M_Q are realized as subroutines.

The structure of the machine M_Q is presented in the Figure 4.6.

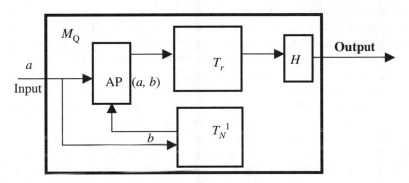

Figure 4.6. The structure of the machine M_Q.

Now we can describe functioning of the machine M_Q.

Given a number a as an input, the machine M_Q works in the following manner. The element a is sent to T_N^1 and to the automaton AP. The automaton T_N^1 produces number 1, which is sent to AP. The automaton AP combines a and 1 into the pair $(1, a)$ and sends it to the machine T_r. The machine T_r transforms the pair $(1, a)$ into the number $r(1, a)$. Then the processor H of the machine M_Q goes to the cell of L_2 that has number $r(1, a)$ and goes from this cell by the connection C. According to the structuring of the memory of M_Q, after going by the connection C, the head comes to the first cell of the tape L_1 when $(1, a)$ belongs to the relation $K(x_1, z)$ and to the second cell of the tape L_1 when $(1, a)$ does not belong to the relation $K(x_1, z)$. After this the processor H writes the content of the cell to which it has come into the output tape of M_Q.

If this result is 1, then either M_Q stops functioning, informing that it has the final result, or, when we want to have an inductive Turing machine that always works without stopping, the machine M_Q continues to function without stopping, but writes nothing else on its output tape.

If the result written in the output tape is 0, then T_N^1 begins the next cycle of the machine M_Q. The automaton T_N^1 produces number 1, which is sent to AP. The automaton AP combines a and 2 into the pair $(2, a)$ and sends it to the machine T_r. The machine T_r transforms the pair $(2, a)$ into the number $r(2, a)$. Then the processor H of the machine M_Q goes to the cell of L_2 that has number $r(2, a)$ and goes from this cell by the connection C. According to the structuring of the memory of M_Q, after going by the connection C, the processor H comes to the first cell of the tape L_1 when $(2, a)$ belongs to the relation $K(x_1, z)$ and to the second cell of the tape L_1 when $(2, a)$ does not belong to the relation $K(x_1, z)$. After this M_Q writes the content of the cell to which the processor H has come into the output tape.

If this result is 1, then either M_Q stops functioning, informing that it has the final result, or, when we want to have an inductive Turing machine that always works without stopping, the machine M_Q continues to function without stopping, but writes nothing else on its output tape. If the result written in the output tape is 0, then T_N^1 begins the next cycle of the machine M_Q.

This process continues and in all possible cases the machine M_Q gives some result. By the definition of the functioning of the machine M_Q, the result that it produces for a is 1 when there is some number x for which the pair (x, a) belongs to $K(x_1, z)$ and the result is 0 when such number x does not exist. In other words, as a is arbitrary natural number, the machine M_Q computes the characteristic function of the relation $Q(z) = \exists x_1 \; K(x_1, z)$. Relation $Q(z)$ is also arbitrary in the class $\boldsymbol{\Sigma_n}$ and M_Q is an inductive Turing machine of order n because its memory is structured by the inductive Turing machine M_H of order $n - 1$, AP is a finite automaton, while T_R^1 and T_N^1 are conventional Turing machines. Consequently, we have demonstrated that any relation from the class $\boldsymbol{\Sigma_n}$ is decidable by some inductive Turing machine of order n.

The same is true for the class $\boldsymbol{\Pi_n}$. Indeed, given a relation $P(z)$ from this class, we know from the definition that $P(z) = \forall x \; K(x, z)$ where the relation $K(x_1, z)$ belongs to the class $\boldsymbol{\Sigma_{n-1}}$. By the inductive assumption, there is an inductive Turing machine M_K of the order $n - 1$ such that given a pair (a, b) of natural numbers, M_K produces 1 when $(a, b) \in K(x_1, z)$ and produces 0 when $(a, b) \notin K(x_1, z)$. This allows us to build an inductive Turing machine M_P of the first order that computes the characteristic function of the relation $P(z)$. This machine has the same structure as the machine M_Q, which is considered above. The only difference is in functioning. As in the case of M_Q, the result of T_R^1 goes to the output tape of M_P. If this result is 0, then either M_P stops functioning, informing that it has the final result, or, when we want to have an inductive Turing machine that always works without stopping, the machine M_P continues to function without stopping, but writes nothing else on its output tape. If the result of T_R^1 is 1, then T_N^1 begins the next cycle of the machine M_P.

This definition of functioning of M_P allows us to show that the machine M_P computes the characteristic function of the relation $P(z) = \forall x \; K(x, z)$. Relation

$P(z)$ is also arbitrary in the class Π_n, and M_P is an inductive Turing machine of order n. Consequently, we have demonstrated that any relation from the class Π_n is decidable by some inductive Turing machine of order n.

The theorem is proved. □

To find the computing power of inductive Turing machines, we use the following result.

Theorem 4.3.9. *If a set X is decidable by some inductive Turing machine, then there is an inductive Turing machine of the same order that computes X.*

The proof is similar to the proof of the statement that all recursively decidable sets are recursively computable or enumerable (cf., for example, (Hopcroft, Motwani, and Ullman, 2001) or (Rogers, 1987)).

Theorems 4.3.8 and 4.3.9 imply the following result.

Corollary 4.3.4. (Superrecursive Computation Theorem). *a) For any arithmetical relation Y, there exists an inductive Turing machine M such that it computes Y.*

b) If Y belongs to the level n of the arithmetical hierarchy, that is $Y \in \Sigma_n \cup \Pi_n$, then there is an inductive Turing machine M of order n such that it computes Y.

Corollary 4.3.5. *Any arithmetical relation is decidable in the class of all inductive Turing machines with inductive memory.*

Corollary 4.3.6. *There is an inductive Turing machine that computes all sentences that have n quantifiers or less and is true in elementary arithmetic.*

Corollary 4.3.7. *There is an inductive Turing machine that determines (decides) if an arbitrary sentence with n quantifiers or less is true in elementary arithmetic or not.*

From the Hierarchy Theorem of Kleene (1955), which states that all inclusions in the arithmetical hierarchy are proper, we deduce the following result.

Corollary 4.3.8. *The hierarchy* **IM** *of inductive Turing machines with the inductively defined memory is strictly increasing, that is, all inclusions* **ITM$_n$** $\subset$ **ITM$_{n+1}$** *are proper.*

It is possible to give a direct proof of this result based on Theorem 4.3.7.

This means that inductively structured memory extends the computability space of inductive Turing machines. In contrast to inductive Turing machines, this nonlinear phenomena of the space extension does not appear in the computability space of Turing machines because a Turing machine with a structured memory that is built by another Turing machine is still equivalent to an ordinary Turing machine. Although as we will see in Chapter 5, such Turing machine can be much more efficient than any Turing machine with the conventional memory in the form of tapes.

Such sets that can be decided by some inductive Turing machine are called inductively decidable.

Corollary 4.3.9. *Any recursively enumerable or computable or acceptable set in inductively decidable.*

In particular, there is an inductive Turing machine that for any Turing machine T and any input symbol decides whether T gives a result or not. There is also an inductive Turing machine that for any Turing machine T decides whether the language of T is empty or not.

It is possible to extend the proof of the Superrecursive Representation Theorem to the arithmetical hierarchy in an arbitrary countable set X and to build inductive Turing machines with the oracles that decide and compute sets from this relative hierarchy.

Let X be some subset of N.

Theorem 4.3.10. (Relative Superrecursive Representation). *a) For any arithmetical in X relation Y, there exists an inductive Turing machine M with the oracle $O(X)$ such that it computes the characteristic function of Y.*

b) If Y belong to the level n of the arithmetical hierarchy in X, that is $Y \in \Sigma_n(X) \cup \Pi_n(X)$, then there is an inductive Turing machine M of order n with the oracle $O(X)$ such that it decides Y.

Theorems 4.3.7 and 4.3.10 imply the following result.

Corollary 4.3.10. (Relative Superrecursive Computation Theorem). *a) For any arithmetical in X relation Y, there exists an inductive Turing machine M with the oracle $O(X)$ such that it computes Y.*

b) If Y belongs to the level n of the arithmetical hierarchy in X, that is, $Y \in \Sigma_n(X) \cup \Pi_n(X)$, then there is an inductive Turing machine M of order n with the oracle $O(X)$ such that it computes Y.

Corollary 4.3.11. *Any relation arithmetical in X is superrecursively decidable.*

All these results show that the computing power of superrecursive algorithms is infinitely larger than the computing power of recursive algorithms. The latter can only compute the first level of the arithmetical hierarchy, while the former can build the whole infinite hierarchy. The same is true for the deciding power of both types of algorithms.

4.3.6 Inductively computable functions

> *Alice was not much surprised at this,*
> *she was getting so well used to queer things happening.*
>
> Lewis Carroll, 1832–1898

It is possible to build an inductively computable function that grows almost everywhere much faster than any recursively computable function. This also shows that inductive Turing machines are more powerful than Turing machines.

Theorem 4.3.11. *For any recursive function* $h(x, z)$, *there is an inductively computable function* $f : N \to N$ *such that for any partial recursive function* $g : N \to N$ *and for almost all n from N for which* $g(n)$ *is defined, we have*

$$f(n) > h(n, g(n)).$$

Proof. To prove the theorem, we use a Gödel enumeration $T_1, T_2, \ldots, T_n, \ldots$ of all Turing machines (cf. (Rogers, 1987)) and build an inductive Turing machine M such that M computes the necessary function $f : N \to X$. This machine M contains a copy of a universal Turing machine U, a copy of a Turing machine that computes the function h and a finite automaton C that given a pair, triple, or n-tuple of words, compare is they are equal or not, that is, C gives 1 when all input words are equal and gives 0 otherwise. We define functioning of M and prove that M has necessary properties defining how M works for an arbitrary input n.

Given input n, M simulates the machines $T_1, T_2, \ldots, T_n$ with the same input n for all of them. Before any of these n Turing machines stops, giving its result, the machine M writes onto its output tape 1 after each odd step of computation and writes onto its output tape 0 after each even step of computation. After the first of these Turing machines stops, giving its result k, the machine M, using a copy of a Turing machine that computes the function h, computes the value $h(n, k) + 1$ and writes it into its output tape. Then M continues writing the same result until the next of these n Turing machines stops, giving its result p. Then the machine M, using copy of a Turing machine that computes the function h, computes the value $h(n, p) + 1$, compares it with the value $h(n, k) + 1$ and writes the larger of the two into its output tape. Then M continues writing the same result until the third of these n Turing machines stops, giving its result q, and so on. In such a way, the machine M computes the function

$$f(n) - \max\{h(n, g_i(n)) + 1; i - 1, 2, \ldots, n\}$$

where $g_i(x)$ is the function computed by the Turing machine T_i and the operation max is defined when, at least, one of the values $h(n, g_i(n))$ is defined.

If a Turing machine T is given, then $T = T_r$ for some r. As a result, $f(n) > h(n, g_r(n))$ for all $n > r$ for which $g_r(n)$ is defined.

The theorem is proved.   □

This result shows that inductive Turing machines can compute functions that grow very fast, much faster than any partial recursive function. At the same time, Corollary 5.3.4 and Theorems 5.3.8 and 5.3.9 demonstrate that inductive Turing machines can compute functions that grow very slowly, slower than any partial recursive function. This explains how inductive computation can extend possibilities of conventional recursive computation.

4.3.7 Universal inductive Turing machines

> *"You see," the Knight went on after a pause,*
> *"it's as well to be provided for everything."*
>
> Lewis Carroll, 1832–1898

Universal Turing machines play a very important role in the theory of recursive algorithms. For example, any universal Turing machine determines what is recursively computable, acceptable or decidable. In addition, the whole theory of Kolmogorov complexity (cf. Section 5.3) is based on the concept of universal Turing machine. Moreover, universal Turing machines influenced computer technology. The structure of a universal Turing machine served as a model for the famous von Neumann architecture for a general-purpose computer (cf. Burgin, 2001). This architecture has determined for decades how computers have been built.

In a similar way, we need universal Turing machines to develop a theory of recursive algorithms and especially, a theory of superrecursive algorithmic complexity (cf. Section 5.3).

Taking a function $\mathbf{c} \colon \mathbf{T} \to \Sigma^+$, we obtain the following result.

Theorem 4.3.12. *For any class* $\mathbf{IT}_n$ *of all inductive Turing machines of order n with the same working alphabet S and alphabet Q of states, there exists a universal inductive Turing machine U of order n.*

Proof. Before beginning the construction of the machine U, let us make some preliminary remarks. First, we remind that a universal inductive Turing machine U of the nth order has to able to simulate any inductive Turing machine T of the nth order with its arbitrary input. That is why we begin the construction of a universal inductive Turing machine U of the nth order with enumeration of all inductive Turing machines of the nth order.

Second, properties of the structured memory allows us to assume that the memory of T is realized in one linear tape and the processor of T contains only one head, while the memory of U contains as many linear tapes and processor of U has as many heads as we need. As it is explained in Section 4.3.1, it is possible to consider any configuration in the structured memory E_T as a single word.

Any inductive Turing machine T of order n can be decomposed into two components: the mainframe BT of T, which consists of the control automaton A, the program (set of rules) R, and the operating device (head) H, and the structured memory E_T of the inductive Turing machine T. The mainframe BT of T is almost the same as the mainframe of any conventional Turing machine. Only the rules can contain description of arbitrary moves specified by ties in the memory. It is possible to consider the mainframe BT as a machine that has everything that has the machine T but the memory of BT is a conventional linear tape, which lacks additional structure. This structure is built by machine T_E, which organizes connections in the memory E_T of T.

Codification of T is constructed by an inductive process. According to the structure of an inductive Turing machine with a structured memory, it is natural to build a code of an inductive Turing machine T of the nth order separated into two parts. The first part is the code of the mainframe BT of T and the second part is the code of the structured memory E_T of the machine T. These parts are separated by $n+5$ symbols "1". The code of the mainframe of T is similar to the code of a conventional Turing machine that is built in Section 2.3. We consider this coding below. As we have seen, for a Turing machine, it is necessary to code only its rules, which constitute the dynamic part of the control device of a Turing machine.

What we are concerned about the structured memory E_T of T is that it may be considered as a linear tape with additional relations. These relations are generated by some Turing machine T_E of order $n - 1$. So, to codify the structure of E_T, we need only to codify the machine T_E. When $n = 1$, T_E is a Turing machine and its code $\mathbf{c}(T)$ is described in Section 2.3. When $n > 1$, we assume that we have the code of T_E. And combine it with the code of the mainframe of T as it is described above.

To apply induction, we note that an inductive Turing machine T of the nth order is stratified into n levels. The first level is the mainframe BT of T. The second level is the mainframe BT_E of the machine T_E, which has order $n - 1$. The third level is the mainframe BT_{2E} of the machine T_{2E} that builds connections in the memory of T_E and has order $n - 2$. The fourth level is the mainframe BT_{3E} of the machine T_{3E} that builds connections in the memory of T_{3E} and has order $n-3$ and so on. Thus, the code of T consists of codes of all these mainframes where the code of the mainframe with a number i is separated from the code of the mainframe with a number $i - 1$ by by $i + 5$ symbols "1".

Now let us codify all inductive Turing machines with a binary alphabet $\Sigma = \{1, 0\}$. To do this, we build a function $\mathbf{c}: \mathbf{IT} \to \Sigma^+$ where $\mathbf{IT}$ is the set of all inductive Turing machines with the alphabet Σ. To represent an inductive Turing machine T as a binary string, we at first assign integers to the states, tape symbols, and directions L, R, and S. If T has the states $q_1, q_2, \ldots, q_k$ for some number k, then the start state is always q_1 and we correspond to each q_i the string 0^i. To the tape symbols 0, 1, and the empty symbol Λ, which denotes an empty cell, we correspond 0, 00, and 000, respectively. If the connection types in the memory E_T of T are $c_1, c_2, \ldots, c_t$ for some natural number t, then we correspond to each c_i the string 0^i.

Once we have established an integer to represent each state, symbol, and direction, we can encode the transition rules or, equivalently, function transition of the transition. When we have a transition rule $q_i, a_j \to q_k, a_l, c_m$, for some natural numbers i, j, k, l, and m, we code this rule by the string $0^i 10^j 10^k 10^l 10^m$. As all of i, j, k, l, and m, are at least one, there are no occurrences of two or more consecutive 1's within the code for one transition rule. Having codes for all transition rules, we write them as one word in which they are separated by couples of 1's. For example, we have the code $0^i 10^j 10^k 10^l 10^m 110^h 10^t 10^r 10^q 10^p$ for two transition rules q_i, $a_j \to q_k, a_l, c_m$ and $q_h, a_t \to q_r, a_q, c_p$.

To simulate input and output operations of the machine T and functioning of the mainframe BT of T, the machine U has several tapes: L_I, L_O, L_1, L_2, L_3, L_4, and L_5. As usually, L_I and L_O are input and output tapes of U, correspondingly. The first tape L_1 is used for storing the word that represents the input w of T. The second tape L_2 is used for storing the word that represents the current state of the control automaton A of T. The third tape L_3 is used for storing the rules for operation of T. The fourth tape L_4 is used for storing the current word that is written in the tape of T. The fifth tape L_5 is used for auxiliary operations of U.

Additional tapes are used for simulating in the active mode the structured memory E_T of the machine T. The structured memory E_T of the machine T is simulated by the structured memory E_U of the machine U. If the machine T has order n, then its memory E_T is $(n - 1)$-inductive, that is, all relation in E_T are constructed by an inductive Turing machine T_E of order $n - 1$. Thus, the machine U has two parts: the first simulates its mainframe B_T of T and the second simulates the structured memory of T. Simulation of the mainframe BT of T does not depend on the order of T.

To simulate input and output operations of the machine T and functioning of the mainframe of T, the machine U has several heads h_c, h_o, h_1, h_2, h_3, h_4, and h_5. The head h_c reads the input of U from its input tape, while the head h_o writes the output of U to its output tape. The first head h_1 is used for reading when it is necessary the input word w of T, which is stored in the first tape L_1. The second head h_2 is used for changing the word that represents the current state of the control automaton A of T. The third head h_3 is used for searching the rules for operation of T. The fourth head h_4 is used for simulating the work of the working head of the machine T, which is changing the word on the tape L_1 that reflects the changing memory content of T. The fifth head h_5 is used for auxiliary operations of U. In addition to this, U has a part that simulates the machine T_E that builds connections in the memory E_T of T.

When $n = 1$, T has recursive memory and it is necessary to simulate only the mainframe BT of T. This is done in the same way as for the mainframe of an inductive Turing machine of arbitrary order. Thus, we consider simulation of T, assuming that its memory is already constructed by the machine U.

Given a word $r(\mathbf{c}(T), w)$ where $r: N^2 \to N$ is one-to-one function that is described in Section 4.3.3, the machine U simulates the machine T working with the word w. Simulation of the machine T is performed in several cycles.

The first cycle:

The machine U examines whether the input v has the form $r(\mathbf{c}(T), w)$, finding, in particular, if the first component of the pair $r^{-1}(v)$ is a legitimate code $\mathbf{c}(T)$ for some inductive Turing machine T. This is possible to do by standard methods in a finite number of steps (cf., for example, (Hopcroft, Motwani, and Ullman, 2001) or (Ebbinghaus et al., 1970). If the answer is *yes*, that is, the input v passes the test, U goes to the second cycle of simulation. If the answer is *no*, U gives no result. It can stop without generating an output or go into an infinite resultless cycle.

The second cycle:

The machine U:

1. rewrites the word w on the first working tape L_1;
2. rewrites the word w to the fourth working tape L_4,
3. writes the initial state q_0 of T on the second working tape of U; and
4. puts its head h_4 into the cell that contains the first symbol a_1 of the word w.

According to the chosen encoding of words, the word 10 is written on the second tape for each 0 of w, and the word 100 is written there for each 1 of w. Note that the empty cells of the simulated tape of M, which are represented by 1000, do not actually appear on the second tape because all cells beyond those used for w hold the empty cell of U. However, U is programmed so that when it looks for a simulated symbol of T and finds its own empty cell, it replaces that blank cell by the sequence 1000 to simulate the empty cell of T.

Then U goes to the third cycle.

The third cycle:

The machine U reads the symbol in the cell where the head h_1 is situated. At the beginning, it is the first symbol a_1 of the word w. If $a_1 = 0$, then U finds the first rule r with one 0 after 1 and one 0 after 11. If $a_1 = 1$, then U finds the first rule r with one 0 after 1 and two 0's after 11.

Then U goes to the fourth cycle.

The fourth cycle:

If the operations for h_1 are prescribed in the rule r that is found in the third cycle, then the machine U performs these operations. They can include:

1. reading with the head h_1 from the first tape L_1;
2. writing with the head h_o the description of an output symbol x of T to the output tape L_o when the rule r prescribes T to write x to its output tape;
3. changing with the head h_2 the content of the second working tape to the number of 0's that go after 11111 in the right part of the rule r;
4. changing with the head h_4 the content of that cell in the fourth working tape: the new content is 0 if there is one 0 that goes after 11 in the right part of the rule r and the new content is 1 if there is two 0's that go after 11 in the right part of the rule r; and
5. moving the head h_4 according to the connection in the corresponding rule; this move simulates in a definite way the corresponding move of the head of T.

The connections between the cells of the tape L_4 are installed by the part of U that simulates the machine T_E. When T has a recursive memory, T_E is a conventional Turing machine, for which there is a universal Turing machine, which performs the same actions.

After executing the prescribed operations, U goes back to the third cycle.

If the machine T has no transition that matches the simulated state and tape symbol, then no transition will be found by U. Thus, the machine T halts in the simulated configuration, and the machine U does likewise.

In such a way, U simulates each step of the machine T, and because the word written in the output tape of T when it works with input word w coincides on each step with the word written in the output tape of T when it works with the input word $(\mathbf{c}(T), w)$, For this reason, U gives the same result as T when T gives a result. In addition, U gives no result when T gives no result. Consequently, U completely simulates functioning of T.

According to its construction, the inductive Turing machine T of nth order is stratified into n levels. Each level has its mainframe, which can be simulated in a similar way as the mainframe of T. As simulation of one mainframe requires seven tapes and heads, the machine U can perform its functions with $7n$ tapes and $7n$ heads.

As T is an arbitrary inductive Turing machine of nth order, Theorem 4.3.12 is proved.   □

This result shows that while the existence of a universal Turing machine, a machine that can compute any recursively computable function, is a pillar of the classical theory of computation, a similar pillar for the theory of inductive computation exists in form of a universal inductive Turing machine, a machine that can compute any inductively computable function. Thus, only universal inductive Turing machine foretold and provides a foundation for the modern general-purpose computer because conventional Turing machines do not offer sufficiently adequate models for modern computers (cf. Section 4.1 and (Burgin, 2001)).

In addition, universal inductive Turing machines allow one to develop invariant theory of algorithmic (Kolmogorov) complexity for inductive computations.

4.4 Grid automata: Interaction and computation

> Alice was puzzled. "In our country,"
> she remarked, "there's only one day at a time."
> The Red Queen said, "That's a poor thin way of doing things.
> Now here, we mostly have days and nights
> two or three at a time, and sometimes
> in the winter we take as many as
> five nights together — for warmth, you know."
>
> Lewis Carroll, 1832–1898

At the end of the twentieth century, computer technology has come to a level when computation and communication are combined into a single process of information processing. We still call this process computation because computation, in some sense, lies at the bottom of communication. For example, communication, as a rule, includes coding and decoding of messages.

New features of information processing are due to the very fast development of information technology, and we need models that reflect these advances and allow us to go further. Now a new approach in information technology appeared. It is called cluster computation. In it computers are combined together as clusters that work as a single system, which is called a cluster computer. The demands of the high performance computing market are causing an outward expansion from localized clusters of computers to clusters of clusters shared between different departments and even between remote sites within the organization. But clusters are relatively limited systems. They tend to have homogenous hardware and software platforms, existing at a single site and having one specific function that they were built to accomplish (Lumb, 2002).

A more advanced level of this approach is grid computing, which, on the other hand, is a natural evolution of the Internet. Grid computing is realized by grid information processing systems or grid arrays. Such system is the organization of computers into interlinked hierarchical networks that can be tapped for processing and communication power (Foster, 2002; Waldrop, 2002). According to California Institute for Telecommunications and Information Technology director Larry Smarr, grid computing has even more growth potential than the Internet explosion of the 1990s. Smarr expects that grid computing will provide an infrastructure to support the entire economy. He foretells a future that features interconnected grids of all sizes, running the gamut from supercomputer clusters to mid-sized nodes of desktops and laptops to PC-based "micronodes." Grid computing projects are initially being established to provide data-crunching power for scientific and academic research, but Ian Foster of Argonne National Laboratory believes that they will also serve as testbeds for commercial applications (Waldrop, 2002). At this time, Grid computing is still a work in progress.

However, as Lumb (2002) writes, grid computing is not a matter of some distant future. It is presently helping many organizations dynamically integrate their disparate, heterogeneous compute resources. Organizations that do not want to wait to reap the benefits of the global Grid have already built their own enterprise grids and partner grids. While classical scheme of computing emphasizes information transformation, grid computing shifts the emphasis, making communication and interaction as important as information transformation.

Thus, we need an efficient model to represent modern computers, networks, and other computing devices that combine in various processes computation and communication. And we have such models: abstract neural networks, cellular automata, Petri nets, iterative arrays, persistent Turing machines and some other models communicate and compute. The advent of the Internet intensified research in this direction. But neither of these models has the unifying property that provides a general context for studies of the huge diversity of contemporary information processing systems. Each particular model represents only some kind of interacting and communicating systems. Abstract neural networks, Petri nets, and cellular automata allow one to study inner interaction in a network of neurons or finite automata, but tell us very little about interaction with the environment. In addition, these models, as well as Internet machines (cf. Section 4.2.7) contain all elements of the same kind. Neural networks are built exclusively of similar neurons. Cellular automata form a network of copies of some finite automaton. Internet machines include only Turing machines as their primitive elements. Petri nets have only identical places and transitions as nodes. Colored Petri nets allow more diversity in element's choice, but their elements are the same places and transitions, only having different names (colors).

Other models of interaction either pay the whole attention to the interaction with the environment, as do persistent Turing machines (Goldin and Wegner, 1988), or as theories of interactive processes (cf. (Milner, 1993), (Hoar, 1978)) focus their attention on local description of interaction, while the global description does not come into the picture.

In contrast to this, global networks, such as the Internet or the future GRID (Foster, 2002; Lumb, 2002; Waldrop, 2002), include a huge diversity of interacting devices. These devices interact not only with one another, but also with a lot of other systems, the main of which are users of computers and staff that provides maintenance of computers, workstations, servers, imbedded devices, and other elements of the network. All this shows that we need a more developed model for modeling, study, and development of such systems and processes. Such model has to be able to represent not only contemporary systems, but provide a theoretical base for study and simulation of future computing systems. A solution to this problem is given by grid arrays and grid automata.

In particular, both approaches of cluster computers and grid computing are synthesized in the concept of grid array. Grid arrays embody the *autonomous distributed computing architecture* and realize the concurrent paradigm of computation in the most complete form. Grid automaton is a theoretical model of an arbitrary grid array. This model allows one to study properties of grid arrays related to their design, functioning, maintenance, and utilization. Flexibility of grid automata makes them an appropriate mathematical model for colonies and other systems of intelligent agents. Grid automata allows one naturally to model and embody swarm intelligence (Kennedy and Eberhart, 2001). To do all this, it is necessary to enhance nodes of a grid automaton with such properties as autonomous functioning, coordination with other nodes, mobility, intelligence, and so on.

4.4.1 Grid computation and grid arrays

> *The next moment soldiers came running through the wood*
> *at first in twos and threes, then ten or twenty together, and*
> *at last in such crowds that they seemed to fill the whole forest.*
>
> Lewis Carroll, 1832–1898

To achieve a better representation of modern information processing systems, we introduce the following concept.

Definition 4.4.1. A *grid array* is a system of information processing systems (computers, networks, imbedded systems, etc.), which are situated in a grid, called nodes, are optionally connected and interact with one another.

The concept of grid array is very general, allowing one to include into the grid people and even nonliving (natural and artificial) systems when they are considered as information processing systems. One grid array can be a node of another grid array. However, the most direct application of the grid approach is to computers and their conglomerates as it is stated above.

Grid arrays, as a rule, have distinct layers. This implies several types of stratification for grid arrays. Physical or hardware stratifications are created trough the hierarchy of physical devices. At the foundation of a grid array of a global network, there are tangible resources to be shared: servers, desktops, network channels, software, storage, software licenses, data, and so on. On the next level of the physical hierarchy, these elements are combined into more complex systems: computers are considered with all their software and databases, local networks and clusters of computers become elements of the new level and so on.

At the same time, in the emerging nomenclature of grid computing, physical resources are treated in terms of the services that they can add to the grid's capabilities. This generates a virtual hierarchy of grid services. In this virtual hierarchy, users and applications get transparent access to the grid services.

Synthesizing these forms of grid arrays, we come to three principal physical layers of a grid array:

1. *Hardware layer* consists of physical elements, components and/or devices, which may be considered with or without their software;
2. *Software layer* consists of program elements (instructions, operators, and so on), components, separate programs, and systems of programs together with those devices that these components and programs utilize in their functioning;
3. *Infware layer* consists of data elements (bits, symbols, and so on), groups of data elements, data bytes, and program data together with those programs that process these data and those devices that these program utilize in their functioning.

Definition 4.4.2. A *data byte* is any standard portion or group of data elements.

For example, the conventional byte is the group of eight bits, where a bit is a binary symbol.

Definition 4.4.2. *Program data* of some program P are all data that are given to this program for processing (input program data), produced by the program (inner program data), and given as the output (output program data).

Input program data may be given to the program as one portion. This is a classical model of computation such, for example, as a Turing machine. In contrast to this, program data may be given to the program as several portions. There are three modes of such data supply:

1. The atemporal mode when portions are given in a sequence without any relation to time. This is also a classical model of computation such as finite automaton.
2. The deterministic temporal mode when portions are given at predetermined intervals or moments of time. An example of such model is given by timed automata (Alur and Dill, 1994).
3. The nondeterministic or random temporal mode when portions are given at random intervals or moments of time. An example of such model is given by interactive automata, such as persistent Turing machines (Goldin and Wegner, 1988) or global Turing machines (van Leeuwen and Wiedermann, 2000a).

It is natural to separate grid arrays into three types:

1. An *actual grid array*, constituting a unified organized system of collaborating information processing systems as nodes.
2. A *virtual grid array*, a set of connected information processing systems as nodes.

3. A *potential grid array*, a set of information processing systems that are connected or have a potency to be connected.

Example 4.4.1. A cluster computer is an actual grid array.

Example 4.4.2. The Internet is a virtual grid array.

Example 4.4.3. All computers in the world form a potential grid array. It shows what a high potential that may be achieved through a synthesis of all computers into one grid array.

Example 4.4.4. Electronic systems of planes, ships, and cars are becoming increasingly complex. Consequently, planes, ships, and cars have extensive networks of electronic devices. For instance, electronic devices control many functions in the car, from anti-lock braking systems (ABS) and fuel injection to the volume control of entertainment systems. Car networks are usually divided into body and power train control branches, and telematics and multimedia subnetworks. Any car or plane network is an actual grid array, which contains several subarrays.

One more characteristic of grid arrays depends on properties of their nodes and connections. According to these properties, we discern four types of grid arrays:

1. A *static grid array* is an array in which devices are nodes of the array and connections between those devices are fixed from the beginning and do not change during functioning.
2. A *grid array with dynamic connections* is an array in which connections between the nodes are eventually changing during functioning.
3. A *grid array with dynamic nodes* is an array in which nodes are eventually changing during functioning.
4. A *grid array with dynamic nodes and connections* is an array in which nodes and connections between the nodes are eventually changing during functioning.

Example 4.4.5. The Internet is a grid array with dynamic nodes and connections.

Example 4.4.6. A computer network with fixed connections in which computers and their software are updated and/or upgraded from time to time is a grid array with dynamic nodes.

Example 4.4.7. A computer is a grid array with dynamic connections as connections between CPU and RAM cells are constantly changing.

4.4.2   Grid automata

> *The parts transfer their own significance to the larger group.*
> Georg Simmel, *Quantitative Aspects of the Group*, 1858–1918

Grid arrays are modeled by grid automata.

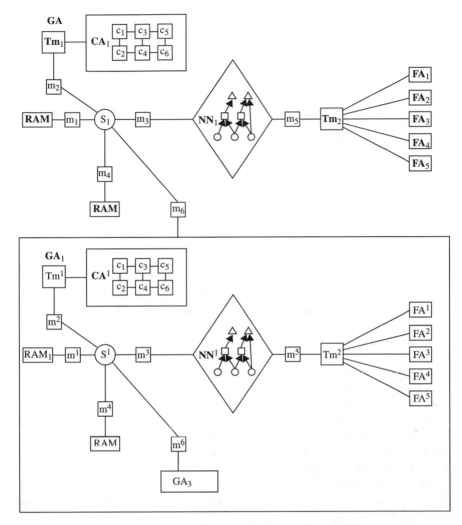

Figure 4.7. GA is a grid automaton; **Tm** is a Turing machine; **RAM** is a random access machine; **S** is a server; **m** is a modem; **NN** is a neural network; **FA** is a finite automaton; **CA** is a cellular automaton.

Definition 4.4.3. A *grid automaton* is a system of automata, which are situated in a grid, are called nodes, are optionally connected and interact with one another.

A grid automaton, one node of which is also a grid automaton similar to the initial is presented in Figure 4.7.

Thus, the difference is that a grid array consists of real (physical) information processing systems and connections between them, while a grid automaton consists of abstract automata as its nodes. Nodes in a grid automaton can be finite automata, Turing machines, vector machines, array machines, random access machines, inductive Turing machines, and so on. Even more, some of the nodes can be also grid automata.

We group connections in grid arrays and grid automata into three main types:

1. Simple connections that are not changing deliberately transmitted data and themselves when the automaton or array is functioning.
2. Transformable connections that may be changed when the automaton or array is functioning.
3. Processing connections that can transform transmitted data.

If we take a global network of computers and represent it as a grid array, connections between the nodes will be processing connections as these connections include different IPS: modems, servers, etc.

Transformable connections are the base of any reconfigurable computer (Casselman, 1993; Thornburg and Casselman, 1994). Reconfigurable computer provides means to change their hardware architecture by the software to suit the application at hand. The main idea is to place algorithm in hardware and to do this on a function by function basis as the application executes. Reconfigurable computers take advantage of parallelism while reducing overhead of load/store, branch operations, and instruction decoding. Reconfigurable devices can also be used either as attached processors or coprocessors. The idea of a reconfigurable computer was suggested by Estrin (cf., (Estrin and Viswanathan, 1962)).

A reconfigurable computer is a potential grid array that: a) contains many virtual grid arrays, and b) is an actual grid array at each computing step.

Grid automata can give many representations of the same grid array. There are three main types of such representations of a grid array G by a grid automaton A:

Hardware representation corresponds nodes of the automaton A to the devices and/or their groups in the array G, where a device may be a separate computer with all its software and infware (for example, databases) or the whole network of computers, servers and other IPS with their resources.

Software representation corresponds nodes of the automaton A to the programs and/or their systems in the array G, where programs are taken with their resources: devices and data that are used by a program.

Infware representation corresponds nodes of the automaton A to the services that are provided by the array G and/or their systems, where services are taken with their resources: devices, programs, and data that are used by a service.

There are different levels of such representations. For example, hardware representation of a grid array G can be on the level of logical elements or gates and on the level of IPS (computers and networks).

A grid automaton G is described by three grid characteristics and three node characteristics.

The grid characteristics are:

1. The *space organization* or structure of the grid automaton G.

This space structure may be in the physical space, reflecting where the corresponding information processing systems (nodes) are situated, it may be the system structure defined by connections between the nodes, or it may be a mathematical structure defined by the geometry of node relations. System structure is so important in grid arrays that in contemporary computers connections between the main components are organized as a specific device, which is called the computer bus. In a computer or on a network, a bus is a transmission path in form of a device or system of devices on which signals are dropped off or picked up at every device attached to the line.

There are three kinds of space organization of a grid automaton: *static structure* that is always the same; *persistent dynamic structure* that eventually changes between different cycles of computation; and *flexible dynamic structure* that eventually changes at any time of computation.

Persistent Turing machines (Goldin and Wegner, 1988) have persistent dynamic structure, while reflexive Turing machines (Burgin, 1992) have flexible dynamic structure.

2. The *topology* of G is determined by the type of the node neighborhood and is usually dependent on the system structure of G.

A natural way to define a neighborhood of a node is to take the set of those nodes with which this node directly interacts. In a grid, these are often, but not always, the nodes that are physically the closest to the node in question.

For example, if each node has only two neighbors (right and left), it defines linear topology in G. When there are four nodes (upper, below, right, and left), the G has two-dimensional rectangular topology.

However, it is possible to have other neighborhoods. For instance, Crutchfield and Mitchell (1995) consider linear cellular automata in which the neighborhood of each cell has the radius $r > 1$. It means that r cells from each side of a given cell directly influence functioning of this cell.

Topology of a grid automaton is very important. For example, changing two-dimensional rectangular topology, which is induced by the structure of the Euclidean plane, to a non-Euclidean topology, which is induced by the structure of the hyperbolic plane, makes possible for cellular automata to achieve much higher efficiency. For example, NP problems are tractable in the space of cellular automata in the hyperbolic plane (Margenstern and Morita, 2001) or polynomial-time cellular automata in the hyperbolic plane accept all PSPACE languages (Iwamoto, Margenstern, Morita, and Worsch, 2002).

3. The *dynamics* of G determines by what rules its nodes exchange information with each other and with the environment of G.

For example, when the interaction of Turing machines in a grid automaton G is determined by a Turing machine, then G is equivalent to a Turing machine (cf. Theorem 4.4.5). At the same time, when the interaction of Turing machines in a grid automaton G is random, then G is much more powerful than any Turing machine (cf. Theorem 4.4.3).

Interaction with the environment separates two classes of grid automata and arrays: *open grid automata or arrays* interact with the environment through definite connections, while *closed grid automata and arrays* have no interaction with the environment. For example, Turing machines are usually considered as closed automata because they begin functioning from some start state and tape configuration, finish functioning in some final state and tape configuration, and do not have any interactions with their environment.

The node characteristics are:

1. The *structure* of the node. For example, one structure determines a finite automaton, while another structure is a Turing machine.
2. The *external dynamics* of the node determines interactions of this node.

According to this characteristic there are three types of nodes: *accepting nodes* that only accept or reject their input; *generating nodes* that only produce some input; and *transducing nodes* that both accept some input and produce some input. Note that nodes with the same external dynamics can work in grids with various dynamics (for example, compare Theorems 4.4.3 and 4.4.4).

3. The *internal dynamics* of the node determines what processes go inside this node.

For example, the internal dynamics of a finite automaton is defined by its transition function, while the internal dynamics of a Turing machine is defined by its rules. Differences in internal dynamics of nodes are very important because a change in producing the output allows us to go from conventional Turing machines to much more powerful inductive Turing machines of the first order (Burgin, 1983).

Cellular automata, neural networks, systolic arrays, iterative arrays, hardware modification machines of Dymod and Cook (1980; 1989), and Petri nets are special kinds of grid automata (cf., for instance, Figure 4.7). In comparison with cellular automata, a grid automaton can contain different kinds of automata as its nodes. For example, finite automata, Turing machines and inductive Turing machines can belong to one and the same grid. In comparison with systolic arrays, connections between different nodes in a grid automaton can be arbitrary like connections in neural networks. In comparison with neural networks and Petri nets, a grid automaton contains, as its nodes, more powerful machines than finite automata. However, grid automata represent much more information processing systems and computational schemes. Thus, an important property of grid automata is a possibility to realize hierarchical structures, that is, a node can be also a grid automaton. In grid automata, interaction and communication becomes as important as computation. This peculiarity results in a variety of types of automata, their functioning modes, and space organization.

Grid automata allow one to represent not only distributed computing systems and schemes (neural networks, cellular automata, systolic arrays, Internet machines, and Petri nets), but also other types of automata and algorithms. Let us consider some examples.

Example 4.4.8. **Turing machine representation**.

A grid automaton that represents a Turing machine T on the highest level has following nodes:

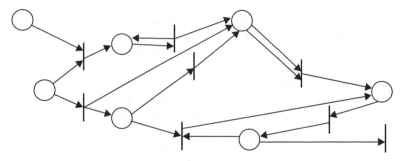

Figure 4.8. A Petri net as a grid automaton with nodes of two types: places ◯ and transitions │.

◇ A finite automaton A performing control operations in T;
◇ A finite automaton h (the head) performing transformation of symbols in T;
◇ A storage automaton c (a cell of the tape of T) performing storage of symbols in the memory (tape) of T.

To build the next layer of a Turing machine representation, we assume that automata in the previous layer are logical circuits, which are built of Boolean elements (cf. Section 2.3). Consequently, the nodes in the grid automaton of the second level representation are logical circuits, while the nodes in the grid automaton of the third level representation are Boolean elements.

This sequence of grid automata forms a physical stratification for a Turing machine T.

Remark 4.4.1. An important property of grid automata is their uniformity. The extent of uniformity is represented by the coefficient of uniformity introduced in (Bratalsky and Burgin, 1986). Cellular automata, systolic and iterative arrays are uniform grid automata. Neural networks are uniform only in nodes as they allow nonregular structure of connections. Petri nets are not completely uniform in nodes as they have two types of nodes: places and transitions. However, their uniformity converges to 1 as the number of elements grows. The same is true for grid automata that represent Turing machines. When the length of the tape of a Turing machine grows, the uniformity of the Turing machine and the uniformity of the representing grid automaton converge to 1. This gives additional supportive evidence to the **principle of asymptotic uniformity** (Bratalsky and Burgin, 1986), which states:

> *When the number of elements of an information processing system grows, its uniformity converges to 1, that is, the system becomes more and more uniform.*

Example 4.4.9. **Partial recursive function representation**.
A partial recursive function is a function that can be generated from the constant zero $0(x) = 0$, the successor function $S(x) = x + 1$, and projections

$P_i(x_1, \ldots, x_n) = x_i$ using any finite number of compositions, primitive recursion, and minimization (cf. Section 3.4). To build grid automata that represent and compute partial recursive functions, we consider such automata S, Z, P_i, B, R, and M that compute correspondingly functions $S(x)$, $0(x)$, $P_i(x_1, \ldots, x_n)$ and perform operations of compositions, primitive recursion, and minimization. In addition, we also take an automaton C that given a number n subtracts 1 from it and checks whether the result is equal to 0. When the result is equal to 0, C opens the gate V_1. When the result is not equal to 0, C opens the gate V_2. If $f(x_1, \ldots, x_n)$ is a partial recursive function, then we correspond to it a grid automaton $G(f)$ in which the type of all input nodes belongs to the set $\{S, Z, P_i; i = 1, 2, \ldots, n\}$ and the type of all other nodes belong to the set $\{C, R, M\}$. Namely, to each function $S(x)$, $0(x)$, $P_i(x_1, \ldots, x_n)$ that belong to a recursive description of $f(x_1, \ldots, x_n)$, a corresponding automaton S, Z, or P_i is related in $G(f)$. These automata are connected to those nodes C, R, and M of the automaton $G(f)$ that perform operations with the functions in the description of $f(x_1, \ldots, x_n)$. Continuing this process, we build the automaton $G(f)$. Simple induction shows that $G(f)$ computes f. The automaton C counts the number of iterations in $G(f)$.

For instance, let us take the function $f(x)$ that is given by the following scheme of recursion:

$$f(1) = 0, \ f(2) = 1, \ f(n) = P_2\big(S(f(n-2)), S(S(f(n-1)))\big), \quad n = 3, 4, \ldots .$$

Then $f(x)$ is represented by the following grid automaton A:

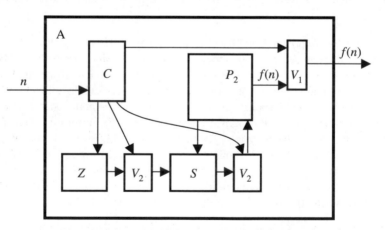

Figure 4.9. The structure of a grid automaton that realizes functional programming mode.

Example 4.4.10. **Persistent Turing machine** (cf. Section 4.2.7) representation:
    On the highest level, the grid automaton that represents a persistent Turing machine consists of two nodes: a Turing machine and the environment. All subsequent

levels are the same as for a Turing machine in Example 4.4.4 plus the nodes of a grid automaton that represent the environment in more detail. The environment of a persistent Turing machine can be always represented by one node. For example, there is one user of the machine and we do not give more detailed model of this user. In other cases, the environment can be a complex system that consists of many subsystems and we represent it by an extensive grid automaton. For example, environment can include many users, embedded devices, external computers and networks.

4.4.3 Properties of grid automata

Safe bind, safe find.

A proverb

While functioning of neural networks, cellular automata, iterative arrays, and systolic arrays is completely synchronous, a grid arrays or automaton has three modes of functioning that represent the dynamics of this array or automaton:

1. The *synchronous mode* when all nodes or automata execute each step of their computation at the same time.
2. The *synchronized mode* when there is a sequence ST of temporal points such that at each point all nodes or automata finish some step of computation and/or begin the next step. The sequence ST is called the *synchronization sequence* for the process.
3. The *asynchronous mode* when different nodes or automata function in their own time.

Proposition 4.4.1. *Any subsequence of a synchronization sequence is a synchronization sequence.*

Lemma 4.4.1. *If X and Y are some sets of nodes of a grid automaton G each of which functions in the synchronous (synchronized) mode and $X \cap Y \neq \emptyset$, then the subgrid $X \cup Y$ of the grid G also functions in the synchronous (synchronized) mode.*

Lemma 4.4.2. *The relations of synchronous and synchronized functioning are equivalence relation for sets of nodes in a grid automaton.*

Temporal modes of functioning can overlap with one another due to the type of their organization. There are three types of spatial organization of temporal modes.

1. *Local organization* when in some neighborhood of each node all nodes function in the same mode (for example, all nodes are synchronized).
2. *Cluster organization* when all nodes from the grid are divided into groups and all nodes in the same group function in the same mode (for example, all of them are synchronized).
3. *Global organization* when all nodes in the grid function in the same mode (for example, the work of all nodes is synchronous).

In a similar way, it is possible to systematize other characteristics of a grid automaton. For example, we have three types of the accepting states:

1. *Local accepting state* of a grid automaton G is a state of one of its nodes.
2. *Cluster accepting state* of a grid automaton G is a set of states of the nodes from a group of nodes from G.
3. *Global accepting state* of a grid automaton G is a set of states of all nodes from G.

Very often a covering of a topological space by a system $\{U_i; i \in I\}$ of open neighborhoods has the following property (**C**):

Any two neighborhoods U_i and U_j from the system $\{U_i; i \in I\}$ have the nonvoid intersection or may be connected by a finite chain $U_0, U_1, \ldots, U_k$ of neighborhoods from $\{U_i; i \in I\}$ such that $U_0 = U_i$, $U_k = U_j$ and $U_t \cap U_{t+1} \neq \emptyset$ for all $t = 1, \ldots, k - 1$.

For grid automata, this property has important implications.

Theorem 4.4.1. *In any grid automaton G, a locally (cluster) synchronous functioning with respect to a system of neighborhoods (groups) with the property (**C**) is globally synchronous.*

Proof. Let G be a grid automaton and U be the set of all nodes (automata) in the system G. By the assumption of Theorem 4.4.1, there is a system $\boldsymbol{E} = \{U_i; i \in I\}$ of neighborhoods in G such that all nodes in each U_i function synchronously and $\cup U_i = U$. As the functioning of G is locally synchronous, then the set $\boldsymbol{V}$ of all subsets V of U such that V functions synchronously is not empty because all U_i. Elements from the set $\boldsymbol{V}$ are ordered by inclusion of sets.

By the Zorn lemma (cf., for example, (Cohn, 1965) or (Kurosh, 1974)), the set $\boldsymbol{V}$ contains, at least, one maximal element W. If $W = U$, then the theorem is proved.

When $W \neq U$, then there is a node d that belongs to W, but does not belong to U. As $\cup U_i = U$, there is some neighborhood U_i that contains d. Let us take some neighborhood U_j that is a subset of W. Then by the condition (**C**) the neighborhoods U_i and U_j from $\{U_i; i \in I\}$ either have the nonvoid intersection or may be connected by a finite chain $U_0, U_1, \ldots, U_k$ of neighborhoods from $\{U_i; i \in I\}$ such that $U_0 = U_i$, $U_k = U_j$ and $U_t \cap U_{t+1} \neq \emptyset$ for all $t = 1, \ldots, k - 1$. In the first case, the functioning of d is synchronized with the functioning of W by Lemma 4.4.1. In the second case, we can use Lemma 4.4.1 several times and prove by induction in the length of the chain that connects U_i and U_j that the functioning of d is synchronized with the functioning of W because by Lemma 4.4.2 synchronous functioning implies a transitive relation on the sets of nodes. Consequently, it is possible to add the node d to the set W and the new system Z will be larger than W. This contradicts our assumption that the system W is maximal in $\boldsymbol{V}$. Thus, $W = U$ and the theorem is proved in the case of local synchronous functioning of G. $\square$

In the case of cluster synchronous functioning of G the proof is similar as all groups of nodes with the synchronous functioning satisfy the condition (C).

Remark 4.4.2. For synchronized functioning of a grid automaton G, the assertion similar to the statement of Theorem 4.4.1 is not true as the following example demonstrates.

Example 4.4.11. Let us consider a grid automaton G with the following system U of devices or nodes $\{d_0; d_{in}; i = 1, \ldots, n; n = 1, 2, 3, \ldots\}$ in the system G. The set $E = \{U_{0n}; U_{in}, i = 1, \ldots, n; n = 1, 2, 3, \ldots\}$ with $U_{0n} = \{d_0, d_{1n}\}$ and $U_{i,n} = \{d_{in}, d_{i+1,n}\}$ for all $i = 1, \ldots, n - 1; n = 1, 2, 3, \ldots$ form the system of neighborhoods in G. Here each set U_{0n} and $U_{i,n}$ is a neighborhood of all its points. The node d_0 performs each operation in a unit of time, that is, d_0 functions in the time $T = \{1, 2, 3, \ldots\}$. The node d_{in} $(i = 1, \ldots, n)$ performs each operation in 2^n units of time T, that is, nodes d_{in} functions in the time $T = \{2^n, 2 \cdot 2^n, 3 \cdot 2^n, \ldots\}$. Thus, the functioning of the grid G is locally synchronized, but a global synchronization sequence for the whole process in G does not exist.

However, additional conditions allow us to prove the assertion similar to the statement of Theorem 4.4.1 for synchronized functioning of grid automata.

Theorem 4.4.2. *In any grid automaton G, a locally (cluster) synchronized functioning with respect to a finite system of neighborhoods (groups) with the property (C) is globally synchronized.*

Proof is similar to the proof of Theorem 4.4.1.

Remark 4.4.3. Theorem 4.4.1 is proved for arbitrary systems, while all real information processing systems are finite. The question is why we need to consider infinite systems. We need to do this because many real grid arrays, being finite, are permanently growing and consequently, do not have the exact number of elements (nodes). This prevents us for building synchronization sequences for the whole functioning of such grid arrays. To consider them as potentially infinite, allows us to ignore their changes. In addition when the number of nodes becomes too large, we encounter phenomena that are peculiar for infinite entities (Kolmogorov, 1961; Burgin, 1997). So, it is reasonable to consider both finite and infinite grid arrays and automata.

When we do not restrict interaction of the nodes in grid automata, these automata become computationally very powerful. For instance, a simple grid automaton that consists of two finite automata with a common or shared infinite tape, similar to the tape of Turing machine, can compute any numerical function $f \colon N \to N$ or word function $h \colon A^* \to A^*$. We prove this for word functions because numbers are usually represented by words.

Let the grid automaton G consists of two finite automata B and C that work with words in the alphabet $\{1, 0\}$ and a common infinite tape L. They function in the following way. After receiving an arbitrary input, the automaton B produces some number of symbols 1 and then stops. The number of 1's is random. After receiving an arbitrary input, the automaton C produces some number of symbols 0 and then stops. The number of 0's is also random. In addition, each of these finite automata at random moments of time writes the obtained symbols in the tape L. Thus, given any word w in the alphabet $\{1, 0\}$, the output in the tape L may be an arbitrary word in the alphabet $\{1, 0\}$. Consequently, any word mapping can be realized by this system.

We have proved the following result.

Theorem 4.4.3. *For any function* $h\colon A^* \to A^*$ *or* $f\colon N \to N$, *there is a grid automaton G with two finite automata nodes that computes this function.*

Corollary 4.4.1. *For any function* $h\colon A^* \to A^*$ *or* $f\colon N \to N$, *there is a grid automaton G with two Turing machine nodes that computes this function.*

Thus, even simple grid automata can go much further than conventional Turing machines and the Church–Turing thesis. However, this contingency has a catch-up. As usually in life, if you get something, it is necessary to pay for it. In this case, much greater computational power of a random grid automaton in comparison with Turing machines results in our inability to direct it and even to choose a necessary process or function for realization. The results of such computation would be scarcely intelligible. When we want to compute what we intend, we inevitably come either to traditional recursive algorithms such as Turing machines or to novel superrecursive algorithms such as inductive Turing machines.

In some cases when the nodes of a grid automaton and their interactions are restricted to some class of automata, it is possible to reduce the whole grid to one automaton from this class.

Theorem 4.4.4. *If all nodes of a grid automaton G with the static structure and finite number of nodes are finite automata, then G is equivalent to a finite automaton.*

Theorem 4.4.5. *If all nodes of a grid automaton G with the static or persistent dynamic structure are recursive automata or algorithms and their interaction is controlled by a recursive automaton or algorithm, then G is a recursive automaton.*

This result means that it is possible to simulate such a grid automaton G by a Turing machine or conventional computer. Only efficiency of simulation will be very low.

Remark 4.4.4. Properties of inductive Turing machines show that for them similar reductions are impossible (Burgin, 2001). This is caused by nonlinear effects in extensions of inductive Turing machines.

4.4.4   Subroutines, autonomous agents, and virtual machines in grid arrays

There is safety in numbers.

A proverb

Grid automata as the most advanced abstract information processing systems have different categories of resources: memory, interface (input and output) devices, control devices, operating devices, software, data or knowledge bases.

Modes of resource utilization yield *interdependence classification* of automata in a grid:

1. *Autonomous automata* with independent resources.
2. *Automata with shared resources*, in which some resource, such as memory, belongs to one node, but some other nodes from the grid can also use it.
3. *Automata with common resources*, for example, common memory or a database, which belong to two or more nodes from the grid.

Each type of automata implies specific styles of exchange in the grid. For example, there are three levels of exchange for autonomous automata:

1. *Data and program exchange* (distributed storage of information).
2. *Task and workspace exchange* (distributed computation and intelligent agent systems).
3. *System exchange* (data, knowledge, tasks, programs, agents are specific systems in such exchange).

Definition 4.4.4. A resource of a node is called *open* when it is accessible for utilization by other nodes.

Example of such resources are: time for execution of its own program; time for execution of some other program; some space in memory (for information storage, for utilization during computation and so on); some software; some device.

Open resources may be used in three modes: for free, for rent, and for lease. There are three techniques to use open resources: to send a task(s) for execution, to send an autonomous agent(s) for work with the resource, and to get direct access to the resource. Each of these techniques has its advantages and shortcomings. The first two techniques are less time and energy consuming for a user, but they are not interactive, that is, they do not allow the user of the resource to make changes in the process of computation until some task is completed. As a result, interaction goes in quanta of actions. A mathematical model for such style of computation is given by persistent Turing machines (Goldin and Wegner, 1988). Agent technology allows interference, but it is mediated as changes are executed only by the agent performing the task.

Access to an open resource can be given in three forms: *autonomous access* when the resource is given exclusively to one user; *shared access* when several users, including the resource owner, can use the same resource; *free access* when anybody can use this resource.

Problems with open resources are numerous. Now many hackers, spammers and others break into people's unprotected computers and use their resources without permission. They can even damage resources — local and remote.

Damaging resources happened for centuries — since the times of the first human society when brute force ruled. However, this state of affairs demonstrated many shortcomings and proved to be an obstacle to progress. Consequently, the rules changed and now society more or less protects its members from damage. Similar situation will come to the world of computers and networks. This means people's assets — computers and networks, software and databases — will be better protected around the world. People will be able to decide for themselves how to use their computing resources, that is, making them open to others or not, and when making them open, people will chose how they open a resource (for free, for rent or for lease). Thus, it will be necessary to be able not only to open resources, but also close definite resources and to regulate access to them.

The open resource technology allows us to change and to extend the concept of a subroutine. Now a subroutine is a specifically organized sequence of instructions in a program for performing a separate task. This allows the subroutine code to be called from multiple places of the program, even from within itself, in which case the form of computation is called recursive. In a grid array or automaton, it is not necessary to have a subroutine as a part of a program P. A virtual subroutine may be a part of a different program, which in turn may be at a different computer or node than the program P that uses this subroutine.

Using agent technology, we can aggregate computers that use different devices and programs from the grid into one virtual computer system. This approach extends the concept of reconfigurable computer, synthesizing it with the idea of cluster computer. In turn, virtual computers may be combined into a virtual network or grid.

Grid automata are sufficiently powerful to model different algorithmic structures. It is evident when a grid automaton contains such structure as a node. However, even with much weaker nodes grid automata are able to simulate advanced types of algorithms. For example, results of (Minsky, 1967) imply the following theorem.

Theorem 4.4.6. *For any finite automaton A, there is a grid automaton G that has only Boolean elements as nodes and simulates A.*

Results of Siegelman and Sontag (1995) imply the following theorem.

Theorem 4.4.7. *For any Turing machine T, there is a grid automaton G that has only neurons with rational weights as nodes and simulates T.*

Theorem 4.4.6 and results from (Trahtenbrot, 1974) imply the following theorem.

Theorem 4.4.8. *For any Turing machine T, there is a grid automaton G that has only Boolean elements as nodes and simulates T.*

The difference between the grid automata from Theorems 4.4.7 and 4.4.8 is that that automaton in the first theorem may have finite number of nodes, while the automaton from the second theorem is potentially infinite.

Definition 4.4.5. A *grid automaton or array G has output exchange* if its nodes can exchange their final outputs.

This condition is not trivial because there are different cases when grid arrays do not satisfy it. For example, due to time limitations, it becomes impossible for a node to send its final result to other nodes. Namely, it is possible that some cluster of computers is set up to solve some very complex problem for a definite period of time, say, for ten days. Each of the computers from the cluster has its own subproblem. Because these subproblems have different complexity, it is likely that some computers from the cluster would not be able to finish their work in ten days, obtaining the final result. As a consequence, they will not be able to send their final results to other computers from the cluster. So, the grid array that corresponds to this cluster lacks the property of output exchange.

Let us consider a network of computers, servers, gadgets, embedded and other electronic devices that is monitoring environment in some region, producing ecological information. This network is a grid array. However, its nodes get, as a rule, only intermediate or partial results. Even if the monitoring period is limited, the nodes cannot exchange their final results because when they obtain these results the network stops functioning. Besides, there are networks that are organized so that the final result of their nodes can go only outside the network. As a result, they do not satisfy the condition from Definition 4.4.4.

Nevertheless, there are many grid arrays and grid automata that model them in the form of networks or clusters that have the output exchange. Actually, all classical models of distributed computation (artificial neural networks, cellular automata, iterative arrays, and Petri nets) have this property. For grid automata with output exchange, we have the following result.

Theorem 4.4.9. *For any inductive Turing machine M of order n, there is a grid automaton G with output exchange that has only inductive Turing machines of the first order as nodes and simulates T.*

All mathematical models of distributed computation, including such popular structures as neural networks, Petri nets, systolic arrays, iterative arrays, and cellular automata, are synthesized in the concept of grid automaton.

In comparison with cellular automata and iterative arrays, a grid automaton can contain different kinds of automata. For example, finite automata, Turing machines and inductive Turing machines can belong to one and the same grid. In comparison with systolic arrays, connections between different nodes in a grid automaton can be arbitrary like connections in neural networks. In comparison with neural networks, a grid automaton contains, as its nodes, more powerful machines than finite automata. Consequently, neural networks, cellular automata, and systolic arrays are kinds of grid automata. While functioning of cellular automata and systolic arrays is completely synchronous, grid arrays have different modes of functioning. Petri nets are grid automata in which there are nodes of two kinds: nodes of the first type only store information (places), while nodes of the second type only transmit and/or produce information (transactions).

It is possible to consider extended grid arrays in which there are nodes that are not technical devices. For example, it is possible to include people that work with the terminals of the grid as nodes of such a generalized grid array. Other dynamical systems may be also nodes in a generalized grid array. For example, the atmosphere of the Earth may be a node of the weather forecasting grid. Moreover, the whole environment of some grid array may be considered as a node in the extended grid array. This is an example of a closed extended grid array. Grid arrays and automata can have a hierarchical structure, containing nodes that are also grid automata or arrays.

As any separate computer or, in more generality, any information processing system can be treated as a degenerate grid array, all previous considerations allow us to formulate the following conjecture.

Any information processing system can be modeled by an extended grid automaton.

As experts predict (Wladawsky-Berger, 2002), the Grid with its universal connectivity and reach, will bring together the qualities of service that people are used to in computing, networking, communication and the qualities of service that we all have gotten used to in utilities like electricity, telephone, our transportation system, all of which tend to be up all the time. The theory of grid automata is aimed at acquiring knowledge on the Grid and using this knowledge for Grid development and advanced safe utilization.

5

Superrecursive Algorithms: Problems of Efficiency

*[Alice] went on. "Would you tell me, please,
which way I have to go from here?"
"That depends a deal on
where you want to get to," said the Cat.*

Lewis Carroll, 1832–1898

Efficiency is a clue problem and a pivotal characteristic of any activity. Inefficient systems are ousted by more efficient systems. Consequently, problems of efficiency are vital to any society and any individual. Many great societies, Roman empire, British empire and others perished because they had become inefficient. However, there are many different criteria of efficiency, and to understand this important and, at the same time, complex phenomenon, it is necessary to use mathematical methods.

Although a high level approach is suggested for mathematical modeling of efficiency and complexity, the aim of this chapter is not a general development of the theory of complexity and its applications to measuring efficiency, but a study of complexity of superrecursive algorithms, mostly, inductive Turing machines, demonstrating that they are essentially more efficient than recursive algorithms.

In this chapter, we consider the following problems:

◇ What are the relations between efficiency and complexity of algorithms (Section 1)?
◇ What is complexity of algorithms and why it is so important (Section 1)?
◇ What is the general situation with mathematical models of complexity, their origin, and problems of classification of measures of complexity (Section 1)?
◇ What resources are necessary for functioning of a given algorithm or program (Section 2)?
◇ How are the minimal resources that are necessary for solving a given problem to be found (Section 3)?

5.1   Measures of computer efficiency, program complexity, and their mathematical models

> *"While you are refreshing yourself,"* said the Queen.
> *"I'll just take the measurements."*
>
> Lewis Carroll, 1832–1898

The mathematical model of efficiency for algorithms and computation is called complexity. However, the term complexity has a wider meaning. Today complexity has become one of the most popular and important notions in science and society. It is a frequent word in present day's scientific literature, in various fields and with diverse meanings, appearing in some contexts as a precise concept, while being a vague idea in others texts. The reason is that people study and create more and more complex systems.

As Wakefield (2001) writes,

> Many companies are increasing profits and efficiency by implementing software based on complexity science, a broad field that includes chaos theory....
>
> Complexity researchers use genetic algorithms, artificial neural networks, and other tools to create models of real world systems ranging from steel production to immune system. A variety of companies such as Bios Group, i2 Technologies, Prediction, and Artificial Life are developing complexity applications for the business world.

This directly refers to computers and computer programming. As stated by the ironic Seventh Law of Computer Programming, *program complexity grows until it exceeds the capabilities of the programmer who must maintain it*. To cope with such situations, we need a developed theory of complexity, which explains why and how complexity emerges and how to solve problems that involve very complex systems. This is especially true for computers, networks, and their software.

At the same time, there is no generally accepted, formalized, and unique definition of complexity. Complexity has proved to be an elusive concept. Different researchers in different fields are bringing new philosophical and theoretical tools to deal with complex phenomena in complex systems. "What is complexity?" is a basic question of Gell-Mann (1994; 1995; 1996). However, after many elaborate considerations and creative insights, he comes to a conclusion that "*a variety of different measures would be required to capture all our intuitive ideas about what is meant by complexity and by its opposite, simplicity.*"

Mathematics develops its own approach to this concept by elaboration of different mathematical structures for measuring complexity and studies these structures by exact methods. Here we are mostly interested in mathematical measures of complexity.

5.1.1   Tractability, efficiency, and complexity

Freedom is the right to do
whatever the laws permit.

Charles Montesquieu, 1689–1755

Problems of algorithmic efficiency were formulated as being among the most important issues from the very beginning of the theory of algorithms and computation. In their seminal works, founders of computer science Gödel (1931; 1934), Church (1936; 1941), Turing (1936), and Post (1936; 1946) discussed what could be *effectively* solved by algorithms, or more exactly, by those mathematical models of algorithms that they had suggested? In fact, some of them went beyond that.

In his Hixon Symposium lecture, von Neumann (1951) voiced the need for a study of computation complexity:

> Throughout all modern logic, the only thing that is important is whether a result can be achieved in a finite number of elementary steps or not. The size of the number of steps that are required, on the other hand, is hardly ever a concern of formal logic. Any finite sequence of correct steps is, as a matter of principle, as good as any other. It is a matter of no consequence whether the number is small or large, or even so large that it couldn't possibly be carried out in a lifetime, or in the presumptive lifetime of the stellar universe as we know it. . . . [On the other hand] in the case of an automaton the thing which matters is not only whether it can reach a certain result in a finite number of steps at all but also how many such steps are needed.

There is evidence that von Neumann discussed these problems with other researchers. For example, (cf. Hemaspaandra and Ogihara, 1998), in a rediscovered 1956 letter to von Neumann, Gödel focused not only on the importance of the number of steps a Turing machine may need to perform a certain task (deciding whether a formula has a proof of a given length), but also used two particular polynomial bounds (linear and quadratic) as examples of efficient computation, in contrast with exhaustive search. However, there is no evidence that Gödel or somebody else at that time ever followed up on the issues that Gödel had raised.

One kind of efficiency is related to the number of problems solved by an algorithm. This characteristic is related to the power of algorithm considered in Section 2.2.

Definition 5.1.1. The more problems can be solved by an algorithm, the more *potentially efficient* is this algorithm.

However, it is important not only to know that it is possible to solve some problem *P* in principle, but also to be able to find a relevant solution in practice. Such problems for which the latter is possible are called *tractable*. Tractability of a problem is a relative property, being dependent on the algorithms that are used for solution. This gives us a definition for pragmatic efficiency of algorithms.

Definition 5.1.2. The more problems are tractable for an algorithm, the more *pragmatically* or *functionally efficient* is this algorithm.

Thus, pragmatic efficiency of an algorithm depends on two parameters: power of the algorithm (cf. Section 2.2) and resources that are used in the process of solution. If an algorithm does not have necessary resources, it cannot solve the problem under consideration. Thus, we come to the concept of resource efficiency of algorithms.

Definition 5.1.3. The fewer resources an algorithm uses for solution of a problem (of problems from some class), the more *resource efficient* is this algorithm with respect to this problem (class of problems).

One more kind of efficiency is related to the quality of solution.

Definition 5.1.4. The better solution for problems gives an algorithm, the more *mission efficient* is this algorithm.

Reliability, exactness, and relevancy are examples of mission efficiency.

Although Definitions 5.1.1–5.1.4 introduce precise concepts, they do not provide mathematical models to study efficiency by mathematical means, which are more powerful than empirical. Thus, we need a mathematical model for efficiency and such model is given by the mathematical concept of complexity of algorithms. Complexity is a mirror reflection of efficiency: the more efficient is an algorithm A for a problem P (problems from K), the less complex is P (are problems from K) for A. Mathematical models of complexity allow one to measure efficiency of various algorithms.

From a general perspective, it seems that complexity is not connected to efficiency. However, if we analyze what does it mean when we say that some system or process is complex, we come to conclusion that it is complex to do something with this system or process: to study it, to describe it, to build it, to control it, and so on. Thus, complexity is always complexity of doing something. Being related to activity and functioning, complexity allows one to represent efficiency in a natural way: when a process has high efficiency, it is simple and when a process has low efficiency, it is complex. For example, we can take time as a measure of efficiency: what is possible to do in one hour is efficient, while what is impossible to do even in a year is inefficient. There is a corresponding measure of computational complexity that estimates time of computation or any other algorithmic process.

These considerations allow us to give an informal definition of complexity. Formal definitions for different kinds of complexity measures of algorithms are studied in Sections 5.2 and 5.3.

Definition 5.1.5. *Complexity* of a system R is the amount of resources necessary for (used by) a process P involving R.

There are different kinds of involvement.

P may be a process in the system R. For example, R is a computer, P is an electrical process in R, and the resource is energy.

P may be a process that is realized by the system R. For example, R is a computer, P is a computational process in R, and the resource is memory.

P may be a process controlled by the system R. For example, R is a program, P is a computational process controlled by R, and the resource is time.

P may be a process that builds the system R. For example, R is a software system, P is the process of its design, and the resource is programmers.

P may be a process that transforms, utilizes, models, and/or predicts behavior of the system R.

In cognitive processes complexity is closely related to information, representing specific kind of information measures.

Processes use different kinds of resources:

Natural resources consumed by a process P: time, space, information, energy, power, minerals, and so on.

Social resources consumed by a process P: people involved, their time, efforts, expertise, knowledge, and so on.

Artificial resources consumed by a process P: system time, system space, data, knowledge, memory, system units, system actions, and so on.

If it is impossible to solve a problem with given resources, we assume that it has infinite complexity. The halting problem (cf. Section 2.5), being restricted to recursive algorithms, may be an example of a problem with infinite complexity since we know that it has no solution.

Pager (1970) defines efficiency of computation as the value that is inversely proportional to complexity of the same computation.

Definition 5.1.5 implies that complexity is always complexity of doing something and although complexity is attributed to a system, it is an essential characteristic of a process and of an algorithm if the process is determined by an algorithm. As we study algorithms here, only measures of algorithm complexity are considered. However, it is possible to extend the constructions of such measures to complexity of arbitrary processes and through processes to arbitrary systems. For instance, if we take some nonalgorithmic process, such as cognition, then it is possible to measure its complexity by the amount of resources this process needs.

It is necessary to have different complexity measures to estimate complexity of different processes, and even one process or system may be characterized by several complexity measures. For example (Gell-Mann, 1995), complexity of working with a model depends on the coarse graining (level of detail) of the description of the entity, on the previous knowledge and understanding of the world that is assumed, on the language employed, on the coding method used for conversion from that language into a string of bits, and on the particular ideal computer chosen as a standard. It is possible to consider these characteristics separately or to build an integral measure.

Some justly assume that the complexity of a problem is a subjective matter. For instance, we can consider a problem related to some system R: to build R, to test R, to increase power of R and so on. Two people having different models or views of the system, different algorithms for dealing with such systems, and even different will to solve the problem in question will have different ideas of the complexity of solving this problem. Waxman (1996) gives the following example. A problem with a car not starting might be very complex for a high-qualified mathematician, but not for the corner mechanic. On the other hand, solving a system of five linear equations with five variables will be simple for the mathematician and very complex for the corner mechanic.

As Grassberger (1990) writes, *"effective complexity and total information emphasize the subjective nature of complexity as a relationship between natural phenomena and an interested observer. Complexity is more in the way that phenomena are observed (that is, the model used) than in complicate elaborations made after the model has been imposed."*

Relativity of complexity is perfectly demonstrated by the following joke.

A Mathematician (**M**) and an Engineer (**E**) attend a lecture by a Physicist. The topic concerns Kaluza–Klein theories involving physical processes that occur in spaces with dimensions of 11, 12 and even higher. The **M** is sitting, clearly enjoying the lecture, while the **E** is frowning and looking generally confused and puzzled. By the end the **E** has a terrible headache. At the end, the **M** comments about the wonderful lecture.

E says, *"How do you understand this stuff?"*

M: *"I just visualize the process."*

E: *"How can you POSSIBLY visualize something that occurs in 11-dimensional space?"*

M: *"Easy, you first visualize it in an n-dimensional space, then let n go to 11."*

Mathematics makes subjective complexity objective introducing various criteria for complexity. For instance, a problem A is complex because its solution demands a huge amount of memory, while a problem B is complex because its solution involves performance of a huge amount of operations. Consequently, the problem A is complex for a computer with small memory, but it is simple for a computer with big memory. At the same time, the problem B is simple for a high performance computer, but is complex for an ordinary computer.

Definitions 5.1.3 and 5.1.5 imply that higher complexity corresponds to lower efficiency and lower complexity corresponds to higher efficiency. This implication looks somewhat inconsistent with our everyday experience. We can see that in many cases complex systems are more efficient than simple systems. For instance, computers becoming more complex work with higher efficiency: they can solve more problems, do this in a better way and in less time, as a rule.

However, this is only an imaginary inconsistency because when we estimate complexity and efficiency of computers, we compare computers by different criteria. Thus, becoming more complex in structure, computers become simpler in interaction, problem solving and decision making. This shows that to model efficiency using complexity measures, we need many different complexity measures and have to choose such measure to correctly reflect the kind of efficiency we want to measure. Efficiency of computation influences what is possible to do in practice from theoretical or ideal processes.

Many properties of systems are related to and depend on complexity. For example, Carlson and Doyle (2002) investigate relations between complexity and robustness in biological, social, economical, and engineering systems. They show a nontrivial interplay of these important properties. Manin (1991) suggests that the development of mathematics, and we would like to add, of science is directed by complexity issues. The reason is that simpler systems are more feasible for cognition. So, cognition goes from simple things to more and more complex ones. In the past, mankind has learned to understand reality mostly through simplification and analysis, ignoring a huge number of factors and details. That is why, for example, physics is more developed than biology: biological systems are much more complex than physical systems.

5.1.2 Structures of algorithmic complexity measures

> *All opinions, properly so called,*
> *are stages on the road to truth.*
>
> Robert Louis Stevenson, 1850–1894

To systematize complexity measures, we consider three types, three classes, and three kinds of complexity measures.

There are **three types** of complexity measures: *static*, *dynamic*, and *processual* complexity measures.

Definition 5.1.6. *Static* complexity measures depend only on an algorithm or program that is measured.

As examples, we can take such measures as the length of an algorithm, which is equal to the number of symbols in its description (Li and Vitanyi, 1996) or numbers of instructions in a program as all computer programs are algorithms.

As it is mentioned above, complexity of software grows with a very high rate. To cope with this situation, the theory of programs has been introduced (Halstead, 1977). The core of this theory is program measurement by means of software metrics (Zuze, 1999). Software metrics are designed and used to evaluate software, software development resources, and/or the software development process.

Software metrics give much more examples of static complexity measures. Let us consider some of them.

Example 5.1.1. The Lines of Code:

In 1974 Wolverton made one of the earliest attempts to formally measure programmer productivity using lines of code (LOC). Studies have shown that the count of the number of lines of code can indicate an estimate of the complexity of understanding of a program (Harrison et al., 1982; Park, 1992; Debnath et al., 2002). The LOC metric is discussed and used till today. It may be computed by counting the number of executable source code lines in a program.

A code line may be considered as an elementary unit of a program or in other words, formally programs are words in the alphabet consisting of code lines. From this perspective, LOC is such a very popular static complexity measure as the length $l(P)$ of a program.

Example 5.1.2. The Length of Program:

In his book *Elements of Software Science* (1977), Halstead considers several software metrics. The first of them is the length of the program $N(P)$. In Software Science, a computer program is considered to be a sequence of tokens, which are divided into groups of operators and operands. The basic metrics in software science are functions of the counts n_1 of the unique operators and n_2 of the unique operands, as well as the total occurrences of operators N_1 and of operands N_2. The simplest of such function is the length of the program $N(P)$ which is defined as

$$N(P) = N_1 + N_2.$$

This metrics can also indicate an estimate of the complexity of writing and storing a program (Halstead, 1977; Christensen et al., 1982).

Example 5.1.3. The Volume of Program:

Another software metrics suggested by Halstead is the volume of the program $V(P)$. It is defined by the formula

$$V(P) = (N_1 + N_2) \cdot \log(n_1 + n_2).$$

Example 5.1.4. The Cyclomatic Number:

The complexity measure that was proposed based on the control flow graph (CFG) model of a program P (McCabe, 1976) is called the *cyclomatic number* $V(P)$ of a program P and is defined by the formula

$$V(P) = e - n + 2,$$

where e is the number of edges and n is the number of nodes in the CFG of a program P.

Definition 5.1.7. *Dynamic* complexity measures depend both on an algorithm or program that is measured and on the input.

As examples, we can take such measures as the time of processing some given data or the volume of memory that is demanded by this algorithm or program.

Definition 5.1.8. *Processual* complexity measures depend on an algorithm or program, its realization, and on the input.

As examples, we can take such measures as the time of processing some given data or the volume of memory that is demanded by this processing.

Proposition 5.1.1. *For deterministic algorithm or program the dynamic and processual complexities coincide.*

However, in nondeterministic cases, one algorithm or program can define different processes, which demand different time or/and memory volume for realization.

Proposition 5.1.2. *Any static complexity measure generates a dynamic complexity measure that is constant for all inputs of a given algorithm and any dynamic complexity measure generates a processual complexity measure that is constant for all realizations of a given algorithm.*

Three classes of complexity measures are: *axiomatic*, *semi-axiomatic*, and *constructive* complexity measures. *Constructive* complexity measures are defined by some construction that allows one to measure consumed resources. Examples are the time of processing, which for an abstract algorithm is taken as the number of steps of this algorithm, or the volume of memory, which for a Turing machine is taken as the number of the cells in its tape that are used in a given computation. *Axiomatic* complexity measures functions, which are defined by axioms, while *semi-axiomatic* complexity measures involve both axioms and concrete constructions. In this setting, the machine-independent computational complexity (Blum, 1967a) and the size of machine (Blum, 1967), that are presented in Section 5.2, are semi-axiomatic complexity measures because they involve definite constructions of models of algorithms (such as Turing machines or partial recursive functions. At the same time, complexity measures from (Burgin, 1982) are axiomatic because they are defined for classes of algorithms specified by axioms (cf. Sections 5.2 and 5.3).

There are **three kinds** of complexity measures: *direct*, *dual*, and *mixed* complexity measures.

Computational complexities, time, space and others, are direct complexity measures. Kolmogorov, communication, and prefix complexities are dual complexity measures.

The *effective measure of complexity* as suggested by Grassberger (1990) is an example of a mixed complexity measure. The value of the effective measure of complexity of a sequence of symbols is defined as the relative memory required to calculate the probability distribution of for the next symbol of the sequence. This value represents the "average usable part of the information about the past which has to be remembered at any time if one wants to be able to reconstruct the sequence X from its shortest encoding."

Another example of a mixed complexity measure is the *logical depth* introduced by Bennet (1985). The logical depth of a string is defined as the computational complexity of the shortest program that produces it. For example if we take time as the computational complexity, then instead of estimating the time taken by the fastest algorithm, first the shortest program has to be found and later its time calculated. The principle of Occam's razor justifies this notion. Really, for a finite sequence of symbols found in nature, it is most likely that it has been produced by the simplest program, and for practical purposes, we can just assume that that is the case (Funes, 1996).

There is an interesting trade-off between different kinds of complexity. For example, in computation, it is possible in some cases to use more memory for program execution, decreasing the time of execution, or to use less memory, paying for with more time. Highly optimized tolerance (HOT) is one recent attempt, in long history of efforts, to develop a general framework for studying complexity in such fields as biology and engineering (Carlson and Doyle, 2002). The main idea of HOT is that higher structural complexity of a system (more complex for construction, modeling or understanding) is aimed at decreasing behavioral or functioning complexity of a system (simpler maintenance, less changes under external influence and so on). This shows how a trade-off between structural and behavioral complexity can inspire the development of systems.

However, here we are not going to develop much further the general theory of complexity. Our goal is to use the general approach developed in (Burgin, 1983; 1985a; 1990; 1992b) for building a theory of complexity for superrecursive algorithms with the main emphasis on inductive Turing machines. This allows us to demonstrate that inductive Turing machines are essentially more efficient than conventional Turing machines and other recursive algorithms.

5.2 Computational complexity: Axiomatic approach and specific measures

Any given program costs more and takes longer.

The First Law of Computer Programming

As Goldreich (2001) writes, computational complexity (also known by the name *Complexity Theory*) is a central field of computer science with a remarkable list of celebrated achievements as well as a very vibrant research activity. The field is concerned with the study of the *intrinsic complexity* of computational tasks, and this study tends to *aim at generality*: it focuses on natural computational resources, and considers the effect of limiting these resources on the class of problems that can be solved.

The first known to the Western reader papers on computational complexity were (Rabin, 1959; 1960; 1963a) and (Hartmanis and Stearns, 1964; 1965). However, earlier Trahtenbrot (1956) and Zeitin studied computational complexity in Russia (cf. Yanovskaya, 1959).

The (half-century) history of computational complexity, although it has become only a part of the Complexity Theory, has witnessed two main research efforts (or directions). The first direction is aimed towards actually establishing concrete lower bounds on the complexity of problems, via an analysis of the evolution (or effect) of the process of computation. Goldreich calls this direction is a "low-level" analysis of computation. Most research in circuit complexity, analysis of computational operations, such as number and matrix multiplication or division, and in proof complexity falls within this category. In contrast, the goal of the second research effort is exploration of the connections among computational problems and constructions, with an interest to individual problems being related only when such problems represent the corresponding class.

5.2.1 Axiomatic complexity measures

Thinking is the desire to get reality by means of ideas.

Jose Ortega Y Gasset, 1883–1955

The first person to introduce an axiomatic approach to complexity measures was Blum (1967; 1967a). He defined static complexity measures in the form of the size of machine (1967) and dynamic complexity measures the form of computational complexity (1967a). These measures are defined by natural axioms for classes of algorithms that are determined by some construction. Far reaching development of these constructions is given in (Burgin, 1982) where both complexity measures and classes of algorithms are defined by axioms.

Let $\mathbf{A} = \{A_i; i \in I\}$ and $\mathbf{B} = \{B_j; j \in J\}$ be some classes of algorithms. Algorithms can be presented in the form of machines or automata or in the form of programs (sets of instructions or arbitrary words as in the case with Turing machines).

Definition 5.2.1. A function $\mathbf{Sc}: I \to N$ is called an *axiomatic static complexity measure* of algorithms from $\mathbf{A}$ (with respect to $\mathbf{B}$) if the following conditions are satisfied:

Totality Axiom (TA). $\mathbf{Sc}(i)$ is a computable in $\mathbf{A}$ (in $\mathbf{B}$) total function.

Inverse Constructibility Axiom (IBA). For all $i \in I$, the value of the inverse function $\mathbf{Sc}^{-1}(i)$ is a finite set with an effective procedure (in $\mathbf{B}$) that builds all its elements.

An effective procedure here means an algorithm. Such an algorithm may be recursive, subrecursive or superrecursive. It gives three types of static complexity measures. The most restricted are the classes of axiomatic static complexity measures that allow to use only subrecursive algorithms for reconstruction of the inverse function $\mathbf{Sc}^{-1}(i)$. When we can use recursive algorithms, the scope of static complexity measures is extended. Superrecursive algorithms extend static complexity measures even more.

For the class $\mathbf{A} = \mathbf{B} = \mathbf{R}$ of all recursive algorithms, the axiomatic static complexity measure of algorithms is the size of machines as introduced by Blum (1967).

Defining size of machines (static complexity measure) for the class $\mathbf{R}$ of all recursive algorithms, Hartmanis and Hopcroft (1971) instead of axiom (IBA) use the following axiom:

Finite Computability Axiom (FCA). There is a recursive function that calculates the number of elements for each set $\mathbf{Sc}^{-1}(i)$.

When the set $\mathbf{A}$ is recursively computable and effective procedure means a recursive algorithm, then FCA is equivalent to IBA.

Example 5.2.1. Let $\mathbf{A}$ consists of algorithms generated by a Turing machine W with two input tapes. One tape is used for data, while the content of the second tape is considered as a program for computation. Each program for W is an algorithm from

A. Then the length $l(p)$ of this program p is a static complexity measure for Turing machines from **A**.

Example 5.2.2. Let **A** consists of all Turing machines. Each Turing machine T has a description $\mathbf{c}(T)$. Then the length $l(\mathbf{c}(T))$ of this description $\mathbf{c}(T)$ is a static complexity measure for Turing machines.

Example 5.2.3. Let **A** consists of all programs, which are written in some programming language (e.g., Java or FORTRAN). Then the length $l(p)$ of a program p as the number of letters (or as the number of words) in p is a very popular static complexity measure.

Such axiomatization of static measures works well for the theory of algorithms. However, applications of complexity theory in such fields as software engineering (Burgin, 1985a; Fenton, 1991), demand more general axioms. Thus, we introduce a more extended classes of *reconstructible* and *computable static measures*, which satisfy axioms TA and IGA or ICA instead of IBA.

Inverse Generability Axiom (IGA). For all $i \in I$, there is an algorithm in **A** (in **B**) that builds all elements from $\mathbf{Sc}^{-1}(i)$.

Computability Axiom (ICA). For $\mathbf{Sc}^{-1}(i)$ is a computable in **A** (in **B**) total function.

All measures defined by Definition 5.2.1 will be called *finite reconstructible static measures*.

Remark 5.2.1. A static measure $\mathbf{Sc}(i)$ may be finite, that is, for all $i \in N$, $\mathbf{Sc}^{-1}(i)$ is a finite set, but it can be noncomputable even in such powerful class as the class **R** of recursive algorithms, in particular, the class of all Turing machines. However, $\mathbf{Sc}(i)$ can be inductively computable (Burgin, 1983).

Software metrics (cf. Section 5.1.2) give different examples of axiomatic static measures of complexity.

Example 5.2.4. When the length of a line is bounded (and this is true for all programming languages as compilers demand this restriction), then the software metrics LOC is a finite reconstructible static complexity measure as it satisfies axioms (a) and (b) from Definition 5.2.1.

Example 5.2.5. Describing a program formally as a sequence of operators and operands, we see that the length of program $N(P)$ (Halstead, 1977) is also a static complexity measure, namely, the length $l(P)$ of a program. For a programming language in which the numbers n_1 of the unique operators and n_2 of the unique operands are finite, $N(P)$ is a finite reconstructible static complexity measure as it satisfies axioms (a) and (b) from Definition 5.2.1. However, some languages (at least, potentially) operate with infinite alphabets of operands, for example, with all real numbers. There are also theoretical models in which there are infinitely many unique operators. In such cases, $N(P)$ is not a finite complexity measure. If the sets of operands and operators are computable, then this measure is also computable.

Example 5.2.6. When it is defined, the volume of the program $V(P)$ (Halstead, 1977) is always a finite reconstructible static complexity measure.

Example 5.2.7. Representing a program formally as a structure of operators and operands, we can demonstrate that the cyclomatic number $V(P)$ (McCabe, 1976) is a static direct complexity measure.

Let $\mathbf{A} = \{A_i; i \in I\}$ be a class of algorithms with the domain $D_\mathbf{A}$. Usually, $D_\mathbf{A}$ is the set X^* of all words in some alphabet X. Each algorithm A_i from $\mathbf{A}$ determines a function f_i such that $A_i(x) = f_i(x)$ for any x from the domain $D_\mathbf{A}$. In particular, $f_i(x)$ is defined if and only if $A_i(x)$ is defined.

Definition 5.2.2. A function $\mathbf{Fc}: D_\mathbf{A} \times I \to N$ is called an *axiomatic computational complexity* of algorithms from $\mathbf{A}$ (with respect to $\mathbf{B}$) if the following conditions are satisfied:

(a) $\mathbf{Fc}(x, i)$ is a computable in $\mathbf{A}$ (in $\mathbf{B}$) function;
(b) for any x from the domain $D_\mathbf{A}$ and any $i \in I$, $\mathbf{Fc}(x, i)$ is defined if and only if $f_i(x)$ is defined.

Examples are given by such popular measures of computational complexity as $\mathbf{T}$ (time) and $\mathbf{S}$ (space).

Definition 5.2.3. *Time complexity* $\mathbf{T}(x, i)$, or $\mathbf{T}(x, Ti)$, of the computation of a Turing machine T_i with the input x is equal to the number of steps that T_i makes until it halts and gives a result.

Remark 5.2.2. This function $\mathbf{T}(x, i)$ gives an efficient model for Turing machines, but in order to be applied adequately to real information processing systems, it need to be either enhanced or based on a correct interpretation. An enhancement example is a correspondence between *model* or *system time*, in which all operations take the same unit of time (e.g., nanosecond) for execution, and *physical time*, in which different operations take different time units for execution. Taking this into account, each type of operation is assigned its own realization time. An example of a correct interpretation is given by a relevant class of functions (for example, functions that are bounded by polynomials) that allows one to stay inside this class for arbitrary transformations of the system time into the physical time.

Definition 5.2.4. *Space complexity* $\mathbf{S}(x, i)$, or $\mathbf{S}(x, T_i)$, of a Turing machine T_i is equal to the number of cells used in the computation of T_i with the input x until it halts and gives a result.

When algorithms work with words in some alphabets, the resources for computation are usually estimated not for separate words, but for sets of words that have the same length n. For example, the most popular definition of time complexity is the following.

Definition 5.2.5. *Time complexity* $\mathbf{T}(n, i)$, or $\mathbf{T}(n, T_i)$, of the computation of a Turing machine T_i is equal to t if whenever T_i is given an input x of length n the number of steps that T_i makes until it halts and gives a result is equal to t, or in other words, $\mathbf{T}(n, T_i)$ is equal to the maximal time that T_i needs to process words of the length n.

In a similar way, space complexity $\mathbf{S}(n, T_i)$ is defined.

Definition 5.2.6. *Space complexity* $\mathbf{S}(n, i)$, or $\mathbf{S}(x, T_i)$ of the computation of a Turing machine T_i is equal to the maximal space (number of cells) that T_i needs processing words of the length n.

This are the, so-called, *worst-case time and space complexity measures*.
Sometimes instead of the worst case, the average case is considered.

Definition 5.2.7. *Average time complexity* $\mathbf{AT}(n, i)$, or $\mathbf{AT}(n, T_i)$, of the computation of a Turing machine T_i is equal to the average number of steps t that T_i makes until it halts and gives a result is equal to t with respect to all inputs x of length n.

Prager (1970) introduces a space-time complexity that combines time and space complexities in one unified measure of complexity. The *space-time complexity measure* is defined as some increasing recursive function from time and space complexity measures.

To give a general definition of dynamic computational complexity, we consider some finite integral operation W (Burgin and Karasik, 1976).

Definition 5.2.8. An *integral operation* W on the set $\mathbf{R}$ of real numbers is a mapping that corresponds a number from $\mathbf{R}$ to a subset of $\mathbf{R}$, and for any $x \in \mathbf{R}$, $W(\{x\}) = x$.

Definition 5.2.9. A *finite integral operation* W on the set $\mathbf{R}$ of real numbers is a mapping that corresponds a number from $\mathbf{R}$ to a finite subset of $\mathbf{R}$, and for any $x \in \mathbf{R}$, $W(\{x\}) = x$.

As a rule, integral operations are partial. That is, they correspond numbers only to some subsets of $\mathbf{R}$.

Examples of integral operation include finite sums, products, minimums, maximums, infinums, supremums, integrals, taking the first element from a given subset, taking the sum of the first and second elements from a given subset, and so on.

Examples of finite integral operation are: sums, products, minimums, maximums, average, weighted average, taking the first element from a given subset, and so on.

Let W be a finite integral operation.

Definition 5.2.10. A function $\mathbf{Fc}\colon D_{\mathbf{A}} \times I \to \mathbf{R}^+$ is called an *axiomatic computational W-complexity* of algorithms from $\mathbf{A}$ (with respect to $\mathbf{B}$) if the following conditions are satisfied:

$$\mathbf{Fc}(n, i) = W\{\mathbf{Fc}(x, i); l(x) = n; i \in I\}$$

where $\mathbf{Fc}(x, i)$ is an axiomatic computational complexity of algorithms from $\mathbf{A}$.

Taking the operation *maximum* as W, we get a worst-case complexity. Taking the operation *average* as W, we get an average complexity.

Starting in the 1960s, computer scientists, unaware of von Neumann's lecture and Gödel's letter and their musings in this direction (cf. Hemaspaandra and Ogihara, 1998; and Section 5.1.1), began to investigate which problems can be *efficiently* solved by computers. The theory of **P** and **NP** and indeed complexity theory itself, sprang from this search for a better understanding of the limits of feasible computation. The notion that polynomial time deterministic algorithms form the right class to represent feasible computation was suggested by Cobham (1964) and Edmonds (1965). One of the main reasons for this was theoretical advantages of the class of all polynomials P. Namely, the growth of polynomials is relatively slow. In addition, the class of all polynomials P is closed under composition (thus allowing subroutine calls in the sense that a polynomial-time machine making subroutine calls to polynomial-time subroutines yields an overall polynomial-time procedure), addition, and multiplication. These features support the claim that P is a reasonable resource bound. The view that P loosely characterizes "feasibility" is now widely accepted.

However, one might argue that an algorithm that runs for $10^{10^{10}} n^{10^{10}}$ steps on inputs of size n is not practical. There are problems for which it is proved that they require high-degree polynomial algorithms. Some of them are artificial like cat-and-mouse games and pebbling problems (Adachi, Iwata, and Kasai, 1984), while others are quite natural like membership problem for permutation groups (Furst, Hopcroft, and Luks, 1980) or robotics configuration space problems (Schwartz and Sharir, 1983). Besides, many natural problems are known to have superpolynomial lower bounds. For example, Meyer and Stockmeyer (1972) and Fischer and Rabin (1974) show, respectively, problems that require exponential space and double exponential nondeterministic time.

It is possible to consider polynomial-time computations in different classes of algorithms. For instance, taking as the base the set of all deterministic Turing machines, we come to the popular class **P** of all problems that have a deterministic polynomial time solution. This class is widely thought to embody the power of reasonable computation. Taking as the base the set of all nondeterministic Turing machines, we come to another popular class **NP** of all problems that have a nondeterministic polynomial time solution.

Nondeterministic algorithms are more efficient, in principle, than deterministic algorithms as nondeterminism allows parallel processing (cf. Section 2.3). Consequently, we have $\mathbf{P} \subseteq \mathbf{NP}$. However, one of the most illustrious problems at the turn of twenty-first century is whether $\mathbf{P} = \mathbf{NP}$. Formally, it is denoted by "$\mathbf{P} = \mathbf{NP}$?".

For the relative case, that is, for computations with an oracle (cf. Section 4.2), we have a similar problem $\mathbf{P}_f = \mathbf{NP}_f$? where $\mathbf{P}_f$ the class of all problems that have a polynomial time solution by a deterministic Turing machine with the advice f and $\mathbf{NP}_f$ the class of all problems that have a polynomial time solution by a nondeterministic Turing machine with the advice f. This problem is solved. Namely, it is proved (Baker, Gill, and Solovey, 1975) that there is an advice f such that $\mathbf{P}_f = \mathbf{NP}_f$ and there is an advice g such that $\mathbf{P}_g \neq \mathbf{NP}_g$. So, both possibilities can be true.

Researchers have not been able to solve the problem "$\mathbf{P} = \mathbf{NP}$?" for a definite period of time (approximately, for thirty years). Besides, the solution is important for Internet and computer security. Consequently, it has attracted attention of many researchers. However, there is little understanding for this and other problems on computational complexity that complexity of a problem depends on data representation for this problem. As an example, we can take the satisfiability problem. It inquires whether for a given Boolean function $f(x)$ where x is a vector of Boolean variables there is a binary vector a such that the value $f(a)$ is equal to 1. Cook (1971) proved that this problem is $\mathbf{NP}$-complete. It means (cf. Definitions 5.2.15–5.2.18) that if it might be possible to solve this problem in a polynomial time, then $\mathbf{P} = \mathbf{NP}$.

At the same time, the theory of Boolean functions implies that if a Boolean function $f(x)$ has a normal disjunctive form, then the satisfiability problem is solved by a deterministic Turing machine time that is linear in number of symbols in $f(x)$. It means that in this case, the satisfiability problem is solved in the lowest nontrivial polynomial time.

However, the main goal here is not to solve separate problems but to build a theory that allows one to efficiently solve a large portion of such problems. This is a more effective way of developing science. For example, after Newton and Leibniz created differential and integral calculus, students began to solve such problems that were previously difficult even for the best mathematicians. So, we proceed to construct and systematize computational complexity measures.

Let us consider the following predicate

$$M(r, i, j, x) = \begin{cases} 1, & \text{when } \mathbf{Fc}(x, i, j) = r; \\ 0, & \text{when } \mathbf{Fc}(x, i, j) \neq r. \end{cases}$$

Definition 5.2.11. A functional complexity measure $\mathbf{Fc}$ of algorithms from $\mathbf{A}$ is called *decidable* in $\mathbf{B}$ if the predicate $M(r, i, j, x)$ is decidable in $\mathbf{B}$.

When $\mathbf{A}$ is the class $\mathbf{T}$ of all deterministic Turing machines, then a decidable in $\mathbf{T}$ functional complexity measure $\mathbf{Fc}$ is an axiomatic computational complexity $\Pi = \{\Phi_i(n); i \in N\}$ in the sense of Blum (1967a). It contains constructive computational complexity measures such as time or space. When $\mathbf{A}$ is the class $\mathbf{T}$ of all deterministic Turing machines, then $\mathbf{Fc}$ is an axiomatic computational complexity in the sense of Burgin (1982).

This axiomatization works well for the theory of algorithms. However, to encompass other useful complexity measures, we need more general axioms. Thus, all measures defined by Definition 5.2.2, will be called *computable* and we introduce two other classes.

Definition 5.2.12. A function $\mathbf{Fc}: D_A \times I \to N$ is called an *axiomatic subcomputational (up-computational) complexity* of algorithms from $\mathbf{A}$ (with respect to $\mathbf{B}$) if the following conditions are satisfied:

Computability Axiom (CA). $\mathbf{Fc}(x, i)$ is a computable in $\mathbf{A}$ (in $\mathbf{B}$) function.

Definability Axiom (CA). For any x from the domain D_A and any $i \in I$, if $\mathbf{Fc}(x, i)$ is defined, then $f_i(x)$ is also defined (if $f_i(x)$ is also defined, then $\mathbf{Fc}(x, i)$ is also defined).

An example of a useful subcomputational complexity is the number $r_T(x)$ of reversions made by the head of a one-tape Turing machine T in the process of computation with the input x (Trahtenbrot, 1974). Really, if the Turing machine T gives a result for the input x, then $r_T(x)$ is defined. However, it is possible that $r_T(x)$ is defined, for instance, equal to zero, the machine T never stops and thus, gives no result.

Let $\mathbf{A} = \{A_i; i \in I\}$ be a class of algorithms having the set X^* of all words in an alphabet X as their domain D_U, $J = \bigcup_{i \in I} I_i$, and $\mathbf{PA} = \{p_{ij}; j \in I_i, i \in I\}$ be the class of processes determined by algorithms from $\mathbf{A}$.

Definition 5.2.13. A function $\mathbf{Pc}: X^* \times I \times J \to N$ is called an *axiomatic processual complexity measure* of algorithms from $\mathbf{U}$ if the following conditions are satisfied:

Computability Axiom (CA). $\mathbf{Pc}(x, i, j)$ is a computable in $\mathbf{U}$ function;

Definability Axiom (CA). For any x from the domain D_U, any $i \in I$ and any $j \in I_i$, $\mathbf{Pc}(x, i, j)$ is defined if and only if the process $p_{ij}(x)$ gives a result.

Proposition 5.1.2 cannot be proved for axiomatic computational complexity because the dynamic measure induced by a static one is always defined, while the axiomatic computational complexity is not defined for those data for which the algorithm give no result. A similar obstacle for axiomatic processual complexity measures does not allow extension of the second part of Proposition 5.1.2. However, we can prove two slightly weaker results.

Proposition 5.2.1. *a) If all algorithms from* $\mathbf{U}$ *give results for all inputs, then any axiomatic static complexity measure generates an axiomatic dynamic complexity measure that is constant for all inputs of a given algorithm.*

b) If all algorithms from $\mathbf{U}$ *are deterministic, then any dynamic complexity measure generates a processual complexity measure that is constant for all realizations of a given algorithm.*

Proposition 5.2.2. *a) Any static complexity measure generates a dynamic complexity measure that is constant for all inputs of a given algorithm when this algorithm gives a result.*

b) Any dynamic complexity measure generates a processual complexity measure that is constant for all realizations of a given algorithm when this realization gives a result.

Remark 5.2.3. In some cases, it is more realistic to consider complexity measures with values not only in integer numbers but in the whole domain of real numbers. Average complexity (time, space, and so on) is an example when we need real numbers, or at least, rational numbers, to reflect properties of real computers and computations.

If **Gc** is a computational complexity measure, **K** is a class of algorithmic problems, and **Q** is a class of algorithms, complete and hard problems for the class **K** relative to the class **Q** are naturally introduced.

Definition 5.2.14. A problem p is called *hard* for the class **K** relative to the class **Q** if any problem from **K** can be reduced to the problem p by some algorithms from **Q**.

Definition 5.2.15. A problem p that is hard for the class **K** relative to the same class **K** is called **K**-*hard*.

It is possible to define the class **Q** as a complexity class. For example, **Q** is the class of all algorithms the dynamic complexity measure **Gc** of which is bounded by a function f or by some function from a class of functions **F**. It gives us the following concept.

Definition 5.2.16. A problem p is called *hard* for the class **K** with respect to the function f (to the class **F**) and the measure **Gc** if any problem from **K** can be reduced to the problem p with complexity **Fc** less than or equal to f (to some function from **F**).

Definition 5.2.17. A problem p from the class **K** is called *complete* for the class **K** relative to the class **Q** if any problem from **K** can be reduced to the problem p by some algorithms from **Q**.

In other words, complete problems for a class are hard problems that belong to the same class.

Definition 5.2.18. A problem p that is *complete* for the class **K** relative to the same class **K** is called **K**-complete.

Definition 5.2.19. A problem p is called *complete* for the class **K** with respect to the function f (to the class **F**) and the measure **Gc** if any problem from **K** can be reduced to the problem p with complexity **Fc** less than or equal to f (to some function from **F**).

It is possible to find many examples of hard and complete problems, for example, in (Balcazar et al., 1988) or (Hopcroft et al., 2001).

Some argue (cf., for example, Edmonds, 1999) that axiomatic complexity measures are too general to exact description of complexity, as they include many other functions that are only formally related to algorithms. To eliminate this discrepancy, we subdivide axiomatic complexity measures into three groups: *abstract*, *general*, and *proper*.

Definition 5.2.20. *Abstract complexity measures* are abstract properties that reflect essential features of resource estimation.

For example, axiomatic computational complexity measures reflect common features of such popular measures as **T** (time) and **S** (space). These features allow one to obtain many properties of complexity measures in an axiomatic setting and then to apply these properties to a variety of proper, semi-axiomatic, and constructive measures.

Definition 5.2.21. *General complexity measures* are abstract properties that reflect common features of resource estimation.

For example, we can chose some computational resource R (time, number of steps, volume of memory, number of changes in the computation direction, number of interactions and so on). Then the amount $Vi(R)$ of the consumed resource R is determined for each step i of computation. Choosing an appropriate integral operation **I** (Burgin and Karasik, 1976), we define the dynamic complexity measure c_R related to the resource R by the formula $c_R(A, x) = \mathbf{I}(V_i(R); i = 1, \ldots, r$ and A makes r steps of computation being applied to x).

Here $V_i(R)$ is the amount of the consumed resource R at the step i for all steps of the computation of the algorithm A with the input data x.

Thus, $c_R(A, x)$ is a general dynamic complexity measure, allowing to obtain a common structure for many popular complexity measures. For example, a natural integral operation that gives relevant results for such measures as time (the measure $\mathbf{T}_A(x)$) or interaction with external sources is summation, while a natural integral operation for such measures as the memory size (the measure $\mathbf{S}_A(x)$) or number of computers (computational cells) involved in performing one step of computation is the operation *max*.

The concept of a computational resource can be also axiomatized by the following construction.

Definition 5.2.22. *Proper complexity measures* are properties of resource estimation.

The most popular examples of proper complexity measures are the length $l(P)$ of a program, computation time $\mathbf{T}_A(x)$ with input data x and memory space $\mathbf{S}_A(x)$ utilized by the computation.

Proper complexity measures are obtained from axiomatic complexity measures. An axiomatic complexity becomes a proper complexity through some interpretation of an axiomatic measure when this interpretation allows one to use it for resource estimation. We consider interpretations related to resources of three types:

1. Resources of the mathematical model (such as time as the number of steps, memory as the number of tape cells and so on).
2. Resources of the modeled system (such as physical time, computer memory, the number of utilized devices and so on).
3. External resources (such as cost of the computation, the number of personnel used to organize computation, the number of programmers to write a program and so on).

Specifications of axiomatic complexity measures are extensively used in software engineering in a form of software metrics (Pressman, 1994). To find complexity of a computer program, it is natural to use direct measures of complexity, which determine resources used by this program. These measures applied to the realizational complexity of a program estimate on computational resources such as time for computation, memory for program storage, memory for program functioning, number and types of computers, number and types of input/output devices, number and types of processors, and so on. Because there are many such measures, it is efficient to study their properties in the axiomatic context. Direct complexity measures allow one to assess functioning of different kind of software, for example, expert systems (Burgin, 1991).

Direct complexity measures applied to the design complexity of a program estimate on computational resources such as time for program development, cost of this process, number and qualification of designers or/and programmers, number of external interfaces, number of encountered problems and other factors of software production. These resources reflect parameters of the process of program development that are evaluated by means of repository software metrics such as cost, schedule, staffing, size, requirements, risk and others (Hefner and Mann, 2002).

Importance of the organizational complexity of a program is underestimated and consequently, such measures are not studied. Utilization of corresponding resources is partially reflected in software metrics such as cost and staffing.

5.2.2 Computational complexity of Turing machines and inductive Turing machines

> *What memory has in common with art is*
> *the knack for selection, the taste for detail....*
>
> Joseph Brodsky, 1940–1996

In Section 4.3, inductive Turing machines with a structured memory were introduced. Now let us consider what advantages can be obtained by utilization of a structured memory for conventional Turing machines.

Theorem 5.2.1. *For any recursively computable function f, there is a Turing machine T with a recursive memory that computes f within a space that is linear in input and output, that is, if n is an arbitrary number, then $\mathbf{S}(n, T) = c_1 \cdot n + c_2 \cdot f(n)$ for some constants c_1 and c_2 and all input values of x.*

Proof. According to the condition of the theorem, there is a Turing machine D that computes the function f, that is, $f = f_D$. To prove the theorem, we build such Turing machine T that computes the function f, using only $c_1 \cdot n + c_2 \cdot m$ cells for some constants c_1 and c_2 and an arbitrary input n where m is the value $f(n)$. According to the results of the theory of Turing machines (cf. Section 2.3), it is possible to assume that T has infinitely many tapes L_i that are enumerated by natural numbers and all cells in each tape are also enumerated. As it is always in the theory of Turing machines, this infinity of tapes is only potential, that is, there are only finite number of finite parts of tapes at any given moment of computation. But it is possible to extend these parts and to add as many new tapes as we need.

However, this is only a conventional Turing machine. Addition of a structured memory allows us to enhance this machine with new abilities. Namely, a structured memory provides a possibility to establish various connections between cells in all tapes of the machine. In addition to conventional connections between consecutive/adjacent cells, three other types b, c, and r of connections are constructed.

The memory of T is defined by the connection relation R that is built by another Turing machine M, and thus, R is a recursive relation. Before the machine T begins to work, the machine M builds all necessary connections in the memory of T.

At first, all connections c are installed, connecting first cells in the tapes. Namely, the first cell of the tape L_i is connected by the link c with the first cell in the tape L_{i+1} for each $i = 1, 2, \ldots$.

Then all connections b are set up, each of which goes b goes from the cell with the number $n + 1$ to the cell with the number n. These connections are built for all tapes L_i and for all $n = 1, 2, \ldots$.

For building connections of the type r, the machine M contains a copy D_0 of the machine D, which is realized as a subprogram. The standard technique for realization of a Turing machine as a subprogram (component or submachine) of another Turing machine is given, for example, by Hopcroft et al. (2001) and Ebbinghaus et al. (1970).

The machine T also contains several submachines, which are realized as subprograms. It has a subprogram C that performs the following actions: if the head H of T is situated in the cell with a number k in some tape L_i of T, the machine C finds or computes k and gives it as the output. To find k, C moves the head to the initial cell of this tape, using connections of the type b, and counts the number of steps.

In addition, T has a subprogram Q that given an input n, brings the head H of T from the first cell in the first tape L_1 of T to the first cell in the nth tape L_n of T. To do this, a counter and connections of the type c are used.

The machine M works in the following manner. When a number n is given to M as its input, D_0 computes the value $f(n)$ if it is defined. Then having the value $f(n)$, the machine M determines the connection r from the first cell of the nth tape L_n of T to cell that has number $f(n)$ in the same tape. This is done for all numbers $n = 1, 2, \ldots$. Then the subprogram C computes the number $f(n)$ and gives it as the output of T.

When $f(n)$ is not defined, the machine T gives no output. In a general case, it is possible to realize this situation as a cycle in which T accesses M for a connection for the next move of the head H. Receiving no answer, T continues to perform this access operation all the time and thus, gives no output because a Turing machine has to stop to give a result.

When a number n is given to T as its input, Q brings the head H of T to the first cell in the nth tape L_n of T. Then when $f(n)$ is defined, the head H uses the connection r to go directly to the cell with the number $f(n)$. The first cell of L_n has exactly one such connection r. After this subprogram C works, giving an output which is also the output of the machine T.

Now let us find how many cells from the memory of the machine T are used by the main program of T and its subprograms. The machine T uses n cells for accepting the input for processing. Then it uses at most n cells for moving the head H from the first cell in the first tape L_1 of T to the first cell in the nth tape L_n of T. Then it uses at most $f(n)$ cells for finding $f(n)$. In addition, the subprogram Q that moves the head H of T uses $k \cdot n$ cells where k is a fixed number for all $n = 1, 2, \ldots$. Besides the counter C uses $t \cdot f(n)$ cells where t is a fixed number for all $n = 1, 2, \ldots$. Thus, we come to the conclusion that the number of cells used in the memory of T is bounded by a linear function $c \cdot n + c_2 \cdot f(n)$ for some constants c_1 and c_2.

Theorem is proved.

It is possible to improve this result, achieving linearity not in n and $f(n)$, but in the lengths $l(n)$ and $l(f(n))$ of numbers in the binary or any other code.

Theorem 5.2.2. *For any recursively computable function f, there is a Turing machine T with a recursive memory that computes f within a space that is linear in input and output, that is, if n is an arbitrary number, then $\mathbf{S}(n, T) = c_1 \cdot l(n) + c_2 \cdot l(f(n))$ for some constants c_1 and c_2 and all input values n.*

The proof is similar to the proof of Theorem 5.2.1, but the head moves at the beginning not to the tape L_n, but to the tape $L_{l(n)}$, and instead of one connection r, there are many connections r_t where the index t takes values in the set of all numbers having the length $f(n)$) in the binary code.

As the space taken by computation in a Turing machine is always less then time (Aho, Hopcroft, and Ullman, 1976). Theorem 5.2.2 implies the following result.

Corollary 5.2.1. *For any recursively computable function f, there is a Turing machine T with a recursive memory that needs linear in input time to compute f, that is, $\mathbf{T}(x, T) = c_1 l(x) + c_2 l(f(x))$ for some constants c_1 and c_2 and all input values of x.*

Both values $l(x)$ and $l(f(x))$ are essential to validate of the results in Theorems 5.2.1, 5.2.2 and Corollary 5.2.1 in Theorems 5.2.1, 5.2.2 and Corollary 5.2.1. Really, there are such functions f, for which $l(x)$ grows to infinity, while f takes only two values 1 and 0 (as in deciding algorithms). In this case, $l(1) = l(0) = 1$. Thus, $l(x)$ plays the main role.

In other cases, $f(x)$ grows much faster than x (e.g., $f(x) = x^x$), implying that the sum $c_1 l(x) + c_2 l(f(x))$ depends mostly on $l(f(x))$.

However, in many important cases the process of computation can be much faster, depending only on input.

Theorem 5.2.3. *If all possible results of computation have bounded length, then for any recursively computable function f, there is a Turing machine T with a recursive memory that needs linear in input time to compute f.*

For conventional Turing machines, the situation is opposite as the following result shows.

Theorem 5.2.4. *For any strictly increasing recursive function f, there is a recursive function g taking values in the set $\{1, 0\}$ such that any computation of g by a Turing machine has time complexity larger than f, that is, $\mathbf{T}(x, T) > f(x)$ for all Turing machines T and all x.*

This result is a direct consequence of the Theorem 11.1 from (Aho, Hopcroft, and Ullman, 1976).

Theorem 5.2.5. *If all possible results of computation have bounded length, then for any recursively computable function f, there is a Turing machine T with a recursive memory such that it needs linear in input time to compute f without reading the result or converting it to a conventional form.*

In a similar way, we can consider complexity of inductive computations. For inductive Turing machines, it is natural to introduce measures of computational complexity resembling those that are used for Turing machines.

Definition 5.2.23. Time complexity $\mathbf{T}(x, M)$ of the computation of an inductive Turing machine M with the input x is equal to the number of steps that M makes till its output tape stops changing.

Thus, similar to Turing machines, time of computation of an inductive Turing machine M indicates what time is necessary for M to obtain a result in the output tape.

A technique that is similar to the one used in the proof of Theorem 5.2.1 makes it possible to obtain the following result.

Theorem 5.2.6. *For any function f computable by an inductive Turing machine of order n, there is an inductive Turing machine M of order $n + 1$ that computes f within a space that is linear in input and output, that is, if n is an arbitrary number, then $\mathbf{T}(n, M) = c_1 \cdot n + c_2 \cdot f(n)$ for some constants c_1 and c_2 and all input values of x.*

Corollary 5.2.2 (Burgin, 1999). *For any recursively computable function f, there is an inductive Turing machine M of the first order that computes f within a space that is linear in input and output, that is, if n is an arbitrary number, then $\mathbf{T}(n, M) = c_1 \cdot n + c_2 \cdot f(n)$ for some constants c_1 and c_2 and all input values of x.*

Corollary 5.2.3. *For any function f computable by an inductive Turing machine of the first order, there is an inductive Turing machine M of the second order that needs linear in input time to compute f, that is, $\mathbf{T}(x, M) = c_1 l(x) + c_2 l(f(x))$ for some constants c_1 and c_2 and all input values of x.*

Theorem 5.2.3 implies the following result.

Corollary 5.2.4. *If all possible results of computation have bounded length, then for any recursively computable function f, there is an inductive Turing machine M of the first order that needs linear in input time to compute f.*

It is possible to define space complexity in the same way, but inductive Turing machine does not stop after it gets its final output. This *a posteriori* functioning may demand a lot of cells in the working tapes. As a result, we come to several versions of space complexity.

Definition 5.2.24. Minimal space complexity $\mathbf{mS}(x, M)$ of an inductive Turing machine M is equal to the number of cells used in the computation of M with the input x until its output tape stops changing.

A technique that is similar to the one used in the proof of Theorem 5.2.1 makes it possible to obtain the following result.

Theorem 5.2.7. *For any function f computable by an inductive Turing machine of order n, there is an inductive Turing machine M of order $n + 1$ that computes f within a space that is linear in input and output, that is, if n is an arbitrary number, then $\mathbf{mS}(n, M) = c_1 \cdot n + c_2 \cdot f(n)$ for some constants c_1 and c_2 and all input values of x.*

Corollary 5.2.5. *For any recursively computable function f, there is an inductive Turing machine M of the first order that computes f within a minimal space that is linear in input and output, that is, if n is an arbitrary number, then $\mathbf{mS}(n, M) = c_1 \cdot n + c_2 \cdot f(n)$ for some constants c_1 and c_2 and all input values of x.*

Definition 5.2.25. Maximal space complexity $\mathbf{MS}(x, M)$ of an inductive Turing machine M is equal to the number of cells used in the computation of M with the input x.

It is possible that M works without stopping but uses only a finite number of cells. For example, this is the case when an inductive Turing machine simulates a conventional Turing machine.

To use a new cell, it is necessary for an inductive Turing machine to perform, at least, one operation. Thus, as in the case of Turing machines, we have the following result.

Proposition 5.2.3. *For any inductive Turing machine M, the following inequalities are true:*

$$\mathbf{T}(x, M) \leq \mathbf{mS}(x, M) \leq \mathbf{MS}(x, M).$$

Maximal space complexity $\mathbf{MS}(x, M)$ is, as a rule, larger than minimal space complexity $\mathbf{mS}(x, M)$. It is even possible that $\mathbf{mS}(x, M)$ is finite, while $\mathbf{MS}(x, M)$ is infinite. However, we have the following result.

Theorem 5.2.8. *For any recursively computable function f, there is an inductive Turing machine M of the first order that computes f within a space that is linear in input and output, that is, if n is an arbitrary number, then $\mathbf{mS}(n, M) = c_1 \cdot l(n) + c_2 \cdot l(f(n))$ for some constants c_1 and c_2 and all input values n.*

These results show that structural memory allows one to essentially increase speed of computations both for Turing machines and inductive Turing machines. It is possible to ask whether an inductive Turing machine of the first order with ordinary memory can speed-up computations in comparison with conventional Turing machines.

There are no general results in this direction. The situation is similar to applications of parallel computations and their comparison to sequential computations. As we know, there are cases when parallel computations are more efficient than sequential computations and there are cases when parallel computations give no advantage. It is possible to show that there are important classes of problems for which inductive computations can essentially improve productivity.

5.3 Dual complexity measures: Axiomatic approach and Kolmogorov complexity

There are two sides to every problem.

A proverb

While computational complexity based on direct complexity measures tells us what resources will take a given algorithm or machine or program to do some computation, dual complexity measures reflect the minimal necessity in the corresponding resources for solving some problem. Usually, the problem that is considered for algorithms is building (computing) some word or making a decision whether a given element belongs to a given set. Additionally, it is assumed that it is possible to choose the means for the necessary computation (construction) from some set of algorithms or machines or programs. There was an attempt to build a universal dual complexity measure, which does depend on a specific class of algorithms. However, this goal has not been achieved. One reason was that it turned out that the original definition was not sufficient for solving some mathematical and practical problems. For example, such universal measure was not appropriate for formalizing the concept of randomness and for developing algorithmic probability theory and information theory. The second reason for constructing relative dual measures was the discovery of superrecursive algorithms. Prior to the discovery, all believed that Turing machines or other models of recursive algorithms give an absolute class for algorithms and computation. Since the discovery, the situation has changed. The third reason was that computer scientists have already used several distinct dual measures. As a result, the universal approach was discarded and dual measures have been introduced and studied for some specific classes of algorithms. Later an axiomatic approach to dual complexity measures has been elaborated.

The first developed form of dual complexity measures was the so-called *Kolmogorov* or *constructive* or *algorithmic complexity*.

5.3.1 Kolmogorov complexity: A general perspective

It is quite a three-pipe problem....

The Adventures of Sherlock Holmes,
Arthur Conan Doyle, 1859–1930

Traditionally the theory of Kolmogorov complexity has been developed top down: from larger classes to smaller classes of algorithms that were more relevant to computational problems. At first, as the history tells us, Kolmogorov complexity $C(x)$ was defined and studied independently for the class of all recursive algorithms by three mathematicians: Solomonoff (1964), Kolmogorov (1965), and Chaitin (1966). However, some authors (cf., for example, Uspensky and Semenov, 1993) do not consider Chaitin an author of Kolmogorov complexity, although, as Chaitin wrote, he submitted his paper was for publication before the paper of Kolmogorov was published.

Another controversy is related to the term *Kolmogorov complexity*. Although the majority of researchers use Kolmogorov complexity for this measure, other authors prefer a different name. For example, Gell-Mann (1994) trying to avoid the conflicting word "complexity" calls it simply *algorithmic information content*. Some call it *algorithmic complexity* or *algorithmic entropy*. Chaitin seems more comfortable calling this concept by the name *randomness*. Lewis (2001) calls this measure *algorithmic* (AC) or *KCS* (Kolmogorov-Chaitin-Solomonoff) *complexity*. In the fundamental treatise *An Introduction to Kolmogorov Complexity and its Applications* (Li and Vitanyi, 1997), both names, Kolmogorov and algorithmic complexity, are used.

The name *algorithmic complexity* looks more neutral, but it may be related to a much broader context. As we have discussed earlier, any complexity measure is connected to algorithms. The name *constructive complexity* better reflects this type of complexity because it is the complexity of computing or constructing a given word. Nevertheless, here we use both names *Kolmogorov* and *algorithmic complexity* due to the existing tradition in this area (cf. (Li and Vitanyi, 1997)).

It is also necessary to remark that it might be reasonable to use different names for the function $C(x)$ relevant to its applications. When we want to know how difficult it might be in computing or constructing some object with recursive algorithms, we consider *Kolmogorov* or constructive complexity as an appropriate name for $C(x)$ will be Kolmogorov or constructive complexity. When the question is how much information we need to build or compute x with recursive algorithms, we consider information (or more exactly, recursive information) content of x as a better name for $C(x)$.

The original Kolmogorov complexity of a word x is taken to be equal to the size of the shortest program (in number of symbols) for a universal Turing machine U that without additional data, computes the string and terminates. To formalize this, we define Kolmogorov complexity for a class **R** of recursive algorithms such that **R** has a universal algorithm. For example, in the class of all Turing machines, a universal Turing machine is a universal algorithm.

Definition 5.3.1. The *Kolmogorov complexity* $C(x)$ of an object (word) x is defined as

$$C(x) = \min\{l(p); U(p) = x\}$$

where $l(p)$ is the length of the word p and U is a universal algorithm in the class **R**.

This measure is called *absolute* Kolmogorov complexity because Kolmogorov complexity has one more form, which called relative.

Definition 5.3.2. The *relative* to a given word y *Kolmogorov complexity* $C(x \mid y)$ of an object (word) x is defined as

$$C(x \mid y) = \min\{l(p); U(p, y) = x\}$$

where $l(p)$ is the length of the word p and U is a universal algorithm in the class **R**.

Kolmogorov complexity $C(x)$ of a word x in some sources is denoted by $K(x)$, while $C(x \mid y)$ is denoted by $K(x \mid y)$.

Absolute Kolmogorov complexity is a particular case of relative Kolmogorov complexity, namely:

$$K(x) = K(x \mid \Lambda) = \min\{l(p); U(p, \Lambda) = x\}.$$

The aim of introduction of Kolmogorov complexity was to ground probability theory and information theory, creating the new approach based on algorithms. This goal was achieved. The new theories became very popular, although they did not substitute either the classical probability theory, which was grounded before by the same Kolmogorov (1950) on the base of measure theory, or Shannon's information theory.

However, an attempt to define in this setting an appropriate concept of randomness was unsuccessful. It turned out that the original definition of Kolmogorov complexity was not relevant for that goal. To get a correct definition of a random infinite sequence, it was necessary to restrict the class of utilized algorithms. That is why Kolmogorov complexity was defined and studied for various classes of subrecursive algorithms. For example, researchers discussed different reasons for restricting power of the device used for computation when when estimating the minimal complexity is estimated.

When Kolmogorov complexity is defined for the class of Turing machines that compute symbols of a word x, we obtain uniform complexity $KR(x)$ studied by Loveland (1969).

When Kolmogorov complexity is defined for the class of prefix functions (see Section 3.4), we obtain prefix complexity $K(x)$ studied by Gasc (1974), Levin (1974), and Chaitin (1975).

When Kolmogorov complexity is defined for the class of monotonous Turing machines, we obtain monotone complexity $Km(x)$ studied by Levin (1973).

When Kolmogorov complexity is defined for the class of Turing machines that have some extra initial information, we obtain conditional Kolmogorov complexity $CD(x)$ studied by Sipser (1983).

Let $t(n)$ and $s(n)$ be some functions of natural number variables.

When Kolmogorov complexity is defined for the class of recursive automata that perform computations with time bounded by some function of a natural variable $t(n)$, we obtain time-bounded Kolmogorov complexity $C^t(x)$ studied by Kolmogorov (1965) and Barzdin (1968).

When Kolmogorov complexity is defined for the class of recursive automata that perform computations with space (that is, the number of used tape cells) bounded by some functions of a natural variable $s(n)$, we obtain space-bounded Kolmogorov complexity $C^s(x)$ studied by Hartmanis (1983).

When Kolmogorov complexity is defined for the class of multitape Turing machines that perform computations with time bounded by some function $t(n)$ and space bounded by some function $s(n)$, we obtain resource-bounded Kolmogorov complexity $C^{t,s}(x)$ studied by Daley (1973).

To deal with this variety of algorithmic complexity measures, we need a much more general approach to Kolmogorov complexity in order to include in this approach complexity of constructive objects represented or constructed by more powerful superrecursive algorithms. It is done on the basis of axiomatic dual complexity measures and utilization of abstract classes of algorithms. These constructions follow and develop in such a way the approach from (Burgin, 1982; 1990; 1991; 1992b).

5.3.2   General dual complexity measures

> *All intellectual improvement arises from leisure.*
>
> Samuel Johnson, 1709–1784

Dual complexity measures are properties of objects that are constructed and processed by algorithms. On the other hand, it is possible to interpret these measures as properties of classes of algorithms. Here we consider only static dual complexity measures for algorithms.

Let $\mathbf{P} = \{A_i; i \in I\}$ be a class of algorithms, A be an algorithm that works with elements from I as inputs and $\mathbf{Sc}: I \to N$ be a static complexity measure of algorithms from a class $\mathbf{P}$. Elements of I are usually treated as programs for the algorithm A. In addition, developing the theory of Kolmogorov complexity, researchers assume for simplicity that I consists of natural numbers in a form of binary sequences. These numbers can be only indices enumerating algorithms from $\mathbf{P}$ or codes of these algorithms (cf., for example, Section 2.3 for algorithm coding).

Definition 5.3.3. The dual to $\mathbf{Sc}$ complexity measure $\mathbf{Sc}_A^o$ of an object (word) x with respect to the algorithm A is the function from the codomain (the set of all outputs) Y of A that is defined as

$$\mathbf{Sc}_A^o(x) = \min\{\mathbf{Sc}(p); p \in I \text{ and } A(p) = x\}.$$

Naturally when there is no such p that $A(p) = x$, the value of $\mathbf{Sc}_A^o$ at x is undefined.

When $\mathbf{Sc}(x)$ measures information in the word or text x, the dual complexity measure $\mathbf{Sc}_A^o(x)$ estimates minimal information necessary to compute or build x by the algorithm A.

If $L_A^o(x)$ is the dual to the length $l(p)$ of program or algorithm description p complexity measure with respect to a algorithm A, then

$$L_A^o(x) = \min\{l(p); p \in I \text{ and } A(p) = x\}.$$

Let M and T be some algorithms.

Proposition 5.3.1. *If $M(x) > T(x)$ for almost all x and the function $f_M(x)$ defined by M is increasing, then $L_T^o(x) > L_M^o(x)$ for almost all x for which both $L_M^o(x)$ and $L_T^o(x)$ are defined.*

The most interesting case is when A is a universal algorithm V for the class $\mathbf{P}$. Let $\mathbf{c} \colon \mathbf{P} \to X^*$ be some coding of algorithms from $\mathbf{P}$.

Definition 5.3.4. An algorithm V is called *universal* for the class $\mathbf{P}$ if for any $A \in \mathbf{P}$ and any x given the pair $(\mathbf{c}(A), x)$ as its input, the result of V is equal to the result of A applied to x.

Examples of universal algorithms are a universal Turing machine (Section 2.3) and a universal inductive Turing machine (Section 4.3).

The dual complexity measure that corresponds to a universal algorithm gives an invariant characteristic of the whole class $\mathbf{P}$.

Definition 5.3.5. The dual to $\mathbf{Sc}$ complexity measure $\mathbf{Sc}_{\mathbf{P}}^{o}$ of an object (word) x with respect to the class $\mathbf{P}$ is defined as

$$\mathbf{Sc}_{\mathbf{P}}^{o}(x) = \min\{\mathbf{Sc}(p); \; p \in \mathrm{I} \text{ and } V(p) = x\}.$$

Naturally when there is no such p that $A(p) = x$, the value of $\mathbf{Sc}_{\mathbf{P}}^{o}$ at x is undefined. Because algorithm V is universal for the class $\mathbf{P}$, this condition is equivalent to the condition that there is no such algorithm A from $\mathbf{P}$ and such p that $A(p) = x$.

In other words, $\mathbf{Sc}_{\mathbf{P}}^{o}(x) = \mathbf{Sc}_{V}^{o}(x)$ for a universal algorithm V for the class $\mathbf{P}$. However, it is possible that has several universal algorithms. In such a case, the function of $\mathbf{Sc}_{\mathbf{P}}^{o}(x)$ is defined not in unique way. Nevertheless, Theorem 5.3.2 shows, the definition of $\mathbf{Sc}_{\mathbf{P}}^{o}(x)$ is invariant with respect to certain transformations.

When $\mathbf{Sc}(x)$ measures information in the word or text x, the dual complexity measure $\mathbf{Sc}_{\mathbf{P}}^{o}(x)$ estimates minimal information necessary to compute or build x by algorithms from the class $\mathbf{P}$.

Remark 5.3.1. If we can chose different algorithms from $\mathbf{P}$ to build the element x, the dual measure with respect to the class $\mathbf{P}$ is defined in a different way.

Definition 5.3.6. The dual to $\mathbf{Sc}$ complexity measure $\mathbf{Sc}_{\mathbf{P}}^{o}$ of an object (word) x with respect to the class $\mathbf{P}$ with selection is defined as

$$\mathbf{Sc}_{\mathrm{SP}}^{o}(x) = \min\{\mathbf{Sc}(p); \; p \in \mathrm{I}, \; A \in \mathbf{P}, \text{ and } A(p) = x\}.$$

Naturally when there is no such algorithm A from $\mathbf{P}$ and such p that $A(p) = x$, the value of $\mathbf{Sc}_{\mathrm{SP}}^{o}$ at x is undefined.

Lemma 5.3.1. $\mathbf{Sc}_{\mathrm{SP}}^{o}(x) \leq \mathbf{Sc}_{\mathbf{P}}^{o}(x) \leq \mathbf{Sc}(x)$.

In general, both functions $\mathbf{Sc}_{\mathrm{SP}}^{o}(x)$ and $\mathbf{Sc}_{\mathbf{P}}^{o}(x)$ are defined for all elements x from the domain $\bigcup_{A \in \mathbf{P}}(A)$. In particular, when all algorithms from $\mathbf{P}$ have a common domain X, then both functions $\mathbf{Sc}_{\mathrm{SP}}^{o}(x)$ and $\mathbf{Sc}_{\mathbf{P}}^{o}(x)$ are defined for all elements x from X. For example, when $\mathbf{P}$ is the set of all partial recursive functions, both functions $\mathbf{Sc}_{\mathrm{SP}}^{o}(x)$ and $\mathbf{Sc}_{\mathbf{P}}^{o}(x)$ are defined for all natural numbers. This is a consequence of the following more general result, which is true for the most interesting cases.

Let us assume that the class $\mathbf{P}$ contains the identical algorithm $E(x) = x$.

Theorem 5.3.1. $Sc_A^o(x)$ *is a total function on* N^+ *(on the set of all words in some alphabet).*

Dual complexity measures are usually interpreted as complexity of problem solution with the help of algorithms from **P**. More exactly, the problem under the consideration is construction or computation of a word x by means of algorithms from **P**.

The complexity of a problem often differs from the complexity of its solution. Simple problems, that is, problems that have short descriptions, may have only complex solutions, that is, they demand long proofs or a lot of computations. Moreover, as it is proved by Juedes and Lutz (1992) many important problems that have hard solutions (those that are P-complete for ESPACE) have low problem complexity, that is, their Kolmogorov complexity or algorithmic information is rather low.

In the theory of algorithms, a lot of dual complexity measures are studied: Kolmogorov complexity (Solomonoff, 1964; Kolmogorov, 1965; Chaitin, 1969), uniform complexity (Loveland, 1969), prefix complexity (Gasc, 1974; Levin, 1974; Chaitin, 1975), monotone complexity (Levin, 1973), process complexity (Schnorr, 1973), conditional Kolmogorov complexity (Sipser,1983), time-bounded Kolmogorov complexity (Kolmogorov,1965; Barzdin,1968), resource-bounded Kolmogorov complexity (Daley,1973), generalized Kolmogorov complexity (Burgin, 1982) and so on.

However, there are other dual complexity measures. As an example of another kind of a dual complexity measure, we can take Boolean circuit complexity, which is also a nonuniform complexity measure (Savage, 1976; Balcazar et al., 1988).

Basic for this measure are Boolean elements or gates. Each Boolean element is a source (of a Boolean variable or a Boolean constant, 1 or 0) or performs a basic Boolean operation: 0-ary operations — "*true*" and "*false*" (which are sometimes also denoted by 1 and 0); the unary operation of negation, denoted by $--$; and the binary operations — "*and*", denoted by $\wedge$ and "*or*", denoted by $\vee$.

Definition 5.3.7. A *Boolean circuit* (cf. Figure 5.1), also called a *combinational machine* or just a *circuit*, is an acyclic graph which nodes are Boolean elements (gates).

There are two direct and two dual complexity measures for Boolean circuits.

Definition 5.3.8. The *cost* or *size* $c(A)$ of a Boolean circuit A is the number of gates it has.

This is a direct static complexity measure of Boolean circuits.
Let f be a Boolean function.

Definition 5.3.9. The Boolean *cost* $c(f)$ of f is the size of the smallest circuit computing f:

$$c(f) = \min\{c(A); A \text{ defines the function equal to } f\}.$$

This is a dual complexity measure.

Definition 5.3.10. The *depth* of a circuit is the length of the longest path in the graph of the circuit.

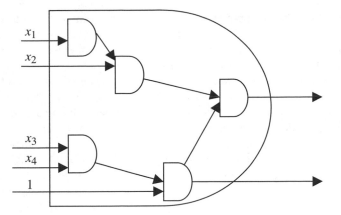

Figure 5.1. A Boolean circuit

This is a direct static reconstructible complexity measure of Boolean circuits.

Definition 5.3.11. The Boolean depth $d(f)$ of f is the depth of the minimal depth circuit computing f:

$$d(f) = \min\{d(A); A \text{ defines the function equal to } f\}.$$

This is a dual complexity measure. Thus, we see that not all dual complexity measures are Kolmogorov complexity or something of its kind. There are other examples of dual complexity measures.

Now due to its applications to problems of cryptography and network security, communication complexity has become one of the most popular types of complexity measures (cf. Hromkovic, 1997; Kushilevitz and Nisan, 1997)). There are different kinds of communication complexity: deterministic, nondeterministic, probabilistic, quantum, one-way, two-way, and so on. The first introduction of communication complexity is attributed to Yao (1979). Communication complexity measures important characteristics of distributed data processing.

Usually communication complexity is considered for the following situation. Two computers (persons) C_1 and C_2 are working together and solving the same problem (cf. Hromkovic, 1997; Kushilevitz and Nisan, 1997)). The problem taken for this purpose is computation of some finite function $f: X_1 \times X_2 \to Y$. As a rule, f is a Boolean function with m variables. At the beginning of the process, the input from X_1 is given to C_1 and the input from X_2 is given to C_2.

These computations, which include communication, are performed by (according to) algorithms P_i that are called communication protocols and describe a distributed computational processes of two computers C_1 and C_2. The goal is for one of them to compute $f(x_1, x_2)$ with the least amount of communication between them. In contrast to computational complexity, here we are not concerned about the number of computational steps or the size of the computer memory used. Communication complexity tries to quantify the amount of communication required for distributed computations.

It is supposed that both computers have unlimited computational resources and thus, they can, for example, always succeed by having C_1 send its whole n-bit string to C_2, allowing C_2 to compute the function, but we are interested in finding better ways of calculating f with less than n bits of communication.

This problem has applications in in many areas. One of them is VLSI circuit design, where it is necessary to minimize energy use by decreasing the amount of electric signals required between the different components during a distributed computation. Communication complexity is used to prove lower bounds for area complexity of general three-dimensional, and multilective VLSI. Another application area comprises Boolean circuits and formulas. The problem is also relevant in the study of data and knowledge bases, in the development of human-computer interaction, and in the optimization of computer networks, reflecting complexity trade-offs for interconnection networks with different topologies. In addition, communication complexity has been useful in proving lower bounds for size of finite automata, time and space complexity of Turing machines, and size of linear programs.

Definition 5.3.12. The *communication complexity* **cc**(P) of a communication protocol P is defined as the length of communicated word or, in other words, the maximal number of bits exchanged during the computational processes defined by P_i for all pairs of inputs. Inputs are taken from some finite sets X_1 and X_2.

This is a direct static complexity measure.

Definition 5.3.13. The *communication complexity* **cc**(f) of a function or problem f is defined as:

$$\mathbf{cc}(f) = \min\{\mathbf{cc}(P);\ P \text{ computes the function } f\}.$$

It is possible to represent any finite function by a table and then to represent this table as a word. In this context, the communication complexity **cc**(f) is a dual complexity measure on the set AP of all protocols.

As the class of all protocols is too broad and vague, it is reasonable to consider a definite class **P** of protocols and consider communication complexity with respect to this class. For instance, it is possible to consider the class **Q** of quantum protocols or the class **A** of conventional protocols.

Definition 5.3.14. The *communication complexity* **cc**(f) of a function f with respect to the class **P** is defined as

$$\mathbf{cc}(f) = \min\{\mathbf{cc}(P);\ P \text{ computes the function } f \text{ and belongs to } \mathbf{P}\}.$$

The communication complexity of a function f is a particular case of the following concept.

Definition 5.3.15. The *communication complexity* **cc**$_\mathbf{P}(q)$ of a problem q with respect to the class **P** is defined as

$$\mathbf{cc}_\mathbf{P}(q) = \min\{\mathbf{cc}(P);\ P \text{ solves the problem } q \text{ and belongs to } \mathbf{P}\}.$$

For a long time, communication complexity has been studied only for one problem — computing a function. Recently, Ambainis, et al. (2003) considered communication complexity for two more kinds of problems – generating a function and sampling. Communication sampling complexity is considered for two classes **P**: conventional and quantum algorithms. There are many other important theoretical and practical collaborative problems for which it might be useful to apply communication complexity.

There are also other approaches leading to dual complexity measures. For example, Gell-Mann (1994) introduced the concept of crude complexity of a system.

Definition 5.3.16. The *crude complexity* of a system R is the length of the shortest message that will describe R, employing language, knowledge, and understanding that both parties, the sender and recipient, share.

This is a complexity measure, which is dual to such direct measure as the length of description and which depends on algorithms of language utilization, understanding and explanation.

Crude complexity looks similar to Kolmogorov complexity, but it is not defined as an exact mathematical structure. However, its vagueness is an advantage as it leads to a much broader concept of dual superrecursive complexity. Really, people use superrecursive algorithms in their everyday life. An example is language learning because as demonstrated Gold (1967), this is a kind of inductive inference or computation. Utilization of superrecursive algorithms implies that the class **P** in the definition 5.3.3 cannot be reduced to recursive algorithms and it is necessary to study and use dual superrecursive complexity measures. In particular, crude complexity is a kind of dual superrecursive complexity measures.

Let $Sc_P^o(x)$ and $Sc_R^o(x)$ be dual to **Sc** complexity measures with respect to classes **P** and **R**, respectively. If $P \subseteq R$, then any algorithm universal for **R** is also universal for **P**. This implies the following results.

Theorem 5.3.2. *If* $P \subseteq R$ *and* $Sc_P^o(x)$ *is defined for* x, *then* $Sc_R^o(x)$ *is defined for* x *and* $Sc_R^o(x) \leq Sc_P^o(x)$.

Corollary 5.3.1. *If* $P \subseteq R$ *and* $Sc_P^o(x)$ *is defined for all* x, *then* $Sc_R^o(x)$ *is defined for all* x *and* $Sc_R^o(x) \leq Sc_P^o(x)$.

Dual complexity measures with respect to the class **P**, that is, determined by a universal algorithm, have invariance properties, defining minimal resources that are necessary in **P** to build or compute objects from Y. The set Y contains such objects that can be computed by algorithms from **P**.

Let **H** and **G** be two sets of functions.

Definition 5.3.17. A function $f(n)$ is called (asymptotically) **H**-optimal in **G** if there is such $h \in H$ that $f(n) \leq h(g(n))$ for any $g \in G$ and (almost) all $n \in N$.

If there is such $h \in \mathbf{H}$ that $f(n) \leq h(g(n))$ for almost all $n \in N$, we denote this relation by $f(x) \preceq_{\mathbf{H}} g(x)$. In the case, when $\mathbf{H}$ consists of such functions that add some constant to the argument, for example, $h(g(n)) = g(n) + c$, we write simply $g(x) \succeq f(x)$ or $f(x) \preceq g(x)$. This relation is basic for the theory of Kolmogorov complexity (Li and Vitanyi, 1997).

Lemma 5.3.2. *Relations* $g(x) \succeq f(x)$ *and* $f(x) \preceq g(x)$ *mean that there is a number* c *such that* $f(x) \leq g(x) + c$ *for all* x.

Lemma 5.3.3. *If* $q(x) \leq f(x)$ *and* $g(x) \leq h(x)$ *for almost all* x *and* $f(x) \preceq g(x)$, *then* $q(x) \preceq h(x)$.

Let $\mathbf{H}(h) = \{h_k(n) = h(h(n) + k), k \in N\}$ and $\mathbf{P}$ be an enumerable in itself set of algorithms that has a universal algorithm U.

Theorem 5.3.3. *(Burgin, 1982). For any axiomatic static complexity measure* $\mathbf{Sc}(p)$ *on* $\mathbf{P}$ *and for some recursively computable function* $h(n)$, *there is an* $\mathbf{H}(h)$-*optimal dual measure* $\mathbf{Sc}^o(x)$.

Definition 5.3.18. $f(n) \preceq_{\mathbf{T}(h)} g(n)$ $(f(n) \preceq^a_{\mathbf{H}(h)} g(n))$ if there is such $h \in \mathbf{H}$ that $f(n) \leq h(g(n))$ for all $n \in N$ (almost all $n \in N$).

Definition 5.3.19. Functions $f(n)$ and $g(n)$ are called (asymptotically) $\mathbf{T}(h)$-equivalent if $f(n) \preceq_{\mathbf{T}(h)} g(n)$ and $g(n) \preceq_{\mathbf{T}(h)} f(n)$ $(f(n) \preceq^a_{\mathbf{T}(h)} g(n)$ and $g(n) \preceq^a_{\mathbf{T}(h)} f(n))$.

Theorem 5.3.4. *(Burgin, 1982). Any two (asymptotically)* $\mathbf{H}(h)$-*optimal functions are (asymptotically)* $\mathbf{H}(h)$-*equivalent.*

This means that optimal dual measures are in some sense invariant.

Theorems 5.3.3 and 5.3.4 imply existence and uniqueness of optimal or invariant measures for many dual complexity measures (Li and Vitanyi, 1997): Kolmogorov complexity, uniform complexity, prefix complexity, monotone complexity, process complexity, conditional Kolmogorov complexity, time-bounded Kolmogorov complexity, space-bounded Kolmogorov complexity, conditional resource-bounded Kolmogorov complexity, time-bounded prefix complexity, resource-bounded Kolmogorov complexity, and so on. We do not need to prove these theorems for each case separately because it is sufficient only to check conditions from theorems 5.3.3 and 5.3.4 and then to apply these theorems.

However, not all properties of optimal dual measures are good. For example, it is proved that Kolmogorov complexity, which is an optimal dual measure for all recursive algorithms, is itself not a recursive function (Li and Vitanyi, 1997), although it can be computed by an inductive Turing machine (Burgin, 1983).

5.3.3 Algorithmic complexity of recursive algorithms

Custom reconciles us to everything.

Edmund Burke, 1729–1797

In the study of dual complexity measures, it is possible to make the following reductions. Algorithms work with words in some alphabet X. We can codify all symbols from X by finite strings consisting of two symbols 1 and 0. This allows us to consider only algorithms that work with words in the alphabet $\{1, 0\}$. In addition, it is practical in some cases interpret such binary words as representations of nonnegative integer numbers and assume that algorithms work with such numbers.

At first, we find some properties of complexity measures $L_{\mathbf{P}}^{o}(x)$ dual to the length $l(p)$ of program or algorithm description p with respect to a general class $\mathbf{P}$ of algorithms that work with words or natural numbers. We assume that $\mathbf{P}$ has a universal algorithm V. Then we have

$$L_{\mathbf{P}}^{o}(x) = \min\{l(p); \ p \in I \text{ and } V(p) = x\}.$$

Theorem 5.3.1 implies following results.

Corollary 5.3.2. $L_{\mathbf{P}}^{o}(x)$ *is a total function on* N^{+} *(on the set of all words in some alphabet).*

The dual to the length of program or algorithm description complexity measure $C_{R}(x)$ with respect to a class $\mathbf{R}$ of recursive algorithms (Turing machines, partial recursive functions, and so on) is called Kolmogorov complexity (Li and Vitanyi, 1996). For simplicity, we consider only such class $\mathbf{R}$ as the class $\mathbf{T}$ of all Turing machines and denote $C_{R}(x)$ by $C(x)$.

Corollary 5.3.3 (Kolmogorov, 1965). $C(x)$ *is a total function on* N^{+} *(on the set of all words in some alphabet).*

Let us suppose that the class $\mathbf{P}$ is infinite and contains only such algorithms that give as the result only one word or one number. In addition, we assume, without loss of generality, that all algorithms from $\mathbf{P}$ are working with natural numbers that are represented by words in the alphabet $\{1, 0\}$.

Lemma 5.3.4. *For any number n there is such number z that for all elements x that are larger than some element z, the values* $L_{\mathbf{P}}^{o}(x)$ *are larger than n.*

Proof. The number of those elements x for which $L_{\mathbf{P}}^{o}(x)$ is less than or equal to a given number n is less than 2^{n+1} because there are at most 2^{n+1} programs having the length less than or equal to n and the universal inductive Turing machine W computes only one word with one program. Consequently, for all elements y that are larger than some element z, the values $L_{\mathbf{P}}^{o}(y)$ are larger than $n - 1$.

Definition 5.3.20. A partial function $f: X^* \to N^{+}$ tends to infinity (we denote it by $f(x) \to \infty$, or $f(x) \to \infty$ when $l(x) \to \infty$) if for any number m from N^{+} there is a number k such that $f(x) > m$ when $l(x) > k$.

Definition 5.3.21. A partial function $f: X^* \to X^*$ tends to infinity (we denote it by $f(x) \to \infty$) if the partial function $l(f(x))$ tends to infinity.

Lemma 5.3.4 implies the following result.

Theorem 5.3.5. $L_P^o(x) \to \infty$ *when* $l(x) \to \infty$.

Proof. Since the number of elements x for which $L_P^o(x)$ is less than or equal to a given number n is finite by Lemma 5.3.3, so as n tends to infinity, the function $L_P^o(x)$ does the same.

Corollary 5.3.4 (Kolmogorov). $C(x) \to \infty$ *when* $l(x) \to \infty$.

Remark 5.3.2. Theorem 5.3.1 implies that Corollary 5.3.3 is also a direct corollary of Theorem 5.3.9.

Let **P** be an enumerable class of recursive or subrecursive algorithms that contains a universal algorithm.

Theorem 5.3.6. $L_P^o(x)$ *is an inductively computable function, namely, it is computable by some inductive Turing machine of the first order.*

It is known (cf. (Li and Vitaniy, 1997)) that the function $C(x)$ is not recursively computable. At the same time, we have the following result implied by Theorem 5.3.6 that shows one more time the advantages of inductive Turing machines.

Corollary 5.3.5 (Burgin, 1982). $C(x)$ *is an inductively computable function, namely, it is computable by some inductive Turing machine of the first order.*

This result also follows from Theorem 4.2.5 and the theorem of Kolmogorov that states that $C(x)$ is a limiting recursive function (cf. Zvonkin and Levin, 1970).

Traditionally (cf., for example, (Li and Vitaniy, 1997)), researchers in Kolmogorov complexity also consider the function $mC(x) = \min\{C(y); y \geq x\}$, which bounds $C(x)$ from below. It has the following properties. It has the following properties.

Theorem 5.3.7. (Kolmogorov). *a)* $mC(x)$ *is a total increasing function;*
    b) $mC(x)$ *is not recursively computable;*
    c) $mC(x)$ *is inductively computable;*
    d) $mC(x) \to \infty$ *when* $l(x) \to \infty$.

Proof. a) Since $C(x)$ is a total function, $mC(x)$ is also a total function. By its definition, $mC(x)$ is increasing. Parts b) and d) are proved in (Li and Vitaniy, 1997). The part c) follows from Corollaries 5.3.3 and 5.3.4.   □

Corollary 5.3.1. *(Kolmogorov, 1965).* $C(x)$ *is not a recursively computable function.*

Moreover, it is possible to prove a stronger result.

Theorem 5.3.8. *(Zvonkin and Levin, 1970). If h is an increasing computable function that is defined in a decidable set V and tends to infinity when $l(x) \to \infty$, then for infinitely many elements x from V, we have $h(x) > C(x)$.*

Noncomputability of Kolmogorov complexity allows one to prove noncomputability of communication complexity $\mathbf{cc}(f)$.

Theorem 5.3.9. (Gupta, 2002). $\mathbf{cc}(f)$ *is not a recursively computable function in a general case.*

To prove this result, we consider two computers (persons) C_1 and C_2 that are solving a problem f and are represented by universal Turing machines. The problem taken for this purpose is computation of x. Thus, we denote the problem by x.

At the beginning of the process, x is given as input to C_1 and nothing is given to C_2, while it is C_2, which has to compute x. That is why the value $\mathbf{cc}(x)$ determines the minimal number of bits that allow C_2 to compute x. By the definition, this number is equal to $C(x)$.

As $C(x)$ is not recursively computable function, $\mathbf{cc}(f)$, which in this case coincides with $C(x)$ is also not recursively computable function.

However, inductive computations realized by inductive Turing machines allow one to compute communication complexity in many interesting cases.

Let us assume that any problem f under consideration can be solved by some recursive algorithm (Turing machine) A and all communication protocols form a recursively enumerable or computable set.

Theorem 5.3.10. $\mathbf{cc}(f)$ *is an inductively computable function, namely, it is computable by some inductive Turing machine of the first order.*

Proof utilizes the Church–Turing thesis and is based on the assumption, which is usually made in studies of communication complexity, that all protocols are recursive algorithms.

Remark 5.3.3. For some classes of distributed computation problems, $\mathbf{cc}(f)$ is a recursively computable function. For example, let us consider two computers (persons) C_1 and C_2 that are solving the problem f and are represented by universal Turing machines. The problem taken for this purpose is computation of x. However, in contrast to the situation in theorem 5.3.9, x is given as input to C_2 and C_2 has to compute x. In this case, $\mathbf{cc}(x)$ is identically equal to 0.

Remark 5.3.4. It is interesting to study computability of the communication complexity $\mathbf{cc}(f)$ in the case when protocols are inductive or other superrecursive algorithms.

5.3.4 Algorithmic complexity of superrecursive algorithms

Constant dropping wears away a stone.

A proverb

The dual to the length of program or algorithm description complexity measure $C(x)$ with respect to a class **SR** of superrecursive algorithms (inductive Turing machines, limit partial recursive functions, grid automata and so on) is called superrecursive Kolmogorov complexity (Burgin, 2000). Here is an explicit definition.

Definition 5.3.22. The superrecursive Kolmogorov complexity $SRC(x)$ of an object or word x is defined as

$$SRC(x) = \min\{l(p); U(p) = x\}$$

where $l(p)$ is the length of word p and U is a universal algorithm in the class **SR**.

For simplicity, we consider only such class **SR** as the class **IT** of all inductive Turing machines of the first order.

Definition 5.3.23. The inductive Kolmogorov complexity $IC(x)$ of an object (word) x is defined as
$$IC(x) = \min\{l(p); U(p) = x\}$$

where $l(p)$ is the length of the word p and U is a universal inductive Turing machine of the first order.

Here we assume, without loss of generality, that all considered inductive Turing machines are working with natural numbers that are represented by words in the alphabet $\{1, 0\}$.

Let T be an inductive Turing machine of the first order.

Definition 5.3.24. The inductive Kolmogorov complexity $IC_T(x)$ of an object or word x with respect to the machine T is defined as

$$IC_T(x) = \min\{l(p); T(p) = x\}.$$

Theorems 5.3.1 and 5.3.4 imply the following result.

Proposition 5.3.2. *$SRC(x)$ is a total function on **N** (on the set of all words in some alphabet).*

Corollary 5.3.2. (Burgin, 1990). *$IC(x)$ is a total function on **N** (on the set of all words in some alphabet).*

As we will see the function $IC(x)$ is essentially smaller than the function $C(x)$. However, $IC(x)$ also tends to infinity as Theorem 5.3.5 implies the following result.

Proposition 5.3.3. *$IC(x) \to \infty$ when $l(x) \to \infty$.*

However, $IC(x)$ grows slower than any total increasing inductively computable function.

Theorem 5.3.11. *If f is a total strictly increasing inductively computable function, then for infinitely many elements x, we have $f(x) > IC(x)$.*

Proof. Let us assume that there is some element z such that for all elements y that are larger than z, we have $f(x) \leq IC(x)$. Because $f(x)$ an inductively computable function, there is an inductive Turing machine T that computes $f(x)$. It is done in the following way. Given a number x, the machine T makes the first step, producing $f_1(x)$ on its output tape. Making the second step, the machine T producing $f_2(x)$ on its output tape. After n steps, T has $f_n(x)$ on its output tape. Since the function is inductively computable, this process stabilizes on some value $f_n(x) = f(x)$, which is the result of computation with the input x. Taking the function $h(m) = \min\{x; f(x) \geq m\}$, we construct an inductive Turing machine M that computes the function $h(x)$.

The inductive Turing machine M contains a copy of the machine T. Utilizing this copy, M finds one after another the values $f_1(1)$, $f_1(2)$, ..., $f_1(m + 1)$ and compares these values to m. Then M writes into the output tape the least x for which the value $f_1(x)$ is larger than or equal to m. Then M finds one after another the values $f_2(1)$, $f_2(2)$, ..., $f_2(m + 1)$ and compares these values to m. Then M writes into the output tape the least x for which the value $f_2(x)$ is larger than or equal to m. This process continues until the output value of M stabilizes. It happens for any number m due to the following reasons. First, $f(x)$ is a total function, so all values $f_i(1)$, $f_i(2)$, ..., $f_i(m+1)$ after some step $i = t$ become equal to $f(1)$, $f(2)$, ..., $f(m + 1)$. Second, $f(x)$ is a strictly increasing function, that is, $f_i(m + 1) > m$. In such a way, the machine M computes $h(m)$. Since m is an arbitrary number, the machine M computes the function $h(x)$.

Since for all elements y that are larger than z, we have $f(y) \leq IC(y)$, there is an element m such that $IC(h(m)) \geq f(h(m))$ and $f(h(m)) \geq m$ as $f(x)$ is a strictly increasing function and $h(m) = \min\{x; f(x) \geq m\}$. By the definition, $IC_T(h(m)) = \min\{l(x); T(x) = h(m)\}$. As $T(m) = h(m)$, we have $IC_T(h(m)) \geq l(m)$. Thus, $l(m) \geq IC_T(h(m)) \geq IC(h(m)) \geq m$. However, it is impossible that $l(m) \geq m$. This contradiction concludes the proof of the theorem. □

We can prove a stronger statement than Theorem 5.3.11. To do this, we assume for simplicity that inductive Turing machines are working with words in some finite alphabet and that all these words are well ordered, that is, any set of words contains the least element. It is possible to find such orderings, for example, in (Li and Vitaniy, 1997).

Theorem 5.3.12. *If h is an increasing inductively computable function that is defined in an inductively decidable set V and tends to infinity when $l(x) \to \infty$, then for infinitely many elements x from V, we have $h(x) > IC(x)$.*

Proof. Let us assume that there is some element z such that for all elements x that are larger than z, we have $h(x) \leq IC(x)$. Because $h(x)$ an inductively computable function, there is an inductive Turing machine T that computes $h(x)$. Taking the function $g(m) = \min\{x; h(x) \geq m$ and $x \in V\}$, we construct an inductive Turing machine M that computes the function $g(x)$.

As V is an inductively decidable set, there is an inductive Turing machine H that given an input x, produces 1 when $x \in V$, and produces 0 when $x \neq V$. It means that H computes the characteristic function $c_V(x)$ of the set V.

The inductive Turing machine M contains a copy of the machine H and a copy of the machine T. Utilizing this copy of T, the machine M computes the value $h_1(1)$ and compares it to m. Utilizing this copy of H, the machine M computes the value $c_{V1}(1)$. If $h_1(1)$ is larger than m and $c_{V1}(1) = 1$, then M writes 1 into the output tape. Otherwise, M writes nothing into the output tape. After this, M finds the values $h_2(1)$ and $h_2(2)$ and compares these values to m. Concurrently, M finds the values $c_{V2}(1)$ and $c_{V2}(2)$. Then M writes into the output tape the least x for which the value $h_1(x)$ is larger than or equal to m and at the same time, $c_{V2}(x) = 1$. This process continues. Making cycle i of the computation, M computes the values $h_i(1)$, $h_i(2), \ldots , h_i(i)$ and compares these values to m. We remind here that $h_i(j)$ is the result of i steps of computation of T with the input j. Concurrently, M computes the values $c_{Vi}(1), c_{Vi}(2), \ldots , c_{Vi}(i)$. Then M writes into the output tape the least x for which the value $h_i(x)$ is larger than or equal to m and at the same time, $c_{Vi}(x) = 1$. Such cycle is repeated until the output value of M stabilizes.

Each value $c_{Vi}(x)$ stabilizes at some step t because $c_V(x)$ is a total inductively computable function. In a similar way, each value $h_i(x)$ stabilizes at some step q because $h(x)$ is an inductively computable function defined for all $x \in V$. Thus, after this step $p = \max\{q, t\}$, the value $h_i(x)$ becomes equal to the value $h(x)$. In addition, there is such a step t when a number n is found for which $h(n) \geq m$. After this step, only such numbers x can go to the output tape of M that belong to V and are less than or equal to n.

This happens for any given number m due to the following reasons. First, $h(x)$ is defined for all elements from V total function, so those values $h_i(1)$, $h_i(2), \ldots , h_i(m + 1)$ for which the argument of h_i belongs to V after some step $i = r$ become equal to $h(1), h(2), \ldots , h(m)$. Second, $h(x)$ is an increasing function that tends to infinity.

This shows that the whole process stabilizes and by the definition of inductive computability, the machine M computes $g(m)$. Since m is an arbitrary number, the machine M computes the function $g(x)$.

To conclude the proof, we repeat the reasoning from the proof of Theorem 5.3.11. Since for all elements y that are larger than z, we have $f(x) \leq IC(x)$, there is an element m such that $IC(g(m)) \geq h(g(m))$ and $h(g(m)) \geq m$ as $h(x)$ is an increasing function and $g(m) = \min\{x; h(x) \geq m\}$. By the definition, $IC_T(g(m)) = \min\{l(x); T(x) = g(m)\}$. As $T(m) = g(m)$, we have $IC_T(g(m)) \leq l(m)$. Thus, $l(m) \geq IC_T(h(m)) \geq IC(h(m)) \geq m$. However, it is impossible that $l(m) \geq m$. This contradiction concludes the proof of the theorem.   □

Remark 5.3.5. Although Theorem 5.3.11 can be deduced from Theorem 5.3.12, we give an independent proof because it demonstrates another technique, which displays essential features of inductive Turing machines.

Corollary 5.3.3. *If h is a total increasing inductively computable function that tends to infinity when $l(x) \to \infty$, then for infinitely many elements x, we have $h(x) > \mathrm{IC}(x)$.*

Corollary 5.3.4. *If h is an increasing inductively computable function that is defined in an recursive set V and tends to infinity when $l(x) \to \infty$, then for infinitely many elements x from V, we have $h(x) > \mathrm{IC}(x)$.*

Since the composition of two increasing functions is an increasing function and the composition of a recursive function and an inductively computable function is an inductively computable function, we have the following result.

Corollary 5.3.5. *If $h(x)$ and $g(x)$ are increasing functions, $h(x)$ is inductively computable and defined in an inductively decidable set V, $g(x)$ is a recursive function, and they both tend to infinity when $l(x) \to \infty$, then for infinitely many elements x from V, we have $g(h(x)) > \mathrm{IC}(x)$.*

Corollary 5.3.6. *The function $\mathrm{IC}(x)$ is not inductively computable. Moreover, no inductively computable function $f(x)$ defined for an infinite inductively decidable set of numbers can coincide with $\mathrm{IC}(x)$ in the whole of its domain of definition.*

In addition to the function $\mathrm{IC}(x)$, we also introduce the function $\mathrm{mIC}(x) = \min\{\mathrm{IC}(y); l(y) \geq l(x)\}$. It has the following properties.

Theorem 5.3.13. *(a) $\mathrm{mIC}(x)$ is a total increasing function;*
(b) $\mathrm{mIC}(x)$ is not inductively computable;
((c) $\mathrm{mIC}(x) \to \infty$ when $l(x) \to \infty$.

Proof. (a) Since $\mathrm{IC}(x)$ is a total function, $\mathrm{mIC}(x)$ is also a total function. By definition, $\mathrm{mC}(x)$ is increasing.

(b) If $\mathrm{mIC}(x)$ is an inductively computable function, then by Theorem 5.3.12, for infinitely many elements x, we have $\mathrm{mIC}(x) > \mathrm{IC}(x)$. However, by the definition of $\mathrm{mIC}(x)$, we have $\mathrm{mIC}(x) \leq \mathrm{IC}(x)$ everywhere. This contradiction completes the proof of the part (b) is proved.

Part (c) follows from Lemma 5.3.1.

Theorem 5.3.13 is proved.   □

Theorems 5.3.6 and 5.3.12 imply the following result.

Theorem 5.3.14. *For any increasing recursive function $h(x)$ that tends to infinity when $l(x) \to \infty$ and any inductively decidable set V, there are infinitely many elements x from V for which $h(\mathrm{C}(x)) > \mathrm{IC}(x)$.*

Corollary 5.3.12. *In any inductively decidable set V, there are infinitely many elements x for which $\mathrm{C}(x) > \mathrm{IC}(x)$.*

Corollary 5.5.13. *In any recursive set* V, *there are infinitely many elements* x *for which* $C(x) > IC(x)$.

Corollary 5.5.14. *In any inductively decidable* (*recursive*) *set* V, *there are infinitely many elements* x *for which* $\ln_2(C(x)) > IC(x)$.

If $\ln_2(C(x)) > IC(x)$, then $C(x) > 2^{IC(x)}$. At the same time, for any natural number k, the inequality $2^n > k \cdot n$ is true almost everywhere. This and Corollary 5.5.6 imply the following result.

Corollary 5.5.15. *For any natural number* k *and in any inductively decidable* (*recursive*) *set* V, *there are infinitely many elements* x *for which* $C(x) > k \cdot IC(x)$.

Corollary 5.5.16. *There are infinitely many elements* x *for which* $C(x) > IC(x)$.

Corollary 5.5.17. *For any natural number* a, *there are infinitely many elements* x *for which* $\ln_a(C(x)) > IC(x)$.

Corollary 5.5.16. *There are infinitely many elements* x *for which* $\ln_2(C(x)) > IC(x)$.

All these results show that, with respect to a natural extension of the Kolmogorov or algorithmic complexity, inductive Turing machines are much more efficient than any kind of recursive algorithms. Informally, it means that in comparison with recursive algorithms, superrecursive programs for solving the same problem are shorter, have lower branching (that is, less instructions of the form IF A THEN B ELSE C), make less reversions and unrestricted transitions (that is, less instructions of the form GO TO X) for infinitely many problems solvable by recursive algorithms.

In addition, connections between communication complexity and Kolmogorov complexity explicated in Section 5.3.3 and results of Theorem 5.3.12 and its corollaries imply that it is possible to decrease communication complexity for many problems if we use inductive computations instead of recursive computations.

Another proof of higher efficiency of inductive Turing machines in comparison with conventional Turing machines and other recursive algorithms is given by their ability to solve such problems that cannot be solved by conventional Turing machines. For instance, Lewis (2001) demonstrates limits of software estimation, using boundaries set by the theory of algorithmic complexity. Inductive Turing machines are able to make many estimations that are inaccessible for conventional Turing machines. It is possible only because inductive Turing machines have lower algorithmic complexity for those problems than conventional Turing machines.

6

Conclusion: Problems of Information Technology and Computer Science Development

> *One never notices what has been done;*
> *one can only see what remains to be done....*
>
> Marie Curie, 1867–1934

Thus, a new definition of algorithms has been introduced that provides a clear distinction between algorithms and their representations. However, for brevity, it is possible to use the name *algorithm* for symbolic representations of algorithms. We consider two main types of representations: physical representation (for example, in the form of electrical charges, currents and so on) and symbolic representations (for instance, in the form of recursive functions, Turing machines, finite automata, flow charts and so on).

The new definition of algorithms has been used as a base for the study of mathematical models of algorithms, representing different modes of:

1) computation: centralized, controlled distributed, and autonomous distributed computation;
2) programming: procedural, functional, and descriptive programming;
3) intelligence modeling: behavioral and structural approaches.

Three types of algorithms and their models have been studied: recursive, subrecursive, and superrecursive. All conventional algorithms and their models are included in recursive and subrecursive types. There are several models/classes of superrecursive algorithms. The emphasis is on inductive Turing machines, which are closer to conventional algorithms and thus, are more feasible for realization. In addition, superrecursive algorithms allow us to explain many peculiarities of computer and network functioning, as well as human thinking and behavior. It has been demonstrated that inductive Turing machines are more powerful (Chapter 4) and more efficient (Chapter 5) than conventional algorithms and their models.

Inductive Turing machines represent a new higher level of mathematical models of algorithms, realizing a transition from terminating computation to intrinsically emerging computation. These properties result in nonlinear growth of computability spaces.

It has many consequences not only for algorithms themselves, but also for scientific study of nature and society. Science, as any other human cognition, builds models of studied phenomena. We can discern three types of models for natural and social processes: qualitative, analytical quantitative, and algorithmic or computational models. However, in many cases, the first two types of models do not give particular results such as numerical values without applying algorithms and performing computation. As a result, investigation of complex phenomena always involves algorithmic/computational models.

Inductive Turing machine as a model for cognitive processes opens new perspectives for cognitive studies and artificial intelligence. They are not only more powerful, but also more efficient for problems of artificial intelligence than Turing machines. Being more relevant for description of human cognition, inductive Turing machines also give more appropriate theoretical representation of modern computers. While Turing machines gave a correct abstract portrayal of computers at the beginning of the "computer era", inductive Turing machines are more adequate as mathematical models for contemporary computers. Really, at the beginning, to obtain a result from a computer, it was necessary to print this result. So, after the result was obtained, it was possible to print it and to shut down the computer. This corresponds to the work of Turing machine and any recursive algorithm, which halts when the result is obtained. Now when a result of computation is displayed on the screen of the monitor, this result exists only when the computer is functioning. If computer stops, the result disappears. This corresponds to the work of inductive Turing machine, which works without stopping to give a result. Only in some cases, it is possible to print the final result and have it as a hard copy. However, this mode is also included in the functioning of inductive Turing machine. Consequently, the recursive model does not represent many cases of real computations, while inductive Turing machines provide such representation. The same is true for many embedded systems, the primary function of which is to interact with ever functioning environment (Heath, 1997). Popular now pervasive computation, which gives people convenient access to relevant information stored on powerful networks and is more efficient in work, is the first step in a practical realization of the computational mode of inductive Turing machine and implementation of the new computing paradigm (Burgin, 1999a; 2000).

Inductive Turing machines and other superrecursive algorithms have many other advantages, providing for a better theoretical frame for functioning of huge and dynamic knowledge- and databases, for computing methods in numerical analysis, for search engines and so on. This type of algorithms is also better suited for modeling different dynamical systems, for which it is usually assumed that they function without stopping.

The results presented in previous chapters allow us to make here several conclusions that might be useful for further development of computer and information science, as well as for accelerating the growth of information technology. We discuss here the following topics:

1. What is the system organization of models of algorithms and automata?
2. How does this system organization of models influence information processing systems?
3. How do information processing systems evolve?

4. What are future trends in the development of computer and information science?

6.1   A systemology for models of algorithms and information processing systems

It takes all sorts to make a world.

A proverb

There is a multitude of different formal computational schemes or mathematical models that are called algorithms or considered as generalizations of algorithms. The problem is how to make an organized system of all these models and approaches. In this way, some try to make a rigid border between algorithms and non-algorithms. From this perspective, experts bounded the scope of algorithms only by such computational schemes that can be modeled by Turing machines. These boundaries are deduced from the Church–Turing thesis. However, some specialists accept as algorithms exclusively such computational schemes that produce only total functions. For others, only finite automata give the correct models of a computer. With introduction of polynomial time computations, many computer scientists use the name *algorithm* only for such computational schemes that give the final result in the polynomial time with respect to the length of their input.

On the other hand, as it is demonstrated in Chapter 4, it is reasonable and even necessary for future development to consider and to use inductive and other super-recursive algorithms that are more powerful than Turing machines. In contrast to the opinion of Teuscher and Sipper (2002), even contemporary computers are able to perform hyper-computations (see Examples 4.2.1–4.2.3 from Section 4.2), which can do much more than computations controlled by recursive algorithms. Thus, any approach to algorithms that tries to restrict algorithms to some mathematical model is not efficient enough for study, design, and utilization of algorithms.

That is why in contrast to restrictions, it is more productive to sanction utilization of different kinds of algorithms and their models, but to organize and systematize them through relevant classifications. In one of such classifications, which is basic for this book, the class of all deterministic Turing machines is taken as the base, and all models of algorithms that are equivalent to this class are called *recursive* algorithms. Consequently, the diversity of different models of algorithms is divided into three categories:

subrecursive algorithms, which have less computing or accepting power than recursive algorithms;

recursive algorithms, and

superrecursive algorithms, which have more computing or accepting power than *recursive* algorithms.

In addition to algorithms in this classification scheme, we also consider algorithmic schemes.

Savage (1976) introduces another classification, which contains three levels of models for IPS and computation:

1. Logic circuits or combinatorial machines.
2. Finite automata or sequential machines.
3. Turing machines.

In this section, we build additional classification that connects algorithms with their environment and IPS that they model. It is based on the structural relativistic approach to the concept of algorithm, as suggested in Chapter 2. Accordingly, the standard question "What is algorithm?" has a fallacy. Namely, it presupposes that there is an absolute concept of algorithm. More relevant is to speak about algorithms relative to some given conditions. We call them *classification conditions* because they allow us to classify and distinct algorithms.

For example, algorithm involves mechanical or cleric operations. However, what is a simple mechanical operation for one IPS, can be a very complex operation for another IPS. Such operation as face recognition is simple for people, but is hard even for the best computers. At the same time, to add 13579 to 8642 is a simple operation even for calculators, but is a complex operation for many people, who cannot do it without a calculator. Because computers can do much more than calculators, an algorithm for a computer is not, as a rule, an algorithm for a calculator.

Classification conditions for algorithms include:

1. a class **P** of information processing systems (IPS), which realize algorithms and may be abstract or real;
2. *actual resources* **R** for functioning or computations of systems from **P**;
3. *potential resources* **ER** for functioning or computations of systems from **P**.

Actual and potential resources may be different in theory and in practice. For example, if we take a resource such as time for a Turing machine, then potential time for a computation is arbitrary, while actual time, according to the contemporary approach, has to be polynomially bounded. For real computers, we have distinctly different conditions. One hour or less is an actual resource for computation, while 10^{10} hours is only a potential resource for computation.

Actual and potential resources separate solvability and tractability of problems.

Definition 6.1.1. A problem is called *solvable* in **P** if it can be solved by some system from **P** given potential resources.

Definition 6.1.2. A problem is called *tractable* in **P** if it can be solved by some system from **P** given actual resources.

Distinction between resources separates not only problems, but also all algorithmic schemes into three classes: actual, potential and ideal algorithms with respect to these conditions.

Definition 6.1.3. *Actual algorithm* is a structure that, given resources from **R**, it allows one to organize in an exact way the functioning of some IPS from **P**.

Definition 6.1.4. *Potential algorithm* is a structure that, given resources from **ER**, it allows one to organize in an exact way the functioning of some IPS from **P**.

If even resources from **ER** do not allow one to organize the functioning of some IPS from **P**, but the scheme describes an abstract information process, then this scheme is an *ideal algorithm or algorithmic scheme*.

For example, as no physical device can work with real numbers with infinite precision whatever resources we are given, recursion with real numbers represents ideal algorithms. As no physical device can do infinite numbers of separate steps whatever resources we are given, infinite time Turing machines are ideal algorithms.

Let us consider some examples. We assume that resources from **R** contain only finite tapes the length of which is bounded by some number n, while resources from **ER** have infinite tapes for Turing machines. The class **P** consists of devices that have the structure of Turing machines. Then all Turing machines are only potential algorithms. Those Turing machines that use the tape with the length less than n are actual algorithms. Such Turing machines are equivalent to finite automata. So, actual algorithms are only finite automata. Inductive Turing machines with a recursive memory are also potential algorithms, while inductive Turing machines of the second order are ideal algorithms.

Another situation is when we have infinite tapes for Turing machines but do not have oracles. In this case, all Turing machines are only actual algorithms, while all Turing machines with oracles for recursively noncomputable functions are ideal algorithms. All algorithmic schemes that work with arbitrary real numbers are also only ideal algorithms.

If we imagine an Algorithmic Universe (cf. Fig. 6.1), we see that its recursive region is a closed system, which entangles depressing incompleteness results, such as Gödel incompleteness theorems. In contrast to this, superrecursive region of the Algorithmic Universe is open. It implies development, creativity, and puts no limits on human endeavor.

An important problem is relation between algorithms and computation. Some reduce computation only to those processes that are realized by some general conventional model, such as Turing machine. This is the mathematical point of view on computation. Engineering approach treats computation as everything that computers can do. Others take such a general view on computation that almost all processes become computations. For example, computation is loosely described in (Goldreich, 2001) as "*a process that modifies an environment via repeated applications of a predetermined rule that depends and affects only a (small) portion of the environment, called the active zone.*" An extreme position of such broad understanding results in, the so-called, algorithmic theories of everything (cf., for example, (Schmidhuber, 2000)).

All this brings us to the important problem of a scientific definition of computation. Suggested by different authors definitions do not solve this problem. For instance, if we take the engineering definition, which looks so natural, we encounter a problem what computer is. Is a human being a computer or not?

Results of this book show that our understanding of computation depends on what kinds of algorithms and their models we consider. In particular, it is necessary to discern actual computations that are realized now from potential computations that only may be realized and from ideal computations that are now only theoretically

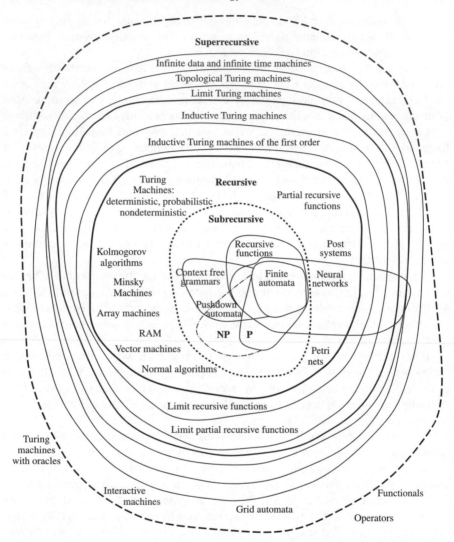

Figure 6.1. Algorithmic universe.

described. Thus, the majority of quantum algorithms and computations described in literature are now only ideal entities. At the same time, many kinds of inductive computations are actual or potential. Building computers and other IPS that realize new kinds of superrecursive algorithms allows one to shift the border between ideal and actual computations.

6.2 Development of information processing systems (IPS)

Nothing ventured, nothing gained.

A proverb

It is reasonable to consider the development of IPS from three aspects: what kind of technical devices they will be, what problems they will be able to solve, and what their role will be as they are placed in society and life of people in general. The latter perspective is well described by Papadopoulos (2002) as three waves. He writes:

> *The first wave was a network of computers that swelled to encompass hundreds of millions of systems, all connected, all continually exchanging data.*
>
> *The second wave, the one we're riding now, could be described as a network of things that embed computers. It's made up of wireless phones, two-way pagers and other handsets, game players, teller machines, and automobiles. In short, billions of potential connections.*
>
> *The third wave is on the way, and even as we create it, we need to prepare ourselves; it's shaping up to be a regular tsunami. I call it a network of things. Trillions of things. Things you'd hardly think of as computers. So-called subIP (Internet Protocol) devices such as light bulbs, environmental sensors and radio-frequency identification tags.*

We need to know how all this will influence our lives. The development of technology changes even the meaning of words. According to Parsons and Oja (1994), before 1940 the term *computer* was referred to a person who performed calculations. Although machines that performed calculations had existed from the 17th century when Blais Pascal (1623–1662) had built first machines for calculation, these machines were called calculators, not computers. The reason was that calculators were able to perform essentially fewer operations than people did. Only the electronic devices with essentially larger abilities in comparison with calculators acquired the name *computer*.

Now computer technology is everywhere: in engineering and design, in banking and accounting, in industrial production and entertainment, at the stock market and universities, and so on and so forth. Many fields of human activities cannot go on without computers and specialized software. Computers have changed and are continuing to change the process of writing and publishing a book or an article. Contemporary communication involves so many IPS that it is difficult even to imagine. IPS help to drive a car and to navigate and control a plane.

However, in future people will become as dependent on computers and other IPS as now they are dependent on their own organism. Embedded IPS will be everywhere and therefore it is so important and even vital to know, understand and foresee how IPS are developing, what they are doing and what they will be capable to do. It is also urgent to be able to utilize this knowledge. Theoretical tools help us to achieve these goals.

To understand the main trends in the development of IPS, we consider three main modes of computation, or in general, of information processing:

◇ *recursive mode* in which an IPS has potentially unbounded (undefined) resources but has to complete computation in finite time;
◇ *subrecursive mode* in which all resources of an IPS are exactly circumscribed;
◇ *superrecursive mode* in which an IPS can have potentially unbounded (unrestricted) resources, including time.

As it we already know, IPS have three layers: hardware, software, and infware. Each of them has static and dynamic characteristics. Consequently, the development of IPS goes along all these lines. Engineers construct more and more efficient hardware based on technological innovations (such as transistors, chips, DNA and so on). Programmers design more and more sophisticated software that can work with more and more advanced information. The advancement of computational and communication processes goes on concurrently with this design. Computer scientists develop new methods and techniques aimed at IPS. Better organization of computations contributes to enhancement of processes, improving their dynamical characteristics.

Natural science, material technology, and mathematics contribute to the development of hardware. Innovations for IPS software come mostly from software technology, computing linguistics and mathematics. Knowledge engineering, information and computer sciences, and mathematics play the leading role for infware enhancement. For instance, an urgent problem is to teach computers to work not only with data, but also with knowledge.

All layers of IPS are interconnected. The development of one of them influences other components. More complex hardware demands highly developed software in the form of an operating system. More sophisticated software is efficient only with a sufficiently advanced hardware. A variety of forms of information representation demand more advanced hardware and software and so on.

IPS are important only because they perform information processing. Information processing is directed by computer and network programs, while each program is an embodiment of some algorithm. Consequently, the theory of algorithms, automata and computation is one of the cornerstones for the development of IPS because any original approach to organization of information processing changes all components and parts of IPS often resulting in a new paradigm for computation as a whole.

It is necessary to admit that now we have much less knowledge about infware than about two other layers of IPS. Being able to build very advanced and sophisticated IPS (computers, local and global networks), people do not understand what information is. Even the best experts on computers do not make correct (if any) distinctions between knowledge and information. Thus, we encounter, for example, statements that "by 2047 almost all information will be in cyberspace — including a large percentage of knowledge and creative works." Actually, as it is demonstrated in the general theory of information (Burgin, 1994; 2002), information, knowledge, and texts are essentially different essences. Texts contain knowledge and both texts and knowledge are carriers of information. In contrast to knowledge, information is an active essence. It relates to knowledge and data as energy relates to substance. The existing misconception demonstrates that the development and correct application of information theory is urgent for a sound advancement of information processing, IPS and their utilization.

Now there are several perspective approaches how to increase the power of computers. We may find distinctions in biological, chemical, physical, and mathematical directions. Three first are applied to hardware only influencing software and infware, while the mathematical approach transforms all three components of IPS. A popular area in the first, biological approach is *membrane computing*, which identifies a new computing model, called P-system, from the natural way biological cells live and function (Berry and Boudol, 1990; Paun, 2002). A P-system is built as a structure of nested compartments surrounded by porous membranes. Initially each membrane contains a number of possible repeated information objects (symbols), which form a certain multiset (cf., for example, (Knuth, 1981)). Once functioning of such system starts, the compartments exchange objects according to specific rules for interaction with other membranes. Each compartment has a number of multiset processing rules. In the simplest case, these processing rules are just multiset rewriting rules. The activity stops when no such rule can be applied any more. The result of such computation equals to the number of objects that reside in a designated compartment called the output membrane. This allows P-systems to achieve high parallelism. There are different kinds of P-systems. In general, P-systems have many similar to Petri nets features and are specific biologically oriented kinds of grid automata.

The second approach is very popular. It is called the molecular computing, the most popular branch of which is the DNA computing (Cho, 2000). Its main idea is to design such molecules that solve computing problems. The third direction is even more popular than the first. It is quantum computing (Deutsch, 2000). Its main idea, which is attributed to Feynman (1982; 1986) and Beniof (1982), is to perform computation on the level of atoms and subatomic particles utilizing a theoretical ability to manufacture, manipulate, and measure quantum states. As there are different quantum process and various models in quantum physics, several approaches to quantum computation have been developed: a quantum Turing machine (Deutsch, 1985; 2000), quantum circuits (Feynman, 1986; Deutsch, 1989), modular functors that represent topological quantum computation (Freedman, 2001; Freedman, Larsen, and Wang, 2002; Freedman, Kitaev, Larsen, and Wang, 2002), and quantum adiabatic computation (Farhi et al., 2000; Kieu, 2002; 2002a; 2003).

It is necessary to remark that some researchers criticize the quantum approach from the theoretical perspective and doubt that quantum computers will work (cf., for example, Schmidhuber, 2002; Levin, 2003).

The most developed mathematical direction that extends the boundaries of contemporary computers is the theory of superrecursive algorithms. It is based on a new paradigm for computation that changes computational procedure. However, it might be interesting to know that analyzing Turing's analysis of the concept of algorithm, Gödel predicts, in a remark published after his death, a necessity of recursive algorithms that realize inductive and topological computations (cf., Blass and Gurevich, 2003). Gödel pointed that Turing disregarded completely the fact that mind, in its use, is not static, but constantly developing, i.e., that we understand abstract terms more and more precisely as we go on using them, and that more and more abstract terms enter the sphere of our understanding. As we know now, this brings us to algorithms that can compute more than the conventional Turing machines.

At the same time, it is known that biologically, chemically and physically based types of computing that are in the process of realization, for example, molecular or standard quantum computation, can do no more than conventional Turing machines theoretically can do. For instance, standard quantum computers are only some kinds of nondeterministic Turing machines (Deutsch, 1985; 2000), while a Turing machine with many tapes and heads model DNA and other molecular computers. DNA and quantum computers will be (when they will be realized) eventually only more efficient. In practical computations, they can solve more real-world problems than Turing machines. However, any modern computer can also solve more real-world problems than Turing machines because these abstract devices are very inefficient. At the same time, superrecursive algorithms can compute what is now considered as noncomputable, both theoretically and practically (cf. Chapter 4). In addition, superrecursive algorithms are more efficient (cf. Chapter 5). Consequently, these mathematical ideas are much more advanced and go much further than physical and chemical innovations.

This situation might be explained by the following metaphor.

Let us imagine time when people have many different cars, but no planes. DNA and quantum computing is like a suggestion to build cars that will have ten times higher speed than any existing car. In contrast to this, theory of superrecursive algorithms suggests building and flying planes.

To see how to go from virtual perspectives to actual reality, we need to consider three questions: how modern computers and networks are related to superrecursive algorithms, what new possibilities open superrecursive computations, and how it is possible to realize technologically these computations.The first two problems are considered in Chapters 4 and 5. To achieve the last but not the least goal, we need, in our case, to develop a new paradigm for computation or, more generally, for information processing. The conventional paradigm is based on our image of computer utilization, which consists of the following stages: 1) formalizing a problem; 2) writing a computer program; 3) obtaining a solution to the problem by program execution. In many cases, the necessary computer program exists and we need only the third stage. After this, we either leave our computer to do something else or you begin to solve another problem.

This process is similar to using a car. We use car to go to some place, then possibly to another and so on. However, at some moment, we park the car we were driving at some place, stop its functioning, and for a definite time do not use it. This is the *Car Paradigm* when some object is utilized only periodically for achieving some goal, but after this it does not function (at least for some time). In particular, this gives us a definite paradigm for computation.

In a very different manner, people use clocks. After buying, they start the clock, and then the clock is functioning until it breaks. People look at the clock from time to time to find what time it is. This is the *Clock Paradigm* when some object is functioning all the time without stopping, while those who utilize it get some results from it from time to time. Recursive algorithms imply that modern computers are utilized according to the *Car Paradigm*, while superrecursive algorithms suggest for computation and computer utilization the new *Clock Paradigm*.

It is also assumed that contemporary computers and their software are utilized according to the *Car Paradigm*. This justifies validity of the Church–Turing thesis for the computer realm. However, an interesting peculiarity of modern computers is that in many cases their computations are organized according to the potential *Clock Paradigm* when obtained results are considered only as intermediate and it is supposed that to get better results it is necessary to make more computational cycles.

Thus, it is possible to ask why the *Clock Paradigm* is new. Really, analysis of the up-to-date utilization of computers shows that even when researchers understand necessity to repeat a cycle of computation, they assume that computation is a finite process. Consequently, computation is represented by recursive algorithms. At the same time, conscious application of the new approach provides for several important benefits. First, it First, it gives a better understanding of the results of a separate part of the whole computational process. Second, it shows how to utilize computers in a better way. Third, it makes possible to use more adequate theoretical models for investigation and development of computation. For example, simulation and monitoring of many processes in industry is better modeled when these processes are treated as potentially infinite.

The Clock Paradigm, which was represented by such theoretical model as inductive Turing machine almost thirty years ago, finds its partial realization now in a form of pervasive computing. Pervasive computing encompasses the dramatically expanding sphere of computers embedded within and intrinsically part of larger devices (Kara, 2000). Pervasive computing provides convenient access to relevant information and applications through a new and powerful class of ubiquitous, intelligent appliances that have the ability to easily function when and where needed. Special networks enable these pervasive computing devices by providing transparent access to e-business and other services. Pervasive computing is useful to business users because it supports global (anywhere), sustainable, and persistent (anytime) environment, improves customer service (loyalty, competition, and differentiation), increases revenue (new channels, markets, and transactions), and decreases costs (efficiency competition, and cycle time).

At the same time, while the pervasive approach is simple, elegant, and mathematically grounded, the implementations are incredibly complex, with critical interdependencies in abundance. To solve these problems in an efficient and reliable way, we need to investigate properties of pervasive computing in the context of inductive computation model. Conventional recursive models may be very misleading for pervasive computing.

The *Car* and *Clock Paradigms* represent two mainstreams in the development of information processing technology. The first one corresponds to the past and present, while the second paradigm is directed into the future. However, this binary classification of paradigms constitutes a part of a more developed paradigmatic system.

Actually, when computers appeared, they have been utilized by many users with the help of programmers and operators. Consequently, computer functioning resembled not an arbitrary car but a taxi. Thus, computation at that time corresponded to the *Taxi Paradigm*.

Elaboration of personal computers changed the situation giving birth to the *One-Person-Car Paradigm*.

At the same time, big computers and, especially, supercomputers are working in the time sharing mode. For example, one computer may be used for running on it several computational processes. It makes possible to share one computer system providing in such a way concurrent facilities to several researchers. This corresponds not to the *Car* but to the *Bus Paradigm* for computation when one IPS is used by many users.

Time sharing on a personal computer corresponds to the *Car Paradigm* in its complete form when a car carries not only a driver, but also some passengers.

Recently, such approach as pervasive computation has been coined to reflect new facilities hidden in the Internet (Ark and Selker, 1999; Kara, 2000). Besides, different embedded systems are spreading very rapidly. Both these issues in connection with the superrecursive model for computation imply one more paradigm for computation. It is called the *Watch Paradigm*. According to this paradigm, computations are going on continuously, while computing device is not fixed at one place as in the *Clock Paradigm*, but moves together with the host system. Such host system may be a user, car, plane, and so on.

Normal functioning of modern computers presupposes that they work without stopping. However, many of them are switched off from time to time. In any case, these devices eventually end computations. At the same time, development of computer technology gave birth to systems that include as their hardware many computers and other electronic devices. As an example, we can take the contemporary World Wide Web. These systems possess many new properties. For instance, who can imagine now that the World Wide Web will stop its functioning even for a short period of time? Thus, the World Wide Web is a system that works according to the *Clock Paradigm*. Consequently, only superrecursive algorithms such as inductive Turing machines can correctly model such systems.

Embedded computing devices that employ superrecursive schema have to work in the Clock Paradigm if the process in their host system is continuous and the system is stationary. For example, embedded systems that regulate temperature in a building will have this feature. At the same time, embedded computing devices that employ superrecursive schema will be working in the Watch Paradigm if their host system moves.

The same is true for ubiquitous computing. According to its main idea, computations are going on continuously, but computing device is not fixed at one place as in the *Clock Paradigm*, It moves together with its owner. Recently, such approach as ubiquitous or nomadic computations has been coined to reflect new facilities hidden in the Internet. Computers will be connected to Internet all the time and will work without stopping. It does not mean that they will function in the same mode all the time, but the whole process will correspond to the Watch Paradigm.

Even Internet computing will be on the lines of the Clock Paradigm because Internet works without stopping and users enter it like people who look at a clock to know time. Only in this case, it will take more efforts to "look" but it will be possible to get much more information. To transform the contemporary Internet into an actual Cyberspace with the Grid as the base for it, we have to provide that it functions without stopping like a clock or nature. Can you imagine that the universe stops functioning? That was not by chance that the great Newton compared nature to a clock. Computers can really transit from a part of our World into our environment only when the new paradigm will be used.

In addition to this, many problems that are now considered undecidable will be repeatedly solved by future computing systems. Those will include such practically important problems as program debugging, software optimization and many others.

According to (Alt, 1997), future methodology for solution of hard problems as such weather forecast and medical control and diagnosis implies that computers that will solve them have to work in the Clock Paradigm. In addition to this, new approach to computation will enable usage of better numerical methods. Consequently, weather forecast and medical control and diagnosis will achieve much higher level of reliability than they have now.

It is worth mentioning that this novel approach of the Clock Paradigm is possible to implement even with the existent hardware, software, and infware. Although, it will be realized only in a partial form. Current situation is reflected in the following metaphor. People have some primitive planes, but do not know how to use them for flying. A new theory explains how to fly on these planes and how to build much more advanced planes. This implies that people will need new skills because flying is different from driving even the best car.

It is necessary to remark that it is claimed that some theoretical models of quantum computation can go beyond the Church–Turing thesis (cf., for example, Kieu, 2002; 2003). At the same time, the corresponding physical model suggests that the halting of a quantum universal Turing machine (computer) is highly problematic. This brings us to necessity of modeling quantum computers with superrecursive algorithms, such as inductive and limit Turing machines.

However advantages of superrecursive algorithms do not imply that they will completely substitute recursive algorithms in future. Where recursive and subrecursive algorithms efficiently solve problems, they will still be used in future. Only in the cases where recursive algorithms are impotent either in principle (as for the halting problem) or due to inefficiency (as for multidimensional optimization), superrecursive algorithms will take their place. We can compare the situation with utilization of calculators (recursive algorithms) and computers (superrecursive algorithms). To add three numbers each of which is less than 1000, we don't need a computer, although it is possible to use a computer as a calculator. Like computers in comparison to calculators, superrecursive algorithms are able to bring new power to people.

For example such important component of e-commerce as customer relationship management and customer analysis is inefficient in the recursive mode. Only the superrecursive mode in form of a topological computation (Burgin, 2001b) can provide necessary efficiency. As Pan and Lee (2003) write, the customer analysis application measures, predicts, and interprets customer behaviors, enabling companies to understand the effectiveness of e-CRM efforts across both unbound and outbound channels. The integrated customer information is used to build a business campaign strategy and assess results. It also builds predictive models to identify the customers most likely to perform a particular activity such as purchase an upgrade from the company. It is necessary to carry on customer analysis constantly or, at least, regularly, working in the Clock Paradigm. Otherwise, a company can miss important changes in the customer behavior. This often results in essential losses to the company.

However, being more advanced and powerful than recursive algorithms and devices, superrecursive algorithms and devices demand more skills and knowledge from their users in the same way as do computers in comparison to calculators. For instance, we may call superrecursive algorithms that direct inductive computations by the name **algorithms for intelligent users**.

To realize the new paradigm to a full extent, we need innovative hardware based on different physical principles (Stewart, 1991), original software implementing superrecursive principles of information processing, and even nonpareil organization of infware.

Being more advanced than recursive algorithms, superrecursive algorithms and devices demand more knowledge and skills from their users. In a similar way, computers demand more knowledge and skills than calculators. Thus, we may call superrecursive algorithms, which direct hyper-computations, *algorithms for intelligent users*.

Now we come to the problem of application of the new paradigm. To utilize its higher possibilities and thus to go beyond the Church–Turing thesis, people and the whole society have to use their creativity. Even more, it will have to be the creativity of special kind that may be called grounded or intelligent creativity. Humans and computers have to be cooperating systems (Norman, 1997). Consequently, creativity will continue to be a clue to the highest achievements, but superrecursive computations will increase these heights to unimaginable extent. Thus, people's creativity multiplied by computing power of superrecursive devices and algorithms will cause the real revolution in information technology and life of people.

At the same time, some aspects of creativity might be programmed (Burgin and Povyakel, 1988). Then programmed creativity times superrecursive computing power will give birth to Artificial Intelligence that will really be on the same or even on the higher level than human intellect.

6.3 From algorithms to programs to technology

All's well that ends well.

A proverb

Laws of the science development, which are discovered in methodology of science (Burgin and Kuznetsov, 1994), show three models for the theory development: extension/unification, intensification, and harmonization/simplification. The first approach is directed to extension of the scope of the theory. This goal is achieved by generalization and abstraction. In mathematics in general and in the theory of algorithms and computation, in particular, the standard methodology for doing this is axiomatization. In mathematics, this approach originated such gemstones as Euclidean geometry, mathematical logic, and topology. In the theory of algorithms, it resulted in creation of the axiomatic theory of programs (Hoare, 1969) and the axiomatic theory of algorithms (Burgin, 1985).

The second direction of the development of a theory is aimed at getting better approximation to reality that the initial theory models. In astronomy, this resulted in the transition from the Ptolemaic celestial system to the Copernicus' celestial system to the Newton's celestial dynamics. In computer science, the development in this direction goes from the *theory of algorithms* to the *theory of programs* (Hoare, 1969; Halstead, 1977) to the *mathematical theory of technology* (Burgin, 1997c; 2002).

To understand this trend better it is important to see how different is functioning of a real computer or network from what any mathematical model in general and a Turing machine (as an abstract, logical device), in particular, reputedly does when it follows instructions. In comparison with instructions of a Turing machine, programming languages provide a diversity of operations for a programmer. Operations involve various devices of computer and demand their interaction. In addition, there are several types of data. As a result, computer programs have to give more instructions to computer and to specify more details than instructions of a Turing machine. The same is true for other models of computation. For example, when a finite automaton represents a computer program, only some aspects of the program are reflected. That is why computer programs give more specified description of computer functioning, and this description is adapted to the needs of computer. Consequently, programs demand a specific theory of programs, which is different from the theory of algorithms and automata.

At the same time, as observes Cleland (2001), instructions of a Turing machine logically predetermine *everything* that is done by the machine — nothing is left open. As a consequence, there is no distinction between *what* an abstract machine does and *how* it does it. This is possible because Turing machine has a simple structure what is surprising for such a powerful device. In contrast to Turing machine instructions, no computer program (considered just *per se*) provides a complete specification of the behavior of the machine implementing it. Even the lowest level programs (machine language programs) depend upon a complicated encoding scheme called the "machine code format" to fix the physical states the machine actually passes through. This gap between the hardware and the software of a concrete computer and even a greater gap between pure functioning of the computer and its utilization by a user demands description of many other operations that lie beyond the scope of a computer program, but might be represented by a technology of computer functioning and utilization. This brings us to necessity of a specific theory of information processing technology.

The theory of programs, which is different from the theory of algorithms and computation and from the theory of programming, is at the very beginning of its development, although it was introduced as a separate discipline by Halstead (1977) 25 years ago. This theory studies programs, their design, development, maintenance, and usability. As such it contains the theory of programming as its part. Consequently, different areas of the theory of programs appeared before the name for this theory was coined. As an example, we can take the axiomatic theory of programs initiated by Hoare (1969). As Hoare writes, "computer programming is an exact science in that all the properties of a program and all the consequences of executing it can, in principle, be found out from the text of the program itself by means of purely deductive reasoning." These assumptions allow Hoare to write down axioms and rules of reasoning about computer programs. The choice of axioms naturally depends on the choice of programming language.

This, looking so practical, but naive approach is, in general, invalid like the Hilbert's belief in possibility to prove consistency of the whole mathematics. As it is demonstrated in Section 2.5, in contrast to this assumption, almost all properties of programs, according to Rice Theorem, are undecidable and thus, cannot be found out from the text of the program itself by means of purely deductive reasoning. Here we see one more time how theory helps practice to understand the real situation.

However, this does not invalidate the axiomatic approach to programming like the Gödel incompleteness theorem does not refute the axiomatic approach in mathematics. It might be useful if properly applied. Although according to the Gödel Theorem (Gödel, 1931), it is impossible to prove by classical inference methods all theorems in mathematics, mathematicians have proved and continue to prove many interesting and useful theorems. In a similar way, although it is impossible to find all properties of programs by standard logical reasoning, axiomatic methods might be useful on different stages of programming, especially, for program verification by means of proving program correctness. A lot of research has been done in this area and different technique has been suggested. At the same time, this formal approach to programming resulted in open controversy in the field of computing (Glass, 2002). On one hand, mathematical approaches to developing software were flourishing in academia. On the other hand, practitioners rarely (if ever) used them, while some of them, as well as some members of academic community challenged utility of logical methods for programming. Now, as Glass (2002) admits, there is a healthier outlet for this disagreement. Academics have toned down their rhetoric, admitting reasons why practitioners avoid the approaches. Practitioners concede the value of the theory, while they continue to challenge its workableness in practice.

Superrecursive algorithms in general and inductive computations, in particular, open new perspectives for formal methods in programming. Namely, it is possible to develop formal logical calculi based not on standard deduction but on inductive inference (Burgin, 1987). Theoretical results (cf. Chapter 4) show that such calculi will be much more powerful than the conventional logic because inductive Turing machines are more powerful than classical deduction. Consequently, these new methods and structures will allow researcher and practitioners to achieve better program design and verification by logical procedures.

The theory of technology is a much younger discipline even in comparison with the theory of programs. Although creation of a theory of technology has been urgent for a long time due to the situation that mankind is developing in the technological direction — more and more technologies are used by people. Technologies become more and more sophisticated and complex. Technology is not only various technical devices, but mostly knowledge about these devices. However, only at the end of the twentieth century, the development of technological knowledge and mathematical achievements made it possible to elaborate mathematical theory of technology. Its starting point is an exact, however, informal definition of technology as a system of knowledge. Basing on this definition, two classes of technologies (the general and specific technologies) are introduced to reflect the situation existing in industry and engineering. Technology represents in an exact form all stages of system life: the analysis of a problem of system development, system analysis, system design, system development, system implementation, system utilization, and system maintenance. This provides for the construction of a general mathematical model of a specific technology as well as for the development of a relevant mathematical apparatus and exact methods for an investigation and design of various technologies (in industry, management, information processing and so on).

The mathematical theory of technology (Burgin, 1997c; 2002) utilizes new mathematical disciplines such as theory of named sets, fuzzy set theory, and theory of structured multidimensional models of systems and processes as well as traditional fields such as algebra, theory of probabilities, and theory of algorithms.

In the mathematical theory of technology, such problems as reliability, equivalence, stability, constructibility, and realizability of technologies are studied. The aim is the development of efficient methods and algorithms of the computer aided design of technologies.

Thus, both theories, the theory of programs and theory of technology are aimed not only at knowledge acquisition, but also at solving practical problems.

Some claim that practice leaves theory behind and theory has only to explain what practice has already gained. It is not so with theory of algorithms. Now chemists are designing only the simplest computational units on the molecular level, while theory of parallel computations, comprising molecular computing, has many advanced results. Physicists are only approaching quantum computers, while theory of quantum computations has many results, demonstrating that it will be much more efficient that contemporary computers. The same even to a greater extent is true for superrecursive algorithms. Now practice has to catch up with the theory and it is urgent to know how to bridge the existing gap. This gives a positive answer to the question of Dijkstra's paper "The end of Computing Science?" (2001). On the contrary, science is even more alive than before. Only many professionals don't know about it.

To conclude, it is worth mentioning that theory of superrecursive computing has an important message for society and not only for technology. It only looks smart to stop and to enjoy after you get something. Theory of superrecursive computing says that it is better to continue to be active. This is the truth of life.

References and Sources for Additional Reading

1. Abiteboul, S., Cluet, S., Milo, T., Mogilevsky, P., Siméon, J., and Zohar, S. (1999). Tools for Data Translation and Integration, *IEEE Data Eng. Bull.*, v. 22, no. 1, pp. 3–8.
2. Abiteboul, S., Papadimitriou, C.H. and Vianu, V. (1994). The Power of Reflective Relational Machines, *LICS*, pp. 230–240.
3. Abiteboul, S., Vardi, M.Y., and Vianu, V. (1992). Fixpoint Logics, Relational Machines, and Computational Complexity. Structure in Complexity Theory Conference, pp. 156–168.
4. Abiteboul, S., Vardi, M.Y., and Vianu, V. (1992). Computing with Infinitary Logic. ICDT, pp. 113–123.
5. Abiteboul, S., Vardi, M.Y., and Vianu, V. (1995). Computing with Infinitary Logic, *Theor. Comput. Sci.*, v. 149, no. 1, pp. 101–128.
6. Abiteboul, S., Vardi, M.Y., and Vianu, V. (1997). Fixpoint logics, relational machines, and computational complexity. *Journal ACM*, v. 44, no. 1, pp. 30–56.
7. Abiteboul, S., and Vianu, V. (1993). Computing on Structures, in *Automata, Languages and Programming*, Proceedings of the Second International Colloquium, ICALP93 (*Lecture Notes in Computer Science*, v. 700, Springer) pp. 606–620.
8. Abiteboul, S. and Vianu, V. (1997). Queries and computation on the web, in *Database Theory - ICDT '97, 6th International Conference* (*Lecture Notes in Computer Science*, v. 1186, Springer) pp. 262–275.
9. Abiteboul, S., and Vianu, V. (2000). Queries and computation on the web. *Theor. Comput. Sci.*, v. 239, no. 2, pp. 231–255.
10. Abrams, M., Page, E.H., and Nance, R.E. (1991). Linking Simulation Model Specification And Parallel Execution Through Unity, *Proceedings of the 1991 Winter Simulation Conference*, Phoenix, Arizona, pp. 223–232.
11. Abramsky, S. and Jung, A. (1994). Domain Theory, in *Handbook of Logic in Computer Science*, v. 3, Clarendon Press.
12. Abramson, F.G. (1971). Effective Computation over the Real Numbers, *Twelfth Annual Symposium on Switching and Automata Theory*, Northridge, Calif.: Institute of Electrical and Electronics Engineers.
13. Adamék, J. (1974). Free algebras and automata realizations in the language of categories, *Comment. Math. Univ. Carolinae*, v. 15, pp. 589–602.
14. Adamék, J. (1975). Automata and categories, finiteness contra minimality, *Lecture Notes in Computer Science*, 32, Springer-Verlag, Berlin, Heidelberg, New York, pp. 160–166.

15. Adámek, J. and Trnková, V. (1990). *Automata and Algebras in Categories*, Kluwer Academic Publishers.

16. Addison, J.W. (1962). The Theory of Hierarchies, in *Proceedings of the 1960 International Congress on Logic, Methodology and Philosophy of Science*, Stanford University Press, Stanford, California, pp. 26–37.

17. Adleman, L. and Blum, M. (1991). Inductive inference and unsolvability, *Journal of Symbolic Logic*, v. 56, no. 3, pp. 891–900.

18. Agerwala, T. (1974). *A Complete Model for Representing the Coordination of Asynchronous Processes*, Hopkins Computer Research Report no. 32, John Hopkins University, Baltimore, MD.

19. Agre, P.E. (1997). *Computation and Human Experience*, Cambridge University Press.

20. Aho, A.V., Hopcroft, J.E., and Ullman, J.D. (1976). *The Design and Analysis of Computer Algorithms*, Reading, MA, Addison-Wesley P.C.

21. Aladyev, V.Z. (1971). *To the Theory of Homogeneous Structures*, Moscow, VINITI (in Russian).

22. Aladyev, V.Z. (1972). Computability in Homogeneous Structures, *Reports of Estonian Acad. of Sci.*, v. 21, no. 1, pp. 79–83.

23. Aladyev, V.Z. (1990). *Homogeneous Structures*, Kiev, Tehnika (in Russian).

24. Alper, J.S., Bridger, M., Earman, J., and Norton, J.D. (2000). What is a Newtonian System? The Failure of Energy and Determinism in Supertasks, *Synthese*, v. 124, no. 2, pp. 281–293.

25. Alefeld, G. and Herberger, J. (1983). *Introduction to Interval Computations*, London, Academic Press.

26. Allen, C. and Dhagat, M. (2001). *LISP Primer*, (http://grimpeur.tamu.edu/ colin/lp/).

27. Alt, F.L. (1997). End-Running Human Intelligence, in *Beyond Calculation: The Next Fifty Years of Computing*, Copernicus, pp. 127–134.

28. Altshuller, G. (1999). *The Innovation Algorithm: TRIZ, systematic innovation, and technical creativity*, Technical Innovation Center, Inc. Worcester, MA.

29. Alur, R. and Dill, D.L. (1994). A theory of timed automata, *Theoretical Computer Science*, v. 126, pp. 183–235.

30. Amari, S. (1967). Theory of adaptive pattern classifiers, *IEEE Transactions*, EC-16, no. 3, pp. 299–307.

31. Ambainis, A., Buhrman, H., Gasarch, W., Kalyanasundaram, B., and Torenvliet, L. (2003). The quantum communication complexity of sampling, *SIAM J. Comput.*, v. 32, no. 6, pp. 1570–1585 (electronic).

32. Ambainis, A., Freivalds, R., and Smith, C.H. (1999). Inductive inference with procrastination: Back to definitions, *Fund. Inform.* v. 40, no. 1, pp. 1–16.

33. Ambainis, A., Schulman, L. J., Ta-Shma, A., Vazirani, U., and Wigderson, A. (2003). The quantum communication complexity of sampling, *SIAM J. Comput.*, v. 32, no. 6, pp. 1570–1585 (electronic).

34. Angluin, D. (1992). Computational Learning Theory: Survey and Selected Bibliography, in *Proc. 24th ACM Symposium on Theory of Computation*, pp. 319–342.

35. Angluin, D. and Smith, C.H. (1983). Inductive inference: Theory and methods, *Comput. Surveys*, v. 15, no. 3, pp. 237–269.

36. Antognetti, P. and Milutinovic, V., eds. (1991). *Neural Networks: Concepts, Applications, and Implementations*, Englewood Cliffs, N.J., Prentice Hall.

37. Apsïtis, K., Arikawa, S., Freivalds, R., Hirowatari, E., and Smith, C.H. (1999). On the inductive inference of recursive real-valued functions. Computability and complexity in analysis, *Theoret. Computer Science*, v. 219, no. 1–2, pp. 3–17.

38. Arbib, M.A. and Manes, E.G. (1974). Machines in a category: An expository introduction, *SIAM Review*, 16, pp. 163–192.
39. Arbib, M.A. and Manes, E.G. (1975). Adjoint machines, state behavior machines and duality, *J. Pure Appl. Algebra*, v. 6, pp. 313–343.
40. Arbib, M.A. and Manes, E.G. (1975a). Fuzzy machines in a category, *Bull. Austral. Math. Soc.*, v. 13, pp. 169–210.
41. Arbib, M.A. and Manes, E.G. (1980). Partially-additive categories and flow-diagram semantics, *J. Algebra*, 62, pp. 203–227.
42. Ark, W.S. and Selker, T. (1999). A look at human interaction with pervasive computers, *IBM Systems Journal*, v. 38, no. 4, pp. 504–507.
43. Asarin, E., Dang, T., and Maler, O. (2002). The d/dt Tool for Verification of Hybrid Systems, in CAV'2002, Lecture Notes in *Computer Science*, v. 2404, pp. 365–370.
44. Asarin, E. and Maler, O. (1998). Achilles and the tortoise climbing up the arithmetical hierarchy, *J. of Computers and Systems Science*, v. 57, pp. 389–398.
45. Asarin E. and Schneider, G. (2002a). Widening the boundary between decidable and undecidable hybrid systems, in *CONCUR'2002, Lecture Notes in Computer Science*, v. 2421, pp. 193–208.
46. Ashby, R.W.R. (1964). *An Introduction to Cybernetics*, Barnes & Noble, London.
47. Ashcraft, M. (1994). *Human Memory and Cognition*, Harper Collins, New York, NY.
48. Atkinson, R.L., Atkinson, R.C., Smith E.E., and Bem, D.J. (1990). *Introduction to Psychology*, Harcourt Brace Jovanovich, Inc., San Diego, New York, Chicago.
49. Atkinson, R.C., and Shiffrin, R.M. (1968). Human Memory: A Proposed System and its Control processes, in *The Psychology of Learning and Motivation*, Academic Press, New York.
50. Attiya, H., and Welch, J. (1998). *Distributed Computing: Fundamentals, Simulations and Advanced Topics*, McGrow-Hill.
51. Ausiello, G., Crescenzi, P., Gambosi, G., Kann, V., Marchetti-Spaccamela, A. and Protasi (1999). *Complexity and Approximation: Combinatorial Optimization Problems and their Approximability Properties*, Springer-Verlag, New York/Berlin/Heidelberg.
52. Ausiello, G. and Protasi, M. (1990). Limiting polynomial approximation of complexity classes, *Int. J. of Foundations of Computer Science*, v. 1, no. 2.
53. Ausiello, G. Protasi, M., and Angelaccio, M. (1991). *A characterization of space complexity classes and subexponential time classes as limiting polynomially decidable sets*, International Computer Science Institute, TR-91-046, Berkeley.
54. Axt, P. (1959). On a Subrecursive Hierarchy and Primitive Recursive Degrees, *Transactions of the American Mathematical Society*, v. 92, pp. 85–105.
55. Babai, L. (1990). E-mail and the Unexpected Power of Interaction, in *Proceedings of the 5th Structure in Complexity Theory Conference*, IEEE Computer Society Press, pp. 30–44.
56. Babai, L. and Fortnov, L. (1991). Arithmetization: A New Method in Structural Complexity Theory, *Computational Complexity*, v. 1, no. 1, pp. 41–66.
57. Babai, L., Frankl, P. and Simon, J. (1986). Complexity Classes in Communication Complexity Theory, in *Proceedings of the 27th Annual ACM Symposium on Theory of Computing (STOC)*, Toronto, pp. 337–347.
58. Bains, S. and Johnson, J. (2000). Noise, physics, and non-Turing computation, *Joint Conference on Information Systems*, Atlantic City, 28 February – 3 March.
59. Baker, M., Buyya, R., and Hyde, D. (2000). *Cluster Computing: A High-Performance Contender*, http://lanl.arXiv.org/.
60. Baker, T., Gill, J., and Solovey, R. (1975). Relativizations of the $P =?NP$ question, *SIAM Journal of Computing*, v. 4, pp. 431–442.

61. Backus, J.W. (1959). The Syntax and Semantics of the Proposed International Algebraic Language of the Zurich ACM-GAMM Conference, *Proc. Internat. Conf on Information Processing*, UNESCO, pp. 125–132.

62. Balcazar, J.L., Diaz, J., and Gabarro, J. (1988). *Structural Complexity*, Springer-Verlag, Berlin, Heidelberg, New York.

63. Ballard, D.H. (1997). *An Introduction to Natural Computation*, MIT Press, Cambridge.

64. Barbin, E., Borowczyk, J., Chabert, J.-L., Guillemot, M., Michel-Pajus, A., Djebbar, A., and Martzloff, J.-C. (1999). *A History of Algorithms: From the Pebble to the Microchip* (Chabert, J.-L., ed.), Springer, New York, Heidelberg, Berlin.

65. Barrett, E., ed. (1988). *Text, Context and Hypertext: Writing with and for the Computer*, Cambridge: MIT Press.

66. Baron, R., and Kalsher, M. (2000). *Psychology*, Allyn and Bacon, Boston, MA.

67. Barr, A. (1983). Artificial intelligence: Cognition as computation, in *The Study of Information: Interdisciplinary Messages*, Wiley-Interscience, New York, pp. 237–265.

68. Barwise, J. (1980). Infinitary logics, in *Modern Logic*: A Survey, Reidel, Dordrecht, The Netherlands, pp. 93–112.

69. Barzdin, J.M. (1968). Complexity of programs which recognize whether natural numbers not exceeding n belong to a recursively enumerable set, *Dokl. Akad. Nauk SSSR*, v. 182, pp. 1249–1252.

70. Barzdin, J.M., Freivalds, R., and Smith, C.H. (1996). Learning with confidence. *STACS 96* (Grenoble, 1996), Lecture Notes in Comput. Sci., 1046, Springer, Berlin, pp. 207–218.

71. Bechtel, W. and Richardson, R.C. (1993). *Discovering Complexity: Decomposition and Localization as Strategies in Scientific Research*, Princeton University Press, Princeton.

72. Bell, G. and Gray, J.N. (1997). The revolution yet to happen, in *Beyond Calculation: The Next Fifty Years of Computing*, Copernicus, pp. 5–32.

73. Benacerraf, P. (1970). Tasks, super-tasks, and the modern eleatics, in *Zeno's Paradoxes*, Indianapolis and New York: Bobbs-Merrill, pp. 103–129.

74. Ben-Hur, A., Siegelman, H.T., and Fisman, S. (2002). A theory of complexity for continuous time systems, *Journal of Complexity*, v. 18, pp. 51–86.

75. Beniof, P.A. (1982). The thermodynamics of computation, *Int. Journal of Theoretical Physics*, v. 21, pp. 905–940.

76. Bennet, C.H. (1985). *Emerging Syntheses in Science*, Pines.

77. Bennett, C.N. and Landauer, R. (1985). On fundamental physical limits of computation, *Scientific American*, pp. 48–56.

78. Bennett, C., and Wiesner, S. (1992). Communication via one- and two-particle Operators on Einstein-Podolsky-Rosen States, *Phys. Rev. Lett.*, v. 69, pp. 2881–2884.

79. Berge, C. (1973). *Graphs and Hypergraphs*, North Holland P.C., Amsterdam, New York.

80. Bernstein, E. and Vazirani, U. (1997). Quantum complexity theory, *SIAM J. Comp.*, v. 26, pp. 1411.

81. Berry, G. and Boudol, G. (1990). The Chemical Abstract Machine, in *POPL'90*, ACM Press, New York, pp. 81–94.

82. Bertalanffy, L. (1976). *General System Theory: Foundations, Development, Applications*, George Brazillier.

83. Black, R. (2000). Proving Church's thesis, *Philos. Math.*, v. 8, no. 3, pp. 244–258.

84. Blake, R.M. (1926). The paradox of temporal process, *Journal of Philosophy*, v. 23, pp. 645–654.

85. Blass, A. and Gurevich, Y. (2002). Algorithms vs. Machines, *Bulletin of the European Association for Theoretical Computer Science*, no. 77, pp. 96–118.

86. Blass, A. and Gurevich, Y. (2003). Algorithms: A Quest for Absolute Definitions, *Bulletin of the European Association for Theoretical Computer Science*, no. 81, pp. 195–225.

87. Blass, A. and Gurevich, Y. (2003). Abstract State Machines Capture Parallel Algorithms, *ACM Transactions on Computation Logic*, v. 4, no. 4, pp. 578–651.

88. Blum, L. (1990). A Theory of Computation and Complexity over the Real Numbers, *Proceedings of the International Congress of Mathematicians*, Kyoto, II, Math. Soc. Japan, Tokyo, pp. 1491–1507.

89. Blum, L. and Blum, M. (1975). Toward a mathematical theory of inductive inference, *Information and Control*, 28, pp. 125–155.

90. Blum, L., Cucker, F., Shub, M., and Smale, S. (1998). *Complexity of Real Computation*, Springer, New York.

91. Blum, L., Shub, M., and Smale, S. (1989). On a theory of computation and complexity over the real numbers: NP-completeness, recursive functions and universal machines, *Bull. Amer. Math. Soc.*, v. 21, no. 1, pp. 1–46.

92. Blum, L., Shub, M., and Smale, S. (1993). The Gödel Incompleteness Theorem and Decidability over a Ring, in *From Topology to Computation*, Proceedings of the Smalefest, Springer-Verlag, pp. 321–339.

93. Blum, M. (1967). On the size of machines, *Information and Control*, v. 11, pp. 257–265.

94. Blum, M. (1967a). A machine-independent theory of complexity of recursive functions, *Journal of the ACM*, v. 14, no. 2, pp. 322–336.

95. Boldi, P., Meyssonnier, C. and Vigna, S. (2001). δ-approximable functions, in *Proceedings of CCA 2000 (Swansea, Wales)*, (Lecture Notes in Computer Science, v. 2064). pp. 187–199.

96. Boldi, P. and Vigna, S. (1998). δ-uniform BSS machines, *Journ. Complexity*, v. 14, no. 2, pp. 234–256.

97. Boolos, G.S. and Jeffrey, R.C. (1989). *Computability and Logic*, Cambridge University Press, Cambridge.

98. Booth, T.L. (1967). *Sequential Machines and Automata Theory*, Wiley, New York.

99. Borodin, A. (1977). On Relating Time and Space to Size and Depth, *SIAM Journal on Computing*, v. 6, pp. 733–744.

100. Bournez, O. (1999). Achilles and the tortoise climbing up the hyper-arithmetical hierarchy, *Theoretical Computer Science*, v. 210, no. 1, pp. 21–71.

101. O.Bournez (1999). *Complexité Algorithmique des Systèmes Dynamiques Continus et Hybrides*. PhD thesis, Ecole Normale Supérieure de Lyon.

102. Branicky, M.S. (1995). Universal computation and other capabilities of hybrid and continuous dynamical systems. *Theoretical Computer Science*, v. 138, no. 1, pp. 67–100.

103. Brandt, U. (1986). The Position of Index Sets of Identifiable Sets in the Arithmetical Hierarchy, *Information and Control*, v. 68, pp. 185–195.

104. Bratalsky, E. and Burgin M. (1986). The principle of asymptotic uniformity in complex system modeling, in *Operation Research and Automated Control Systems*, Kiev, pp. 115–122 (in Russian).

105. Brattka V., and Hertling, P. (1998). Feasible Real Random Access Machines, *J. Complexity*, 14, no. 4, pp.490–526.

106. Bremermann, H. (1974). Algorithms, complexity, transcomputability and the analysis of systems, in *Proceedings of the Fifth Congress of the Deutsche Gesellschaft fur Kybernetik*, Munich.

107. Bringsjord, S. (1992). *What Robots Can and Can't Be*, Kluwer, Dordrecht, The Netherlands.

108. Bringsjord, S. (1995). Computation, among other things, is beneath us, *Minds and Machines*, v. 4, pp. 469–488.
109. Bringsjord, S. (1997). An argument for the uncomputability of infinitary mathematical expertise, in *Expertise in Context*, P. Feltovich, K. Ford and P. Hayes, eds., AAAI Press, Menlo Park, CA, pp. 475–497.
110. Bringsjord, S., Bello, P., and Ferrucci, D. (2001). Creativity, the Turing Test, and the (Better). Lovelace Test, *Minds and Machines*, v. 11, pp. 3–28.
111. Bringsjord, S. and M. Zenzen, (1997). Cognition is not computation: The argument from irreversibility?, *Synthese*, v. 113, pp. 285–320.
112. Bringsjord, S. and Zenzen, M. (2003). *Superminds: People Harness Hypercomputation, and More*, Kluwer Academic Publishers, Dordrecht, The Netherlands.
113. Büchi, J.R. (1960). Weak second order arithmetic and finite automata, *Z. Math. Logic and Grudl. Math.*, v. 6, no. 1, pp. 66–92.
114. Buchi, J.R. (1962). On a Decision Method in Restricted Second Order Arithmetic, in *Proceedings of the* 1960 *International Congress on Logic*, Stanford University Press, Stanford, California, pp. 1–11.
115. Budach, L. and Hoehnke, H.-J. (1975). *Automata und Funktoren*, Akademic Verlag, Berlin.
116. Buhrman, H., Cleve, R. and Wigderson, A. (1998). Quantum vs. Classical Communication and Computation, in *Proceedings of the* 30th *Annual ACM Symposium on Theory of Computing (STOC)*, Dallas, Texas, pp. 63–68.
117. Burgin, M.S. (1980). Functional equivalence of operators and parallel computations, *Programming and Computer Software*, v. 6, no. 6, pp. 283–294.
118. Burgin, M.S. (1982). Generalized Kolmogorov complexity and duality in theory of computations, *Notices of the Academy of Sciences of the USSR*, v. 264, no. 2, pp. 19–23 (translated from Russian, v. 25, no. 3).
119. Burgin, M. (1982a). Products of operators of multidimensional structured model of systems, *Mathematical Social Sciences*, no. 2, pp. 335–343.
120. Burgin, M.S. (1983). Inductive Turing machines, *Notices of the Academy of Sciences of the USSR*, v. 270, no. 6, pp. 1289–1293 (translated from Russian, v. 27, no. 3).
121. Burgin, M.S. (1985). Algorithms and algorithmic problems, *Programming*, no. 4, pp. 3–14 (*Programming and Computer Software*, v. 11, no. 4). (translated from Russian).
122. Burgin, M.S. (1985a). Computational complexity measures and program quality, in *Reliability and Quality of Software*, Kiev, 1985 (in Russian).
123. Burgin, M.S. (1986). Quantifiers in the theory of properties, in *Non-standard Semantics of Non-classical Logics*, Moscow, pp. 99–107 (in Russian).
124. Burgin, M.S. (1987). The notion of algorithm and the Turing-Church thesis, in *Proceedings of the VIII International Congress on Logic, Methodology and Philosophy of Science*, Moscow, v. 5, part 1, pp. 138–140.
125. Burgin, M.S. (1988). Arithmetic hierarchy and inductive Turing machines, *Notices of the Academy of Sciences of the USSR*, v. 299, no. 3, pp. 390–393 (translated from Russian).
126. Burgin, M.S. (1990). Generalized Kolmogorov complexity and other dual complexity measures, *Cybernetics*, no. 4, pp. 21–29 (translated from Russian).
127. Burgin, M.S. (1991). Dual complexity measures and expert system functioning, *Problem Solving in Intellectual Computer Media*, Kiev, pp. 118—126 (in Russian).
128. Burgin, M.S. (1992). Universal limit Turing machines, *Notices of the Russian Academy of Sciences*, v. 325, no. 4, pp. 654–658 (translated from Russian).
129. Burgin, M.S. (1992a). Infinite in finite or metaphysics and dialectics of scientific abstractions, *Philosophical and Sociological Thought*, no. 8, pp. 21–32 (in Russian and Ukrainian).

130. Burgin, M.S. (1992b). Complexity measures in the axiomatic theory of algorithms, in *Methods of Design of Applied Intellectual Program Systems*, Kiev, pp. 60–67 (in Russian).

131. Burgin, M.S. (1992c). Reflexive calculi and logic of expert systems, in *Creative Processes Modeling by Means of Knowledge Bases*, Sofia, pp. 139–160.

132. Burgin, M. (1993). Procedures of sociological measurements, in *Catastrophe, Chaos, and Self-Organization in Social Systems*, Koblenz, pp. 125–129.

133. Burgin, M. (1994). Evaluation of scientific activity in the dynamic information theory, *Science and Science of Science*, no. 1–2, pp. 124–131.

134. Burgin, M. (1995). Neoclassical analysis: Fuzzy continuity and convergence, *Fuzzy Sets and Systems*, v. 75, pp. 291–299.

135. Burgin M. (1995a). Logical Tools for Inconsistent Knowledge Systems, *Information: Theories & Applications*, v. 3, no. 10, pp. 13–19.

136. Burgin, M. (1996). Flow-charts in programming: Arguments pro et contra, *Control Systems and Machines*, no. 4–5, pp. 19–29 (in Russian).

137. Burgin, M.S. (1997). *Fundamental Structures of Knowledge and Information*, Kiev, AIS (in Russian).

138. Burgin, M. (1997a). *Non-Diophantine Arithmetics or Is It Possible that* $2 + 2$ *Is Not Equal to* 4, Ukrainian Academy of Information Sciences, Kiev (in Russian).

139. Burgin, M. (1997b). Time as a factor of science development, *Science and Science of Science*, no. 1/2, pp. 45–59.

140. Burgin, M. (1997c). Mathematical theory of technology, in *Methodological and Theoretical Problems of Mathematics and Informatics (Computer Science)*, Kiev, pp. 91–100 (in Russian).

141. Burgin, M. (1998). *On the Essence and Nature of Mathematics*, Ukrainian Academy of Information Sciences, Kiev (in Russian).

142. Burgin, M. (1999). Super-recursive algorithms as a tool for high performance computing, *Proceedings of the High Performance Computing Symposium*, San Diego, pp. 224–228.

143. Burgin, M. (1999a). A new paradigm for computation and its application to global networks, huge databases, and embedded systems, *Data-Link*, ACM, 44, pp. 1–2.

144. Burgin, M. (2000). Theory of super-recursive algorithms as a source of a new paradigm for computer simulation, in *Proceedings of the Business and Industry Simulation Symposium*, Washington, pp. 70–75.

145. Burgin, M. (2001). How we know what technology can do, *Communications of the ACM*, v. 44, no. 11, pp. 82–88.

146. Burgin, M. (2001a). Mathematical models for computer simulation, in *Proceedings of the Business and Industry Simulation Symposium*, SCS, Seattle, Washington, pp. 111–118.

147. Burgin, M. (2001b). Topological Algorithms, in Proceedings of the ISCA 16th International Conference *Computers and their Applications*, ISCA, Seattle, Washington, pp. 61–64.

148. Burgin, M. (2001c). *Diophantine and Non-Diophantine Arithmetics: Operations with Numbers in Science and Everyday Life*, LANL, Preprint Mathematics GM/0108149, 27 p. (electronic edition: http://arXiv.org).

149. Burgin, M. (2001d). *Mathematical Models for Artificial Intelligence*, Elsevier, Preprint 0107066, 15 p. (electronic edition: http://www.mathpreprints.com/math/Preprint/).

150. Burgin, M. (2001e). Uncertainty and imprecision in analytical context: Fuzzy limits and Fuzzy derivatives, *International Journal of Uncertainty, Fuzziness and Knowledge-Based Systems*, 9 pp. 563–685.

151. Burgin, M. (2001f). *Continuity and Connectedness in Discontinuous Topology*, University of California, Los Angeles, Mathematics Report Series, MRS Report 01-06, 28 p.

152. Burgin, M. (2002). Information, organization, and system functioning, in *Proceedings of the 6th World Multiconference on Systemics, Cybernetics and Informatics*, v. 2, Orlando, Florida, pp. 155–160.

153. Burgin, M. (2002a). *Elements of the System Theory of Time*, LANL, Preprint in Physics 0207055, 21 pp. (electronic edition: http://arXiv.org).

154. Burgin, M. (2002b). Mathematical models for simulating technological processes, in *Proceedings of the Business and Industry Simulation Symposium*, Society for Modeling and Simulation International, San Diego, California, pp. 165–168.

155. Burgin M. (2002c). *The Rise and Fall of the Church-Turing Thesis*, LANL, Preprint in Computer Science CC/0207055 (electronic edition: http://arXiv.org).

156. Burgin M. (2003). Nonlinear Phenomena in Spaces of Algorithms, *International Journal of Computer Mathematics*, v. 80, no. 12, pp. 1449–1476.

157. Burgin, M.S. and Borodyanskii, Yu. M. (1991). Infinite processes and super-recursive algorithms, *Notices of the Academy of Sciences of the USSR*, v. 321, no. 5, pp. 800–803 (translated from Russian).

158. Burgin, M.S. and Borodyanskii, Yu. M. (1993). Alphabetic operators and algorithms, *Cybernetics and System Analysis*, no. 3, pp. 42–57 (translated from Russian: v. 29, no. 3, pp. 338–349).

159. Burgin, M.S. and Borodyanskii, Yu. M. (1994). Operations with trans-recursive operators, *Cybernetics and System Analysis*, no. 4, pp. 3–11 (translated from Russian: v. 30, No.4, pp. 473–478).

160. Burgin, M.S. and Borodyanskii, Yu. M. (1994a). Problems of artificial intelligence and trans-recursive operators, *Notices of the National Academy of Sciences of Ukraine*, no. 11–12, pp. 29–34 (in Ukrainian).

161. Burgin, M. and Karasik, A. (1975). A study of an abstract model of computers, *Programming and Computer Software*, no. 1, pp. 72–82.

162. Burgin, M. and Karasik, A. (1976). Operators of multidimensional structured model of parallel computations, *Automation and Remote Control*, v. 37, no. 8, pp. 1295–1300.

163. Burgin, M. and Karasik, A. (1976a). On a construction of matrix operators, *Problems of Radio-Electronics*, no. 8, pp. 9–25 (in Russian).

164. Burgin, M. and Kuznetsov, V. (1994). *Introduction to the Modern Exact Methodology of Science*, Moscow (in Russian).

165. Burgin, M., Liu, D., and Karplus, W. (2001). The Problem of Time Scales in Computer Visualization, in *Computational Science*, Lecture Notes in Computer Science, v. 2074, part II, 2001, pp.728–737.

166. Burgin, M., Liu, D., and Karplus, W. (2001a). *Visualization in Human-Computer Interaction*, UCLA, Computer Science Department, Report CSD - 010010, Los Angeles, July, 2001, 108 p.

167. Burgin, M.S. and Povyakel, N.I. (1988). Programmable and Reflexive Components of Creativity, in *Activity: Philosophical and Psychological Aspects*, Simferopol, 37–38 (in Russian).

168. Burgin, M. and Ya. Shmidskii, (1996). Is it possible to compute non-computable or why programmers need the theory of algorithms, *Computers & Software*, no. 5, pp. 4–8 (in Russian).

169. Burgin, M. and Westman, J. (2000). Fuzzy calculus approach to computer simulation, *Proceedings of the Business and Industry Simulation Symposium*, Washington, pp. 41–46.

170. Burks, A.W. (ed.). (1970). *Essays on Cellular Automata*, Urbana, Univ of Illinois Press.
171. Burks, A.W. and Wright, J.B. (1953). Theory of logical nets, *Proceedings IRE*, v. 41, no. 10, pp. 1357–1365.
172. Buss, S.L., Kechris, A.S., Pillay, A, and Shore, R.A. (2001). The prospects for mathematical logic in the Twenty First Century, *Bulletin of Symbolic Logic*, no. 2, pp. 169–196.
173. Campagnolo, M.L., Moore, C., and Costa, J.F. (2000). An analog characterization of the subrecursive functions, in *Proc. of the 4th Conference on Real Numbers and Computers*, Odense University, pp. 91–109.
174. Cantor, G. (1895). Beiträge zur Begrundung der transfiniten Mengenlehre, *Math. Ann.*, bd. 46, s. 481–512.
175. Cardelli, L. (1997). Global computation, *ACM Sigplan Notices*, 32-1, pp. 66–68.
176. Carlson, J.M. and Doyle, J. (2002). Complexity and robustness, *Proc. Nat. Acad. Science of the USA*, v. 99, no. 1, pp. 2538–2545.
177. Casselman, S. (1993). Virtual Computing and the Virtual Computer, IEEE Workshop on FPGAs for Custom Computing Machines (FCCM93), pp. 43–48.
178. Casselman, S., Thornburg, M., and Schewel, J. Hardware Object programming on the EVC1: A Reconfigurable Computer, http://www.vcc.com/Papers/hotevc.pdf
179. Chadzelek, T. (1998). *Analytische Maschinen*, Dissertation an der Universität des Saarlandes.
180. Chadzelek, T., and Hotz, G. (1997). Analytic machines. Technical Report 12/97, Sonderforschungsbereich 124 (VLSI-Entwurfsmethoden und Parallelität), Universität des Saarlandes.
181. Chadzelek, T. and Hotz, G. (1999). Analytic machines, *Theoretical Computer Science*, v. 219, no. 1/2, pp. 151–167.
182. Chaitin, G.J. (1966). On the length of programs for computing finite binary sequences, *J. Association for Computing Machinery*, v. 13, no. 4, pp. 547–569.
183. Chaitin, G.J. (1969). On the simplicity and speed of programs for computing definite sets of natural numbers, *J. Association for Computing Machinery*, v. 16, pp. 407–412.
184. Chaitin, G.J. (1975). A theory of program size formally identical to information theory, *J. Association for Computing Machinery*, v. 22, no. 3, pp. 329–340.
185. Chaitin, G. (1975a). Randomness and mathematical proof, *Scientific American*, v. 232, no. 5 pp. 47–52.
186. Chaitin, G.J. (1977). Algorithmic information theory, *IBM Journal of Research and Development*, v. 21, no. 4, 350–359.
187. Chaitin, G.J. (1987). *Algorithmic Information Theory*, Cambridge University Press, Cambridge.
188. Chaitin, G.J. (1994.). Randomness and Complexity in Pure Mathematics, *International Journal of Bifurcation and Chaos*, v. 4, pp. 3–15.
189. Chaitin, G.J. (1999). *The Unknowable*, Springer-Verlag, Berlin, Heidelberg, New York.
190. Chalmers, D.J. (1997). A Computational Foundation for the Study of Cognition, http://www.u.arizona.edu/chalmers/papers/computation.html.
191. Chandra, A.K. and Stockmeyer, L.J. (1976). Alternation, in *Proceedings of the 23rd Symposium on Foundations of Computer Science*, pp. 98–108.
192. Chandy, K.M. and Misra, J. (1988). *Parallel Program Design: A Foundation*, Reading, MA: Addison Wesley.
193. Cherniak, C. (1998). Undebuggability and cognitive science, *Communications of the ACM*, v. 31, no. 4, pp. 402–412.
194. Cho, A. (2000). Hairpins trigger an automatic solution, *Science*, v. 288, no. 5469.

195. Chomsky, N. (1956). Three models for the description of language, *IRE Trans. On Information Theory*, v. 2, no. 3, pp. 113–124.
196. Chow, H.A., Alnuweiri, H., and Casselman, S. (1995). FPGA-Based Transformable Computers for Fast Digital Signal Processing, 3rd Canadian Workshop on Field-Programmable Devices (FPD95), pp. 25–31.
197. Christensen, K., Fitsos, G.P., and Smith, C.P. (1981). A perspective on software science, *IBM Systems Journal*, v. 20, no. 4, pp. 372–387.
198. Church, A. (1932/33). A set of postulates for the foundations of logic, *Annals of Mathematics*, v. 33, pp. 346–366; v. 34, pp. 839–864.
199. Church, A. (1936). An unsolvable problem of elementary number theory, *The American Journal of Mathematics*, v. 58, pp. 345–363.
200. Church, A. (1941). The Calculi of Lambda-Conversion, *Annals of Mathematics Studies*, no. 6, Princeton University Press.
201. Church, A. (1957). Application of Recursive Arithmetic to the Problem of Circuit Synthesis, in *Summaries of Talks presented at the Summer Institute of Symbolic Logic at Cornell University*, 1, pp. 3–50.
202. Cleland, C.E. (1993). Is the Church-Turing Thesis true?, *Minds and Machines*, v. 3, pp. 283–312.
203. Cleland, C.E. (1995). Effective procedures and computable functions, *Minds and Machines*, v. 5, pp. 9–23.
204. Cleland, C.E. (2001). Recipes, algorithms, and programs, *Minds and Machines*, v. 11, pp. 219–237.
205. Cleland, C.E. (2002). *On Effective Procedures, Minds and Machines*, v. 12, pp. 159–179.
206. Clote, P. (1999). Computational models and function algebras, in *Handbook of Computability Theory*, Elsevier, pp. 589–681.
207. Cobham, A. (1964). The Intrinsic Computational Difficulty of Functions, in *Proceedings of the 1964 International Congress on Logic, Methodology and Philosophy of Science*, Amsterdam, pp. 24–30.
208. Codd, E.F. (1968). *Cellular Automata*, New York, Academic.
209. Cohn, P.M. (1965). *Universal Algebra*, Harper & Row, New York, Evanston, London.
210. Colwell, R.R. (2001). Closing the circle of information technology, *Commun. ACM*, v. 44, no. 3, pp. 31–32.
211. Cook, S. (1971). The complexity of theorem-proving procedures, in *Proceedings of the 3rd ACM Symposium on Theory of Computing*, pp. 151–158.
212. Copeland, J. (1996). What is Computation? *Synthese*, vol. 108, pp.335–359.
213. Copeland, J. (1997). The Broad Conception of Computation, *American Behavioral Scientist*, vol. 40, pp. 690–716.
214. Copeland, J. (1997a). The Church-Turing thesis, in *Stanford Encyclopedia of Philosophy*, Center for the Study of Language and Information (CSLI), Stanford University, Stanford, CA.
215. Copeland, J. (1998). Super Turing-Machines, *Complexity*, vol. 4, pp. 30–32.
216. Copeland, J. (1998a). Turing's O-Machines, Penrose, Searle, and the Brain, *Analysis*, vol. 58, pp.128–138.
217. Copeland, J. and Proudfoot, D. (1996). On Alan Turing's Anticipation of Connectionism, *Synthese*, vol. 108, pp. 361–377.
218. Copeland, J. and Sylvan, R. (1999). Beyond the Universal Turing Machine, *Australasian Journal of Philosophy*, v. 77, pp. 46–66.
219. Cormen, T.H., Leiserson, C.E., Rivest, R.L., and Stein, C. (2001). *Introduction to Algorithms*, The MIT Press, Cambridge, Massachusetts.

220. da Costa, N.C.A. and Doria, F.A. (1991). Classical physics and Penrose's Thesis, *Foundations of Physics Letters*, 4, pp. 363–374.

221. da Costa, N.C.A. and Doria, F.A. (1996). H-Computation, in *Advances in Artificial Intelligence, 13th Brazilian Symposium on Artificial Intelligence*, SBIA '96, Lecture Notes in Computer Science, Vol. 1159, Springer.

222. Cover, T. and Thomas, J. (1991). *Elements of Information Theory*, Wiley Interscience, New York.

223. Crutchfield, J.P. and Mitchell, M. (1995). *The Evolution of Emergent Computation*, Proceedings of the National Academy of Sciences, USA, v. 92 (23). 10742–10746.

224. Cucker, F., Montaña, J.L, and Pardo, L. (1995). Models for parallel computation with real numbers, *Number-theoretic and Algebraic Methods in Computer Science* (Moscow, 1993), World Sci. Publishing, River Edge, NJ, pp. 53–63.

225. Cutland, N.J. (1980). *Computability: An Introduction to Recursive Function Theory*, Cambridge University Press.

226. Daley, R.P. (1973). Minimal-program complexity of sequences with restricted resources, *Information and Control*, v. 23, pp. 301–312.

227. Davies, E.B. (2001). Building infinite machines, *British Journal of Philosophy*, v. 52, pp. 571–582.

228. Davis, M. (ed.), (1965). *The Undecidable*, Raven Press, Hewlett, New York.

229. Davis, M. (1982). Why Gödel didn't have Church's thesis, *Information and Control*, v. 54, no. 1–2, pp. 3–24.

230. Davis, M. (1988). Influences of mathematical logic on computer science, in *The Universal Turing Machine: A Half-century Survey*, Oxford Sci. Publ., Oxford Univ. Press, New York, pp. 315–326.

231. Davis, M. and Weyuker, E. (1983). *Computability, Complexity and Languages*, Academic Press, Orlando.

232. M. Davis, (2003). The myth of hypercomputation; manuscript to appear in: *Alan Turing: Life and Legacy of a Great Thinker*, Springer-Verlag, Berlin.

233. M. Davis, R. Sigal, and E. Weyuker, (1994). *Computability, Complexity, and Languages: Fundamentals of Theoretical Computer Science*, Academic Press, New York.

234. Davis, M. and Weyuker, E. (1983). *Computability, Complexity and Languages*, Academic Press, Orlando.

235. De Bakker, J.W. and de Roever, W.P. (1973). A Calculus for Recursive Program Schemata, Automata, Languages and Programming, Proc. Symp., IRIA.

236. Debnath, N.C. and Burgin, M. (2003). Software Metrics from the Algorithmic Perspective, in Proceedings of the ISCA 18th International Conference *Computers and their Applications*, International Society for Computers and their Applications, Honolulu, Hawaii, pp. 279–282.

237. Debnath, N.C., Lee, H.K., Lee, R.Y., and Smari, W.W. (2002). Metrics Comparison for Java Programs, in Proceedings of the ISCA 17th International Conference *Computers and their Applications*, International Society for Computers and their Applications, San Francisco, California.

238. Deutsch, D. (1985). Quantum theory, the Church-Turing principle, and the universal quantum Turing machine, *Proc. Roy. Soc.*, Ser. A, v. 400, pp. 97–117.

239. Deutsch, D. (1989). Quantum computational networks, *Proc. Roy. Soc.*, Ser. A, v. 425, pp. 73–90.

240. Deutsch, D., Ekert, A., and Lupacchini, R. (2000). Machines, logic and quantum physics, *Bulletin of Symbolic Logic*, v. 6, no. 3, pp. 265–283.

241. Dietrich, E. (1990). Computationalism, *Social Epistemology*, v. 4, no. 2, pp. 135–154.

242. Dijkstra, E.W. (2001). The end of computing science? *Communications of the ACM*, v. 44, no. 3, p. 92.

243. Dix, A., Finlay, J. Abowd, G. and Beale, R. (1998). *Human-Computer Interaction*, Prentice Hall.

244. Doyle, J. (1982). *What is Church's Thesis?* An Outline, Laboratory for Computer Science, MIT.

245. Doyle, J. (2002). What is Church's Thesis? An outline, *Minds and Machines*, v. 12, pp. 519–520.

246. Dreyfus, H.L. (1979). *What Computers Can't Do – The Limits of Artificial Intelligence*, Harper and Row.

247. Dymond, P.W. and Cook, S.A. (1980). Complexity Theory of Parallel Time and Hardware, in *Proc. 21st IEEE Symp. on Foundations of Computer Science*, pp. 360–371.

248. Dreyfus, H.L. (1979). *What Computers Can't Do – The Limits of Artificial Intelligence*, Harper and Row.

249. Dymond, P.W. and Cook, S.A. (1989). Complexity theory of parallel time and hardware, *Information and Computation*, v. 80, pp. 205–226.

250. Earman, J. (1995). *Bangs, Crunches, Whimpers, and Shrieks: Singularities and Acausalities in Relativistic Spacetimes*, The Clarendon Press/Oxford University Press, New York.

251. Earman, J., and Norton, J.D. (1993). Forever is a day: Supertasks in Pitowski and Malament-Hogarth spacetimes, *Philosophy of Science*, v. 60, no. 1, pp.22–42

252. Earman, J. and Norton, J.D. (1996). Infinite Pains: The Trouble with Supertasks, in *The Benacerraf and his Critics*, Blackwell Publ., Cambridge, MA, pp. 231–261.

253. Ebbinghaus, H.-D., Jacobs, K., Mahn, F.-K., and Hermes, H. (1970). *Turing Machines and Recursive Functions*, Springer-Verlag, Berlin, Heidelberg, New York.

254. Edalat, A. (1997). Domains for computation in mathematics, physics and exact real arithmetic, *Bull. Symbolic Logic*, v. 3, no. 4, pp. 401–452.

255. Edalat, A. and Sünderhauf, P. (1998). A domain-theoretic approach to real number computation, *Theoretical Computer Science*, v. 210, no. 1, pp. 73–98.

256. Edmonds, B. (1995). What is Complexity? The philosophy of complexity per se with application to some examples in evolution. In *The Evolution of Complexity*, Kluwer, Dordrecht.

257. Edmonds, B. (1999). *Syntactic Measures of Complexity*, CPM Report no. 99-55, University of Manchester, Manchester, UK.

258. Edmonds, J. (1965). Paths, trees, and flowers, *Canadian Journal of Mathematics*, v. 17, no. 3, pp. 449–467.

259. Elgot, C.C. and Robinson, A. (1964). Random-access stored-program machines, an approach to programming languages, *J. Assoc. Comput. Mach.*, 11, pp. 365–399.

260. Elgot, C.C. and Rutledge, J.D. (1964). 17. xr/v RUTLEDOE, RS-machines with almost blank tape, *J. ACM*, no. 11, pp. 313–337.

261. Elman, J.L. (1990). Finding structure in time, *Cognitive Science*, 14, pp. 179–211.

262. Engeler, E. (1993). *Algorithmic Properties of Structures*, World Scientific.

263. Estrin, G. and Viswanathan, C.R. (1962). Organization of a "fixed-plus-variable" structure computer for computation of eigenvalues and eigenvectors of real symmetric matrices, *Journal of the ACM*, v. 9, no. 1, pp. 41–60.

264. Evey, J. (1963). The Theory and Applications of Pushdown-stem Machines, Doctoral thesis, Rep. NSF-10, Harvard University, Cambridge, Mass.

265. Ewing, T. and Tentner, A. (1999). A scalable architecture for modeling and simulation of intelligent transportation systems, in *Proceedings of the High Performance Computing Symposium*, San Diego, pp. 170–174.

266. Farhi, E., Goldstone, J. Gutmann, S. and Sipser, M. (2000). Quantum computation by adiabatic evolution, E-print http://arxiv.org/pdf/quant-ph/0001106.

267. Feferman, S. (1992). Turing's 'Oracle': From absolute to relative computability – and back, in *The Space of Mathematics*, Walter de Gruyter, Berlin, pp. 314–348.

268. Feferman, S. (1996). Computation on abstract data types. The extensional approach, with an application to streams, *Annals of Pure and Applied Logic*, 81, pp. 75–113.

269. Fenton, N.E. (1991). *Software Metrics: A Rigorous Approach*, Chapman & Hall.

270. Fetzer, J. (1994). Mental algorithms: Are minds computational systems?, Pragmatics and Cognition, v. 2, no. 1, pp. 1–29.

271. Feynman, R.P. (1982). Simulating physics with computers, *Int. Journal of Theoretical Physics*, v. 21, pp. 467–488.

272. Feynman, R.P. (1986). Quantum mechanical computers, *Foundations of Physics*, v. 16 (6), pp. 507–531.

273. Feys, R. (1965). *Modal Logics*, Paris.

274. Fischer, M. and Rabin, M. (1974). Super-exponential complexity of Presburger arithmetic, in *Complexity of Computation*, SIAM-AMS Proceedings.

275. Fischer, P.C. (1963). On computability by certain classes of restricted Turing machines, in *Proc. Fourth Ann. Symp. Switching Circuit Theory and Logical Design*, Chicago, pp. 23–32.

276. Fischer, P.C. (1965). Multi-tape and infinite-state automata, *Communications of the ACM*, v. 8, no. 12, pp. 799–805.

277. Flavin, C. (1991). Understanding fault tolerant computer systems, *Communications of the ACM*, v. 34, no. 2.

278. Forrest, S. (ed.). (1991). *Emergent Computation*, MIT Press, Cambridge, Massachusetts, London, England.

279. Forrest, S. (1991). Emergent Computation: Self-Organization, Collective and Cooperative Phenomena in Natural and Artificial Computing Networks, in *Emergent Computation*, MIT Press, Cambridge, Massachusetts, London, England, pp. 1–11.

280. Foster, I. (2002). The grid: A new infrastructure for 21st century science, *Physics Today* (http://www.Physics Today.org).

281. Foster, I. and Kesselman, C. (1998). *The Grid: Blueprint for a New Computing Infrastructure*, Morgan-Kaufmann Publ., San Francisco.

282. Fraenkel, A.A. and Bar-Hillel, Y. (1958). *Foundations of Set Theory*, North Holland P.C., Amsterdam.

283. Franklin, S. and Garzon, M. (1991). Neural Computability, in *Progress in Neural Networks*, vol. 1, Norwood, N.J., Ablex.

284. Free Online Dictionary of Computing, http://foldoc.doc.ic.ac.uk/.

285. Freedman, M.H. (2001). Quantum Computation and the Localization of Modular Functors, *Foundations Comput. Math.*, v. 1, no. 2, pp. 183–204.

286. Freedman, M.H., Kitaev, A., Larsen, M.J., and Wang, Z. (2002). Topological quantum computation, *Bull. Amer. Math. Soc.*, v. 40, no. 1, pp. 31–38.

287. Freedman, M.H., Larsen, M.J., and Wang, Z. (2002). A modular functor which is universal for quantum computation, *Comm. Math. Phys.*, v. 227, no. 3, pp. 605–622.

288. Frege, G. (1893/1903). *Grundgesetze der Arithmetik*, Begriffschriftlich Abgeleitet, Viena.

289. Freund, R. (1983). Real functions and numbers defined by Turing machines, *Theoretical Computer Science*, v. 23, no. 3, pp. 287–304.

290. Freyvald, R.V. (1974). Functions and functionals computable in limit, in *Transactions of Latvijas Vlasts Univ. Zinatn. Raksti*, v. 210, pp. 6–19 (in Russian).

291. Friedman, H. (1971). Algorithmic procedures, generalized Turing algorithm, and elementary recursion theory, in *Logic Colloquium* 1969, North-Holland, pp. 361–390.
292. Fukushima, K. (1975). Cognitron: a self-organizing multilayered neural network, *Biological Cybernetics*, v. 20, pp. 121–136.
293. Funes, P. (1996). Complexity measures for complex systems and complex objects, http://www.cs.brandeis.edu/ pablo/complex.maker.html
294. Furst, M., Hopcroft, J., and Luks, E. (1980). Polynomial-time algorithms for permutation groups, in *Proceeding of the 21st IEEE Symposium on Foundations of Computer Science*, pp. 36–41.
295. Gakwaya, J.-S. (1996). Extended Grzegorczyk hierarchy in the BSS Model of Computability, Technical Report NC-TR-97-041.
296. Gakwaya, J.-S. (1997). A Survey of the Grzegorczyk hierarchy and its Extension through the BSS Model of Computability, Technical Report NC-TR-97-041.
297. Gandy, R. (1980). Church's thesis and principles for mechanisms, in Proceedings of the Kleene Symposium (Univ. Wisconsin, Madison, Wis., 1978), *Stud. Logic Foundations Math.*, v. 101, pp. 123–148.
298. Gandy, R. (1988). The Confluence of Ideas, in 1936, in *The Universal Turing Machine: A Half-Century Survey*, Oxford University Press, Oxford, pp. 55–112.
299. Garson, M. and Franklin, S. (1989). Neural computability, in *Proc. Third International Joint Conference on Neural Networks*, v. 2, pp. 631–637.
300. Gasarch, W. and Smith, C.H. (1997). A survey of inductive inference with an emphasis on queries, *Complexity, Logic, and Recursion Theory*, Lecture Notes in Pure and Appl. Math., 187, Dekker, New York, pp. 225–260.
301. Gasc, P. (1974). On a symmetry of algorithmic information, *Soviet Math. Dokl.*, v. 218, no. 6, pp. 1265–1267.
302. Gell-Mann, M. (1994). *The Quark and the Jaguar*, W.H. Freeman.
303. Gell-Mann, M. (1995). Remarks on simplicity and complexity, *Complexity*, v. 1, no. 1, pp. 16–19.
304. Gell-Mann, M. and Lloyd, S. (1996). Information measures, effective complexity and total information, *Complexity*, v. 2, no. 5, pp. 16–19.
305. Geroch, R. and Hartle, J.B. (1986). Computability and physical theories, *Foundations of Physics*, v. 16, pp. 533–550.
306. Gibbs, W.W. (2002). Autonomic computing, *Scientific American*.
307. Gill, L. (2002). What supercomputers can and cannot do – yet, *NewsFactor Network*, June 17, 2002.
308. Giles, C.L. and Jenkins, B.K. (1986). Complexity Implications of Optical Parallel Computing, *Proceedings Twentieth Annual Asilomar Conference on Signals, Systems, and Computers*, Pacific Grove, CA.
309. Ginsburg, S. (1962). *An Introduction to Mathematical Machine Theory*, Addison-Wesley, Reading, Mass.
310. Ginsburg, S. and Spanier, E.H. (1964). Mappings of languages by two-tape devices, in *Proc. Fifth Ann. Syrup. Switching Circuit Theory and Logical Design*, Princeton, pp. 57–67.
311. Glass, R.L. (2002). The proof of correctness wars, *Communications ACM*, v. 45, no. 8, pp. 19–21.
312. Glass, R.L. (2002). Sorting out software complexity, *Communications ACM*, v. 45, no. 11, pp. 19–21.
313. Gleick, J. (1989). *Chaos: Making a New Science*, Cardinal, London.
314. Glushkov, V.M., Zeitlin, G.E., and Yushchenko, E.L. (1974). *Algebra, Languages, Programming*, Kiev, Naukova Dumka, (in Russian).

315. Gödel, K. (1931). Über formal unentscheidbare Sätze der Principia Mathematics und verwandter System I, *Monatshefte für Mathematik und Physik*, b. 38, s. 173–198.
316. Gödel, K. (1933). Eine Interpretation des intuitionistischen Aussagen kalkuls, *Ergebn. Math. Koll.*, v. 4, pp. 39–40.
317. Gödel, K. (1934). On undecidable propositions of formal mathematical systems, Lectures given at the Institute for Advanced Studies, Princeton, in *The Undecidable* (Davis, M., ed.). Raven Press, 1965, pp. 39–71.
318. Gold, E.M. (1965). Limiting recursion, *Journal of Symbolic Logic*, v. 30, no. 1, pp. 28–46.
319. Gold, E.M. (1967). Language identification in the limit, *Information and Control*, v. 10, pp. 447–474.
320. Goldin, D. and Wegner, P. (1988). *Persistent Turing Machines*, Brown University Technical Report.
321. Goldreich, O. (2001). Computational complexity, in *Mathematics Unlimited: 2001 and Beyond*, Springer, New York, pp. 507–524.
322. Goldschlager, L. and Lister, A. (1998). *Computer Science: A Modern Introduction*, Prentice Hall, New York, London, Toronto, Technical Report.
323. Goldwasser, S., Micali, S., and Rackoff, C. (1985). The Knowledge Complexity of Interactive Proofs, in *Proc. 17th ACM Symp. on Theory of Computing*, pp. 291–305.
324. Grassberger, P. (1990). Information and Complexity Measures in Dynamical Systems, *Information Dynamics*, Plenum Press.
325. Gray, J. (1985). *Why do Computers Stop and What We Can Do about It?* Tandem Computers Technical Report TR85.7, Cupertino, CA.
326. Gray, P. (1994). *Psychology*, Worth Publishers, New York.
327. Greenleaf, N. (1995). Computability and data types, *Journal of Computing in Small Colleges*, v. 11, no. 7, pp. 219–223.
328. Grossberg, S. (1987). Competitive learning: from interactive activation to adaptive resonance, *Cognitive Science*, v. 11, pp. 23–63.
329. Grossberg, S. (1988). Nonlinear neural networks: Principles, mechanisms, and architectures, *Neural Networks*, v. 1, pp. 17–61.
330. Grzegorczyk, A. (1953). Some classes of recursive functions, *Rosprawy Mate.*, IV.
331. Grzegorczyk, A. (1955). Computable functionals, *Fund. Math.*, v. 42, pp. 168–202.
332. Grzegorczyk, A. (1957). On the definition of computable real continuous functions, *Fundamenta Matematicae*, v. 44, pp. 61–71.
333. Guillame, M. (1978). Axiomatique et Logique, in *Abrégé d'histoire des Mathématiques 1700-1900*, Paris, Hermann.
334. Gupta, V.S. (2002). Communication complexity for file synchronization is undecidable, *ACM SIGACT News*, v. 33, no. 3, pp. 110–112.
335. Gupta, V., Jagadeesan, R., and Saraswat, V.A. (1999). Computing with continuous change, *Science of Computer Programming*, v. 30, no. 1/2, pp. 3–49.
336. Gurevich, Y. (1994). Evolving Algebras, in *IFIP 1994 World Computer Congress, Volume I: Technology and Foundations*, Elsevier, Amsterdam, pp. 423—427.
337. Gurevich, Y. (2000). Sequential abstract state machines capture sequential algorithms, *ACM Transactions on Computational Logic*, v. 1.
338. Gurevich, Y. (2001). Logician in the land of OS: Abstract State Machines at Microsoft, *Sixteenth Annual IEEE Symposium on Logic in Computer Science*, IEEE Computer Society, pp.129–136.
339. Gurevich, Y. and Spielmann, M. (1997). Recursive abstract state machines, *J. of Universal Computer Science*, v. 3, no. 4, pp. 233—246.

340. Halmos, P.R. (1962). *Algebraic Logic*, New York.
341. Halpern, J.Y., et al. (2001). On the unusual effectiveness of logic in computer science, *Bulletin of Symbolic Logic*, v. 7, no. 2, pp. 213–236.
342. Halstead, M.H. (1977). *Elements of Software Science*, New York, Elsevier.
343. Ham, F.M. and Kostanic, I. (2001). *Principles of Neurocomputing for Science and Engineering*, New York, McGraw-Hill.
344. Hamkins, J.D. and Lewis, A. (2000). Infinite time Turing machines, *Journal of Symbolic Logic*, v. 65, no. 3, pp. 567–604.
345. Hamming, R.W. (1962). *Numerical Methods for Scientists and Engineers*, New York.
346. Harel, D. (1987). *The Science of Computing*, Addison-Wesley, Boston, San Francisco, New York.
347. Harel, D. (2000). *Computers Ltd: What They Really Can't Do*, Oxford University Press.
348. Harrington, L.A. (1973). *Contributions to Recursion Theory on Higher Types*, Thesis, Cambridge, Mass., MIT.
349. Harrison, W.A., Magel, K.I., Kluczny, R., and Dekock, A. (1982). Applying software complexity metrics to program maintenance, *Computer*, v. 15, no. 9, pp. 65–79.
350. Hartley, R. and Szu, H. (1987). A comparison of the computational power of neural networks, in *Proc. IEEE Conference on Neural Networks*, pp. 17–22.
351. Hartmanis, J. (1983). Generalized Kolmogorov Complexity and the Structure of Feasible Computations, in *24 IEEE Symposium on Foundations of Computer Science*, pp. 439–435.
352. Hartmanis, J. and Hopcroft, J.E. (1971). An overview of the theory of computational complexity, *J. Association for Computing Machinery*, v. 18, no. 3, pp. 444–475.
353. Hartmanis, J. and Stearns, R.E. (1964). Computational complexity of recursive sequences, *IEEE Proc. 5th Annual Symp. on Switching Circuit Theory and Logical Design*, pp. 82–90.
354. Hartmanis, J. and Stearns, R.E. (1965). On the computational complexity of algorithms, *Transactions American Mathematical Society*, no. 5, pp. 285–306.
355. Harvey, I. and Bossomaier, T. (1997). Time out of joint: attractors in asynchronous random boolean networks, in *Fourth European Conference on Artificial Life*, MIT Press, Cambridge, Mass.
356. Hayashi S. and Nakata, M. (2000). Towards limit computable mathematics, in *Types for Proofs and Programs*, International Workshop, TYPES 2000, Durham, UK, December 2000, Lecture Notes in Computer Science, v. 2277, pp. 125–144.
357. Haykin, S. (1994). *Neural Networks: A Comprehensive Foundation*, New York, Macmillan.
358. Heath, S. (1997). *Heath, Embedded Systems Design*, Butterworth-Heinemann.
359. Hebb, D.O. (1949). *The Organization of Behavior*, John Wiley.
360. Hefner, R. and Mann B. (2002). The Evolution of a Software Measurement Program, in *Proceedings of the 6th World Multiconference on Systemics, Cybernetics and Informatics*, v. 7, Orlando, Florida, pp. 307–312.
361. van Heijenoort, J. (1967). *From Frege to to Gödel*, Cambridge University Press, Cambridge, Mass.
362. Hemaspaandra, L.A. and Ogihara, M. (1998). *The Complexity Theory Companion*, Springer, New York.
363. Hemmerling, A. (1999). On approximate and algebraic computability over the real numbers, *Theoretical Computer Science*, v. 219, pp. 185–323.
364. Hennie, F. C. (1961). *Iterative Arrays of Logical Circuits*, MIT Press, Cambridge, Mass.
365. Herbrand, J. (1932). Sur la non-contradiction de l'arithmetique, *Journal fur die reine und angewandte Mathematik*, 166, pp. 1–8.

366. Herman, F. and Margenstern, M. (2003). A universal cellular automaton in the hyperbolic plane, *Theoretical Computer Science*, v. 296, no. 2, pp. 327–364.

367. Higgins, P.J. (1963). Algebras with a scheme of operators, *Math. Nachrichten*, v. 27, no. 1–2, pp. 115–132.

368. Higgins, P.J. (1973). *Gropoids and Categories*, North-Holland.

369. Hintikka, Ja. and Mutanen, A. (1998). An Alternative Concept of Computability, in *Language, Truth, and Logic in Mathematics*, Dordrecht, pp. 174–188.

370. Hoare, C.A.R. (1969). An axiomatic basis for computer programming, *Communications of ACM*, v. 12, pp. 576–580, 583.

371. Hoare, C.A.R. (1978). Communicating sequential processes, *Communications of ACM*, v. 21, pp. 666–677.

372. Hoare, C.A.R. (1985). *Communicating Sequential Processes*, Prentice-Hall.

373. Hogarth, M.L. (1992). Does general relativity allows an observer to view an eternity in a finite time? *Foundations of Physics Letters*, v. 5, pp. 173–181.

374. Hogarth, M.L. (1994). Non-Turing computers and non-Turing computability, *PSA 1994*, v.1, East Lansing, Philosophy of Science Association, pp. 126–138.

375. Hogg T. (1999). Quantum search heuristics, *IEEE Intelligent Systems*, July–August, pp. 12–14.

376. Hopcroft, J.E., Motwani, R., and Ullman, J.D. (2001). *Introduction to Automata Theory, Languages, and Computation*, Addison Wesley, Boston, San Francisco, New York.

377. Horowitz, E. and Sahni, S. (1978). *Fundamentals of Computer Algorithms*, Computer Science Press, Inc.

378. Hotz, G., Schieffer, B., and Vierke, G. (1995). Analytic Machines, TR95- 025, *Electronic Colloquium on Computational Complexity*, http://www.eccc.uni-trier.de/eccc.

379. Hromkovic, J. (1997). *Communication Complexity and Parallel Computing*, Springer, New York.

380. Hughes, J. (1989). Why functional programming matters, *Computer Journal*, v. 32, No. 2.

381. Hyotyniemi, H. (1996). Turing machines are Recurrent Neural Networks, in *Proceedings of SteP'96*, Finnish Artificial Intelligence Society, Helsinki, pp. 13–24.

382. Interview: IBM's Wladawsky-Berger Explains Grid Computing (April 4, 2002). (electronic edition: http://www.gridcomputingplanet.com/opinions

383. Ivanov, L.L. (1986). *Algebraic Recursion Theory*, Ellis Horwood Series: Mathematics and its Applications, Ellis Horwood Ltd., Chichester; Halsted Press [John Wiley & Sons, Inc], New York.

384. Iwamoto, C., Margenstern, M., Morita, K., and Worsch, T. (2002). Polynomial-Time Cellular Automata in the Hyperbolic Plane Accept Exactly the PSPACE Languages, *Proc. 6th World Multi-Conference on Systemics, Cybernetics and Informatics*, pp. 411–416.

385. Juedes, D.W. and Lutz, J.H. (1992). Kolmogorov Complexity, Complexity Cores and the Distribution of Hardness, in *Kolmogorov Complexity and Computational Complexity*, Springer-Verlag, Berlin, Heidelberg, New York.

386. Kalmar, L. (1943). Egyzzerü pelda eldönthetetlen aritmetikai problemara, *Matemat. es Fizikai Lapok*, v. 50, no. 1, pp. 1–23.

387. Kalmar, L. (1959). An argument against the plausibility of Church's thesis, in *Constructivity in Mathematics*, Studies in Logic and the Foundations of Mathematics, North-Holland Publishing Co., Amsterdam, pp. 72–80.

388. Kampis, G. (1988). Information, Computation and Complexity, in *Nature, Cognition and Systems*, Carvallo, M.E. (ed.), Kluwer, Dordrecht, pp. 313–320.

389. Kaneko, K. and Tsuda, I. (1994). Constructive complexity and artificial reality: An introduction, *Physica D*, 75, pp. 1–10.
390. Kara, D. (2000). Pervasive computing era, *Software Magazine*, April, http://www.findarticles.com/cf-dls/m0SMG/mag.jhtml.
391. Karp, R.M. and Lipton, R. (1982). Turing machines that take advice, *Enseignment Mathematique*, 28, pp. 191–209.
392. Karplus, W.J. (1992). *The Heavens are Falling: The Scientific Prediction of Catastrophes in Our Times*, Plenum Press, New York.
393. Karpunin, G.A. and Shaposhnikov, I.G. (2000). Crossed homomorphisms of finite multi-base universal algebras with binary operations, *Discrete Math. Appl.*, v. 10, no. 2, pp. 183–202.
394. Kedrov, F. (1980). *Ernest Rutherford*, Moscow, Znanie (in Russian).
395. Kelly J.L. (1957). *General Topology*, Princeton, New York: Van Nostrand Co.
396. Kelly, K.T. and Schulte, O. (1997). Church's Thesis and Hume's Problem, in *Logic and Scientific Methods*, Dordrecht: Kluwer, pp. 383–398.
397. Kelly, K.T., Schulte, O., and Juhl, C. (1997a). Learning theory and the philosophy of science, *Philosophy of Science*, v. 64, pp. 245–267.
398. Kennedy, J. and Eberhart, R.C. (2001). *Swarm Intelligence*, Morgan Kaufmann Publishers.
399. Kieu, T.D. (2002). Quantum hypercomputation, *Minds and Machines*, v. 12, pp. 541–561.
400. Kieu, T. (2002a). *Computing the noncomputable*, LANL, Preprint in quant-ph/0203034 (electronic edition: http://arXiv.org).
401. Kieu, T.D. Computing the noncomputable, *Contemporary Physics*, v. 44 (2003). pp. 51–77.
402. Kieu, T.D. and Danos, M. (2001). A no-go theorem for halting a universal quantum computer, *Acta Phys. Hung.*, NS-H, v. 14, pp. 217–225.
403. Kleene, S.C., (1935). A theory of positive integers in formal logic, *American Journal of Mathematics*, v. 57, pp. 153–173, 219–244.
404. Kleene, S.C. (1936). General recursive functions of natural numbers, *Mathematische Annalen*, v. 112, no. 5, pp. 727–729.
405. Kleene, S.C. (1936a). λ-definability and recursiveness, *Duke Math.*, v. 2, pp. 340–353.
406. Kleene, S.C. (1955). Arithmetical predicates and function quantifiers, *Trans. of the American Math. Society*, v. 79, pp. 312–340.
407. Kleene, S.C. (1956). Representation of events in nerve nets, *Automata Studies*, Princeton University Press, Princeton, NJ, pp. 3–41.
408. Kleene, S.C. (1959). Recursive functionals and quantifiers of finite type, I, *Trans. of the American Math. Society*, v. 91, pp. 1–52.
409. Kleene, S. (1960). Constructive and Non-constructive Operations, in *Proceedings of the International Congress of Mathematicians*, Edinburgh, 1958, Cambridge.
410. Kleene, S.C, (1963). Recursive functionals and quantifiers of finite type, II, *Trans. of the American Math. Society*, v. 108, pp. 106–142.
411. Kleene, S.C. (1987). Reflections on Church's thesis, *Notre Dame Journal of Formal Logic*, v. 28, no. 4, pp. 490–498.
412. Kline, M. (1967). *Mathematics for Nonmathematicians*, New York, Dover Publications, Inc.
413. Klir, G.J. (1985). Complexity: Some general observations, *Systems Research*, 2, pp. 131–140.
414. Klir, G.J. (1984). The Many Faces of Complexity, in *The Science and Praxis of Complexity*, United Nations University, Tokyo, pp. 81–98.

415. Knuth, D. (1973). *The Art of Computer Programming*, v. 1: *Fundamental Algorithms*, Addison-Wesley.
416. Knuth, D. (1981). *The Art of Computer Programming*, v. 2: *Seminumerical Algorithms*, Addison-Wesley.
417. Ko, K.-I. (1991). *Complexity Theory of Real Functions*, Birkhäuser, Boston, MA.
418. Kogge, P. (1981). *The Architecture of Pipeline Computers*, McGraw Hill.
419. Kohonen, T. (1982). Self-organized formation of topologically correct feature maps, *Biological Cybernetics*, v. 43, pp. 59–69.
420. Kohonen, T. (1984). *Self-organization and Associative Memory*, Springer Verlag.
421. Kolmogorov, A.N. (1950). *Foundations of the Theory of Probability*, Chelsea.
422. Kolmogorov, A.N. (1953). On the concept of algorithm, *Russian Mathematical Surveys*, v. 8, no. 4, pp. 175–176.
423. Kolmogorov, A.N. (1961). Automata and Life, in *Knowledge is Power*, no. 10 and No. 11.
424. Kolmogorov, A.N. (1965). Three approaches to the definition of the quantity of information, *Problems of Information Transmission*, no. 1, pp. 3–11.
425. Kolmogorov, A.N. (1968). Logical basis for information theory and probability theory, *IEEE Trans. Inform. Theory*, vol. IT-14, pp. 662–664.
426. Kolmogorov, A.N. and Uspensky V.A. (1958). On the definition of algorithms, *Russian Math. Surveys*, v. 13, pp. 3–28.
427. Komar, A. (1964). Undecidability of macroscopically distinguishable states in quantum field theory, *Physical Review*, 2nd series, 133B, pp. 542–544.
428. Kosovsky, N.K. (1981). *Elements of Mathematical Logic and Its Application to the Theory of Subrecursive Algorithms*, LSU Publ.
429. Kozen, D. (1976). On Parallelism of Turing Machines, in *Proceedings of the 23rd Symposium on Foundations of Computer Science*, pp. 89–97.
430. Krinitsky, N.A. (1977). *Algorithms Around Us*, Moscow, Nauka (in Russian).
431. Kugel, P. (1977). Induction, pure and simple, *Information and Control*, v. 33, pp. 276–336.
432. Kugel, P. (1986). Thinking may be more than computing, *Cognition*, v. 22 pp. 137–196.
433. Kugel, P. (2002). Intelligence requires more than computing … and Turing said so, *Minds and Machines*, v. 12, no. 4, pp. 563–579.
434. Kung, H.T. and Leiserson, C.E. (1979). Systolic arrays (for VLSI), in *Sparse Matrix Proc.* (Society for Industrial and Applied Mathematics), pp. 256–282.
435. Kurosh, A.G. (1974). *General Algebra*, Moscow, Nauka (in Russian).
436. Kushilevitz, E. and Nisan, N. (1997). *Communication Complexity*, Cambridge University Press, Cambridge.
437. Lacombe, D. (1955). Extension de la notion de fonction recursive aux fonctions d'une ou plusieurs variables reelles, *I. C. R. Acad. Sci. Paris*, v. 240, pp. 2478–2480.
438. LaForte, G., Hayes, P. and Ford, K. (1998). Why Gödel's theorem cannot refute computationalism, *Artificial Intelligence*, v. 104, pp. 265–286.
439. Landow, G. (1992). *Hypertext: The Convergence of Contemporary Critical Theory and Technology*, Baltimore: Johns Hopkins University Press.
440. Langton, C.G. (1984). Self-reproduction in cellular automata, *Physica D*, v. 10, pp. 135–144.
441. Langton, C.G. (1989). Artificial Life, in *Artificial Life*, pp. 1–47.
442. Leblanc, L. (1962). Nonhomogeniuos polyadic algebras, *Proceedings of the American Mathematical Society*, v. 13, no. 1, pp. 59–65.

443. de Leeuw, K., Moore, E.F., Shannon, C.E. and Shapiro, N. (1956). Computability by probabilistic machines, *Automata Studies*, Princeton University Press, Princeton, NJ, pp. 183–212.

444. Lehman, M., et al. (1985). *Program Evolution*, Academic Press, London.

445. Levin, L.A. (1973). On the notion of a random sequence, *Soviet Math. Dokl.*, v. 14, no. 5, pp. 1413–1416.

446. Levin, L.A. (1974). Laws of information (nongrowth). and aspects of the foundation of probability theory, *Problems of Information Transmission*, v. 10, no. 3, pp. 206–210.

447. Levin, L.A. (1976). Various measures of complexity for finite objects (axiomatic description), *Soviet Math. Dokl.*, v. 17, pp. 522–526.

448. Levin, L.A. (2003). The tale of one-way function, *Problems of Information Transmission*, v. 39, no. 1, pp. 92–103.

449. Lewis, J.P. (2001). Limits to software estimation, *Software Engineering Notes*, v. 26, no. 4, pp. 54–59.

450. Li, M. and Vitanyi, P. (1992). Philosophical issues in Kolmogorov complexity, *Lecture Notes in Computer Science*, v. 623, pp. 1–15.

451. Li, M. and Vitanyi, P. (1997). *An Introduction to Kolmogorov Complexity and its Applications*, Springer-Verlag, New York.

452. Li, W. (1992). An open logic system, *Science in China* (*Scientia Sinica*). (series A), v. 10, pp. 1103–1113 (in Chinese).

453. Li, W., Ma, S., Sui, Y., and Xu, K. (2001). *A Logical Framework for Convergent Infinite Computations*, Preprint cs.LO/0105020 (electronic edition: http://arXiv.org).

454. Littlewood, J.E. (1953). *Miscellany*, Methuen, London.

455. Loveland, D.W. (1969). A variant of the Kolmogorov concept of complexity, *Information and Control*, v. 15, pp. 510–526.

456. Lucas, J.R. (1963). Minds, machines, and Gödel, *Philosophy*, v. 36, pp. 112–127.

457. Lucas, J.R. (1967). Minds, machines, and Gödel, in *Minds and Machines*, Prentice-Hall, Englewood Cliffs, NJ, pp. 43–59.

458. Lucas, J.R. (1996). Minds, machines, and Gödel: A retrospect, in *Machines and Thought: The Legacy of Alan Turing*, Oxford University Press, Oxford, pp. 103–124.

459. Luna, F. (1996). Computable learning, neural networks, and institutions, università di Venezia, *Note di Lavoro*, v. 2.

460. Luna, F. (1997). Learning in a Computable Setting. Applications of Gold's Inductive Inference Model, in *Computational Approaches to Economic Problems* (H. Amman et al., eds.), Kluwer Academic Press, pp. 271–288.

461. Lumb, I. (2002). *The Emperor's New Grid*? (electronic edition: http://www.gridcomputingplanet.com/opinions).

462. MacLennan, B.J. (1990). *Field Computation: A Theoretical Framework for Massively Parallel Analog Computation*, Technical Report CS-90-100, Computer Science Department, University of Tennessee, Knoxville.

463. MacLennan, B.J. (1994). Image and Symbol: Continuous Computation and the Emergence of the Discrete, in *Intelligence and Neural Networks: Steps Toward Principled Integration*, New York, Academic Press, pp. 207–240.

464. MacLennan, B.J. (1999). Field computation in natural and artificial intelligence, *Information Sciences*, v. 119, pp. 73–89.

465. MacLennan, B.J. (2001). *Transcending Turing Computability*, Technical Report UT-CS-01-473, November 12.

466. Machlin, R. and Stout Q.F. (1991). The Complex Behavior of Machines, in *Emergent Computation*, MIT Press, Cambridge, MA, London, England, 1991, pp. 85–98.

467. Malcev, A.I. (1965). *Algorithms and Recursive Functions*, Nauka, Moscow (in Russian).
468. Manes, E.G. and Arbib, M.A. (1986). *Algebraic Approaches to Program Semantics*, Springer-Verlag, New York.
469. Manin, Yu I. (1991). *Course in Mathematical Logic*, Springer-Verlag, New York.
470. Manna, Z. (1974). *Mathematical Theory of Computability*, McGraw Hill, Inc., New York.
471. Manna, Z. and Vuillemin, J. (1972). Fixed point approach to the theory of computation, *Communications of the ACM*, v. 15, no. 7, pp. 528–536.
472. Margenstern, M. (2000). New tools for cellular automata of the hyperbolic plane, *Journal of Universal Computer Science*, v. 6, no. 12, pp. 1226–1252.
473. Margenstern, M. (2002). Cellular automata in the hyperbolic plane: A survey, *Romanian Journal of Information Science*, v. 5, no. 1/2, pp. 155–179.
474. Margenstern, M. (2002a). Tiling the hyperbolic plane with a single pentagonal tile, *Journal of Universal Computer Science*, v. 8, no. 2, pp. 297–316.
475. Margenstern, M. (2003). Cellular automata and combinatoric tilings in hyperbolic spaces, *Lecture Notes in Computer Science*, 2731, pp. 48–72.
476. Margenstern, M. (2003). A Combinatorial Approach to Hyperbolic Geometry as a New Perspective for Computer Science and Technology, in Proceedings of the ISCA 18th International Conference *Computers and their Applications*, International Society for Computers and their Applications, Honolulu, Hawaii, pp. 468–471.
477. Margenstern, M. and Morita, K. (2001). NP problems are tractable in the space of cellular automata in the hyperbolic plane, *Theoretical Computer Science*, v. 259, pp. 99–128.
478. Markov, A.A. (1951). Theory of algorithms, *Transactions of the Mathematical Institute of the Academy of Sciences of the USSR*, v. 38 (in Russian).
479. Markov, A.A. (1954). Theory of algorithms, *Transactions of the Mathematical Institute of the Academy of Sciences of the USSR*, v. 42 (in Russian).
480. Martin, E. and Osherson, D.N. (2001). Induction by enumeration, *Information and Computation*, v. 171, no. 1, pp. 50–68.
481. Martin, J.C. (1991). *Introduction to Languages and the Theory of Computation*, McGraw-Hill, New York.
482. Matherat, P. and Jaekel, M.-T. (2001). *Concurrent computing machines and physical space-time*, LANL, Preprint in Computer Science DC/0112020, 15 p. (electronic edition: http://arXiv.org).
483. Mathienssen, A. (1978). A heterogeneous algebraic approach to some problems in automata theory, many-valued logics and other topics, in *Contr. to General Algebra*, Proc. Klagenfurt Conf.
484. McCabe, T.J. (1976). A complexity measure, *IEEE Transaction on Software Engineering*, SE-2, pp. 308–320.
485. McCulloch, W.S. and Pitts, E. (1943). A logical calculus of the ideas immanent in nervous activity, *Bulletin of Mathematical Biophysics*, v. 5, pp. 115–133.
486. Mealy, G.H. (1953). A method for synthesizing sequential circuits, *Bell System Techn. J.*, v. 34, pp. 1045–1079.
487. Meer, K. (1992). A note on a $P \neq NP$ result for a restricted class of real machines, *Journal of Complexity*, v. 8, pp. 451–453.
488. Meer, K. (1993). Real number models under various sets of operations, *Journal of Complexity*, v. 9, pp. 366–372.
489. Mendelson, E. (1990). Second thoughts about Church's thesis and mathematical proofs, *Journal of Philosophy*, v. 87, no. 5, pp. 225–233.
490. Mendelson, E. (1963). On some recent criticism of Church's thesis, *Notre Dame J. Formal Logic*, v. 4, pp. 201–205.

491. Mesarovic, M.D. and Takahara, Y. (1975). *General Systems Theory: Mathematical Foundations*, Academic Press, New York, London, San Francisco.
492. Meyer, A. and Stockmeyer, L. (1972). The Equivalence Problem for Regular Expressions with Squaring Requires Exponential Space, in *Proc. of the 13th IEEE Symposium on Switching and Automata Theory*, pp. 125–129.
493. Meyers, B.A. (1998). A brief history of human computer interaction technology, *ACM Interactions*, v. 5, no. 2, pp. 44–54.
494. Michel, A.N. and Derong Liu (2002). *Qualitative Analysis and Synthesis of Recurrent Neural Networks*, New York, Marcel Dekker.
495. Milne, G.J. (1985). Circal and the representation of communication, concurrency and time, *ACM Transactions on Programming Languages and Systems*, v. 7, pp. 270–298.
496. Milne, G.J. and Milner, A.J.R.G. (1979). Concurrent processes and their syntax, *Journal of ACM*, v. 26, no. 2, pp. 302–321.
497. Milner, A.J.R.G. (1973). Processes: A mathematical model of computing agents, *Proc. Logic Colloquium '73*, IFIP, North Holland, pp. 157–174.
498. Milner, A.J.R.G. (1979). Flow graphs and flow algebras, *Journal of ACM*, v. 26, no. 4, pp. 794–818.
499. Milner, A.J.R.G. (1980). A calculus of communicating systems, *Lecture Notes in Computer Science*, v. 92.
500. Milner, A.J.R.G. (1983). Calculi for synchrony and asynchrony, *Journal of Theoretical Computer Science*, v. 25, pp. 267–310.
501. Milner, A.J.R.G. (1986). *A Calculus of Communicating Systems*, Report ECS-LFCS-86-7, Computer Science Department, University of Edinburgh.
502. Milner, M. (1989). *Communication and Concurrency*, Prentice Hall, New York, London, Toronto.
503. Minsky, M. (1967). *Computation: Finite and Infinite Machines*, Prentice-Hall, New York, London, Toronto.
504. Minsky, M. (1986). *The Society of Mind*, Simon and Schuster, New York.
505. Minsky, M. (1998). The mind, artificial intelligence and emotions, *Brain & Mind Magazine*, no. 7, September/November.
506. Minsky, M. and Papert, S. (1969). *Perceptrons*, MIT Press.
507. Moll, R.N., Arbib, M.A., and Kfoury, A.J. (1988). *An Introduction to Formal Language Theory*, (with contributions by J. Pustejovsky), Springer-Verlag.
508. Moore, E.F. (1956). Gedanken-experiments on sequential machines, in *Automata Studies*, Princeton University Press, Princeton, NJ, pp. 129–153.
509. Moore, A.W. (1990). *The Infinite*, Routledge, New York, NY.
510. Moore, R.E. (1966). *Interval Analysis*, Prentice Hall, New York.
511. Moore, C. (1990). Unpredictability and undecidability in dynamical systems, *Physical Review Letters*, v. 64, pp. 2354–2357.
512. Moore, C. (1996). Recursion theory on the reals and continuous-time computation: Real numbers and computers, *Theoretical Computer Science*, 162, no. 1, pp. 23–44.
513. Morita, K., Margenstern, M., and Imai, K. (1999). Universality of reversible hexagonal cellular automata, *Theoretical Informatics and Applications*, 33, pp. 535–550.
514. Moschovakis, Ya. (1974). *Elementary Induction on Abstract Structures*, Amsterdam, North-Holland.
515. Moschovakis, Ya. (2001). What is an Algorithm?, in *Mathematics Unlimited: 2001 and Beyond*, Springer, New York.
516. Mostowski, A. (1957). On computable sequences, *Fundamenta Mathematicae*, v. XLIV, pp. 12–36.

517. Myhill, J. (1960). Linear bounded automata, *Wadd Tech. Notes*, Wright-Patterson AFB, Ohio, pp. 60–165.

518. Naur, P., et al. (1960). Report on the algorithmic language ALGOL 60, *Communications of the ACM*, v. 3, no. 5, pp. 299–314.

519. Naughton, T.J. (2000). Continuous-space model of computation is Turing universal, in *Critical Technologies for the Future of Computing*, Proceedings of SPIE, v. 4109, pp. 121–128.

520. Naughton, T.J. and Woods, (2001). On the computational power of a continuous-space optical model of computation," *Third International Conference on Machines, Computations, and Universality*, Lecture Notes in Computer Science v. 2055, pp. 288–299.

521. Nelson, R.J. (1987). Church's thesis and cognitive science, *Notre Dame J. of Formal Logic*, v. 28, no. 4, 581–614.

522. Nerode, A. and Kohn, W. (1993). *Models for hybrid systems: Automata, topologies, stability*, TR. 93–11, Mathematical Sciences Institute, Cornell Univ.

523. von Neumann, J. (1951). The general and logical theory of automata, in *Cerebral Mechanisms in Behavior*, The Hixon Symposium, Wiley, New York, pp. 1–31.

524. von Neumann, J. (1963). *Collected Works*, MacMillan.

525. von Neumann, J. (1966). *Theory of Self-Reproducing Automata*, 1949 University of Illinois Lectures on the Theory and Organization of Complicated Automata, edited and completed by Arthur W. Burks. Urbana, University of Illinois Press.

526. Nielsen, J. (1990). *Hypertext and Hypermedia*, New York: Academic Press.

527. Nobili R. and Pesavento, U. (1996). Generalised von Neumann's automata I: A revisitation, in *Artificial Worlds and Urban Studies*, DAEST Pubblication, Convegni 1, Venezia.

528. Norman, D.A. (1997). Why it's Good that Computers don't Work like the Brain, in *Beyond Calculation: The Next Fifty Years of Computing*, Copernicus, pp. 105–116.

529. Odifreddi, P. (2000). *Classical Recursion Theory II*. Elsevier.

530. Orponen, P. (1997). On the computational power of continuous time neural networks, in *Proc. SOFSEM'97, the 17th Seminar on Current Trends in Theory and Practice of Informatics*, Lecture Notes in Computer Science, Springer-Verlag, pp. 86–103.

531. Orponen, P. (1997a). A survey of continuous-time computation theory, in *Advances in Algorithms, Languages, and Complexity*, Kluwer Academic Publishers, Dordrecht, pp. 209–224.

532. Osherson, D.N., Stob, M., and Weinstein, S. (1972). A universal method of scientific inquiry, *Machine Learning*, v. 9, pp. 261–271.

533. Osherson, D.N., Stob, M., and Weinstein, S. (1982). Learning strategies. *Information and Control*, v. 53, no. 1/2, pp. 32–51.

534. Osherson, D.N., Stob, M., and Weinstein, S. (1983). Formal Theories of Language Acquisition: Practical and Theoretical Perspectives, *IJCAI*, pp. 566–572.

535. Osherson, D.N., Stob, M., and Weinstein, S. (1986). Aggregating inductive expertise, *Information and Control*, v. 70, no. 1, pp. 69–95.

536. Osherson, D.N., Stob, M., and Weinstein, S. (1988). Synthesizing inductive expertise, *Information and Computation*, v. 77, no. 2, pp. 138–161.

537. Osherson, D.N., Stob, M., and Weinstein, S. (1989). On Approximate Truth, *COLT*, pp. 88–101.

538. Osherson, D.N., Stob, M., and Weinstein, S. (1990). A Mechanical Method of Successful Scientific Inquiry, *COLT*, pp. 187–201.

539. Osherson, D.N., Stob, M., and Weinstein, S. (1991). A universal inductive inference machine, *Journal of Symbolic Logic*, v. 56, no. 2, pp. 661–672.

540. Overton, R. (2000). Of catenanes and porphyrins, *Business*, v. 5, no. 21, pp. 238.

541. Pager, D. (1970). On the efficiency of algorithms, *Journal of the ACM*, v. 17, no. 4, pp. 708–714.

542. Pan, S.L. and Lee, J.-N. (2003). Using ECRM for a unified view of the customer, *Communications of the ACM*, v. 46, no. 4, pp. 95–99.

543. Papadopoulos, G. (2002). Why PC design must change, *ZDNet*.

544. Park, R.E. (1992). *Software Size Measurement: A Framework for Counting Source Statements*, Software Engineering Institute, Pittsburgh, SEI-92-TR-020, 220 p.

545. Parker, D.B. (1982). Learning-logic, Technical Report 581-64, Office of Technology Licensing, Stanford University.

546. Parsons J.J. and Oja D. (1994). *New Perspectives on Computer Concepts*, Course Technology, Inc., Cambridge, MA.

547. Paul, B. (2002). Complexity – the enemy of integration, *Managing Information Strategies*, http://www.misweb.com.

548. Paun, G. (2002). *Membrane Computing: An Introduction*, Springer, Berlin.

549. Peichl, T. and Vollmer, H. (2001). Finite automata with generalized acceptance rule, *Discrete Mathematics and Theoretical Computer Science*, v. 4, pp. 179–194.

550. Penrose, R. (1989). *The Emperor's New Mind*, Oxford University Press, Oxford.

551. Penrose, R. (1994). *Shadows of the Mind: A Search for the Missing Science of Consciousness*, Oxford University Press, Oxford.

552. Peterson, J.L. (1981). *Petri Net Theory and the Modeling of Systems*, Prentice-Hall, Inc., Englewood Cliffs, N.J.

553. Petri, C. (1962). *Kommunikation mit Automaten*, Ph.D. Dissertation, University of Bonn, Bonn, Germany.

554. Pitowsky, I. (1990). The physical Church thesis and physical computational complexity, *Iyyun*, v. 39, pp. 81–99.

555. Platek, R. (1966). *Foundations of Recursion Theory*, Thesis, Stanford, California, Stanford University.

556. Plotkin B.I. (1991). *Universal Algebra, Algebraic Logic, and Databases*, Moscow, Nauka (in Russian).

557. Pollack, J.B. (1987). *On Connectionist Models of Natural Languages*, Ph.D. Thesis, Computer Science Department, University of Illinois, Urbana.

558. Post, E. (1936). Finite combinatory process, *Journal of Symbolic Logic*, v.1, pp. 103–105.

559. Post, E. (1946). A variant of recursively unsolvable problem, *Bull. of the AMS*, v. 52, no. 4, pp. 264–268.

560. Pour-El, M.B. (1974). Abstract computability and its relation to the general-purpose analog computer, *Trans. Amer. Math. Society*, v. 199, pp. 1–28.

561. Pour-El, M.B. (1999). The structure of computability in analysis and physical theory: An extension of Church's thesis, *Handbook of Computability Theory, Stud. Logic Found. Math.*, 140, North-Holland, Amsterdam, pp. 449–471.

562. Pour-El, M.B. and Richards, I. (1979). A computable ordinary differential equation which possesses no computable solution, *Annals of Mathematical Logic*, v. 17, pp. 61–90.

563. Pour-El, M.B. and Richards, I. (1981). The wave equation with computable initial data such that its unique solution is not computable, *Advances in Mathematics*, v. 39, pp. 215–239.

564. Pour-El, M.B. and Richards, J.I. (1989). *Computability in Analysis and Physics, Perspectives in Mathematical Logic*, Springer-Verlag, Berlin.

565. Prasse, M. and Rittgen, P. (1998). Why Church's thesis still holds. Some notes on Peter Wegner's tracts on interaction and computability, *The Computer Journal*, 41, pp. 357–362.

566. Prather, R.E. (1977). Structured Turing machines, *Information and Control*, v. 35, no. 2, pp. 159–171.

567. Prather, R.E. (1984). An axiomatic theory of software complexity research, *The Computer Journal*, v. 27, pp. 340–346.

568. Pratt, V., Rabin, M., and Stockmeyer, L.J. (1974). A Characterization of the Power of Vector Machines, in *6th ACM Symposium on the Theory of Computing*, pp. 122–134.

569. Pressman, R.S. (1994). *Software Engineering: A Practitioner's Approach*, McGraw Hill.

570. Pribram, K.H. (2002). Brain and quantum holography: Recent ruminations, in Yasue, Kunio, Mari Jibu and Tarcisio Della Senta (eds.), *No Matter, Never Mind*, Proceedings of Toward a Science of Consciousness: Fundamental approaches, Tokyo 1999, John Benjamins Pub Co., pp. 1–11.

571. Prigogine, I. (1980). *Form Being to Becoming: Time and Complexity in the Physical Systems*, San Francisco, Freeman and Co.

572. Putnam, H. (1960). Minds and machines, in *Dimensions of Mind*, New York University Press.

573. Putnam, H. (1965). Trial and error predicates and the solution to a problem of Mostowski, *Journal of Symbolic Logic*, 30, no. 1, pp. 49–57.

574. Pylyshyn, Z.W. (1984). *Computation and Cognition: Toward a Foundation for Cognitive Science*, MIT Press.

575. Rabhi, F. and Lapalme, G. (1999). *Algorithms: A Functional Programming Approach*, Addison-Wesley.

576. Rabin, M.O. (1959). Speed of Computation of Functions and Classification of Recursive Sets, Proc. 3rd Conference of Scientific Societies, Jerusalem, pp. 1–2.

577. Rabin, M.O. (1960). *Degree of Difficulty of Computing a Function and a Partial Ordering of Recursive Sets*, Tech. Report 2, Hebrew University, Jerusalem.

578. Rabin, M.O. (1963). Probabilistic automata, *Information and Control*, v. 6, no. 3, pp. 230–244.

579. Rabin, M.O. (1963a). Real-time computation. *Israel Journal of Math.*, v. 1, pp. 203–211.

580. Rabin, M.O. (1969). Decidability of second-order theories and automata on infinite trees, *Transactions of the AMS*, v. 141, pp. 1–35.

581. Rabin, M.O. and Scott, D. (1959). Finite automata and their decision problems, *IBM Journal of Research and Development*, v. 3, pp. 114–125.

582. Rapaport, W. (1998). How minds can be computational systems, *Journal of Experimental and Theoretical Artificial Intelligence*, v. 10, pp. 403–419.

583. Rapaport, W. (1988). Syntactic semantics: Foundations of computational natural-language understanding, in *Aspects of Artificial Intelligence*, Kluwer, Dordrecht, pp. 81–131.

584. Rees, M. (1997). *Before the Beginning: Our Universe and Others*, Simon & Schuster, London.

585. Rice, H.G. (1951). Recursive real numbers, *Proceedings of the AMS*, v. 5, pp. 784–791.

586. Riguet, J. (1965). Programmation et Theorie des Categories, in *Symbolic Languages and Data*, New York, London, Gordon & Breach, pp. 83–98.

587. Riordan, J. and Shannon, C.E. (1942). The number of two-terminal series-parallel networks, *J. Math. Physics*, v. 21, pp. 83–93.

588. Ritchie, R.W. (1963). Classes of predictably computable functions. *Transactions Amer. Math. Soc.*, v. 106, pp.139–173.

589. Robinson, A. (1966). *Nonstandard Analysis*, Amsterdam, North Holland.
590. Robinson, J. (1950). General recursive functions, *Proceedings of the American Mathematical Society*, v. 1, pp. 703–718.
591. Rogers, H. (1987). *Theory of Recursive Functions and Effective Computability*, MIT Press, Cambridge Massachusetts.
592. Romera, M.E. (2000). *Using Finite Automata To Represent Mental Models*, A Thesis Presented to the Faculty of the Department of Psychology, San Jose State University.
593. Rose, H.E. (1984). *Subrecursion: Functions and Hierarchies*, Clarendon Press.
594. Rosen, K.H. (1999). *Discrete Mathematics and its Applications*, McGraw-Hill.
595. Rosenblatt, F. (1962). *Principles of Neurodynamics*, Spartan Books.
596. Rosolini, G. (1987). Categories and Effective Computations, in *Proceedings on Category Theory and Computer Science*, Edinburgh, UK, Lecture Notes in Computer Science, v. 283, pp. 1–11.
597. Rubel, L.A. (1988). Some mathematical limitations of the general-purpose analog computer, *Advances in Applied Mathematics*, v. 9, pp. 22–34.
598. Rubel, L.A. (1989). Digital simulation of analog computation and Church's thesis, *Journal of Symbolic Logic*, v. 54, pp. 1011–1017.
599. Rubel, L.A. (1993). The extended analog computer, *Advances in Applied Mathematics*, v. 14, pp. 39–50.
600. Rumelhart, D.E., Hinton, G.E., and Williams, R.J. (1986). Learning representations by back-propagating errors, *Nature*, 323, pp. 533–536.
601. Rydeheard, D.E. (1988). *Computational Category Theory*, Prentice Hall.
602. Sacks, G.E. (1990). *Higher Recursion Theory*, Springer-Verlag, New York.
603. Sambrook, T. and Whiten, A. (1997). On the nature of complexity, *Cognitive and Behavioural Science, Theory and Psychology*, v. 7, pp. 191–213.
604. Savage, J.E. (1976). *The Complexity of Computing*, John Wiley & Sons, New York, London, Sydney.
605. Scarpellini, B. (1963). Zwei unentscheidbare Probleme der Analysis, *Z. Math. Logik Grundlagen Math.*, v. 9, pp. 265–289.
606. Schmidhuber, J. (2000). *Algorithmic Theories of Everything*, Technical Report IDSIA-20-00, Lugano.
607. Schmidhuber, J. (2002). Hierarchies of generalized Kolmogorov complexities and nonenumerable universal measures computable in the limit, *International Journal of Foundations of Computer Science*, v. 3, no. 4, pp. 587–612.
608. Schmidhuber, J. (2002a). Speed Prior and Optimal Simulation of the Future, in *Proceedings of the Business and Industry Simulation Symposium*, Society for Modeling and Simulation International, San Diego, California, pp. 40–45.
609. Schneider, M. and Gersting, J. (1995). *An Invitation to Computer Science*, West Publishing Company, New York.
610. Schnorr, C.-P. (1973). Process complexity and effective random tests, Fourth Annual ACM Symposium on the Theory of Computing (Denver, Colo., 1972), *J. Comput. System Sci.*, v. 7, pp. 376–388.
611. Schöning, U. (1988). Complexity theory and interaction, in *The Universal Turing Machine – A Half-Century Survey*, Oxford University Press, Oxford, pp. 561–580.
612. Schubert, L.K. (1974). Iterated limiting recursion and the program minimization problem, *Journal of the ACM*, v. 21, no. 3, pp. 436–445.
613. Schwartz, J. and Sharir, M. (1983). On the piano movers' problem III: Coordinating the motion of several independent bodies, *Int. Journal of Robotics Research*, v. 2, no. 3, pp. 46–75.

614. Scolem, T. (1962). A Theorem on Recursively Enumerable Sets, *International Congress of Mathematicians*, P. II.
615. Scott, D. (1971). Outline of a Mathematical Theory of Computation, in *Proc. Princeton Conference on Information Science*.
616. Scott, D. and Strachey, C. (1971). Towards a Mathematical Semantics for Computer Languages, in *Proc. Symposium on Computers and Automata*, Polytechnic Institute of Brooklyn, New York.
617. Searle, J.R. (1980). Minds, brains and programs, *Behavioral and Brain Sciences*, v. 3, pp. 417–57.
618. Searle, J.R. (1990). Is the brain a digital computer? *Proceedings and Addresses of the American Philosophical Association*, v. 64, pp. 21–37.
619. Seife, C. (2001). The quandary of quantum information, *Science*, v. 293, no. 5537, pp. 2026–2027.
620. Seig, W. (1997). Step by recursive step: Church's analysis of effective calculability, *Bulletin of Symbolic Logic*, v. 3, no. 2, pp. 154–180.
621. Shagrir, O. (2002). Effective computation by humans and machines, *Minds and Machines*, v. 12, pp. 221–240.
622. Shannon, C. (1938). A symbolic analysis of relay and switching circuits, *Transactions AIEE*, v. 57, pp. 713–723.
623. Shannon, C. (1941). Mathematical theory of the differential analyzer, *J. Math. Physics*, MIT, v. 20, pp. 337–354.
624. Shannon, C. (1949). The synthesis of two-terminal switching circuits, *Transactions Bell Systems Technical Journal*, v. 28, pp. 59–98.
625. Shapiro, S. (1981). Understanding Church's thesis, *Journal of Philosophical Logic*, v. 10, pp. 353–365.
626. Shaposhnikov, I.G. (1999). Congruences of finite multibase universal algebras, *Discrete Math. Appl.*, v. 9, no. 4, pp. 403–418.
627. Shepherdson, J.C. and Sturgis, H.E. (1963). Computability of recursive functions, *Journal of the ACM*, v. 10, no. 2, pp. 217–255.
628. Shoenfield, J.R. (1967). *Mathematical Logic*, Addison-Wesley, Reading, Mass.
629. Shönhage, A. (1980). Storage modification machines, *SIAM Journal on Computing*, v. 9, pp. 490–508.
630. Sieg, W. (1994). Mechanical procedures and mathematical experience, in *Mathematics and Mind*, Oxford University Press, Oxford.
631. Sieg, W. (1997). Step by recursive step: Church's analysis of effective calculability, *Bull. Symb. Logic*, v. 3, no. 2, pp. 154–180.
632. Siegelman, H.T. (1995). Computation beyond the Turing limit, *Science*, 268 (5210), pp. 545–548.
633. Siegelmann, H.T. (1999). *Neural Networks and Analog Computation: Beyond the Turing Limit*, Birkhäuser, Berlin.
634. Siegelman, H.T. and Fishman, S. (1998). Analog computation with dynamical systems, *Physica D*, v. 120, pp. 214–235.
635. Siegelman, H.T. and Sontag, E.D. (1991). Turing computability with neural nets, *Applied Mathematics Letters*, 4 (6), pp. 77–80.
636. Siegelman, H.T. and Sontag, E.D. (1994). Analog computation via neural networks, *Theoretical Computer Science*, 131, pp. 331–360.
637. Siegelman, H.T. and Sontag, E.D. (1995). On analog computational power of neural networks, *Journal of Computer and System Sciences*, 50 (1), pp. 132–150.
638. Silverman, R.E. (11/14/2000). Tech-Project Inefficiencies Found in Corporate Study, *Wall Street Journal*, p. B18.

639. Simon, H.A. (1969). The architecture of complexity, in *The Sciences of the Artificial*, Cambridge, MA, MIT Press, pp. 192–229.
640. Sipser, M. (1984). A topological view of some problems in complexity theory, in *Theory of Algorithms*, pp. 387–391.
641. Skordev, D. (1974). A certain generalization of the theory of recursive functions, *Dokl. Akad. Nauk USSR*, v. 219, pp. 1079–1082 (translated from Russian).
642. Skordev, D.G. (1976). The concept of search computability from the point of view of the theory of combinatory spaces, *Serdica*, v. 2, no. 4, pp. 343–349 (in Russian).
643. Skordev, D. (1982). An algebraic treatment of flow diagrams and its application to generalized recursion theory, *Universal Algebra and Applications*, Banach Center Publ., v. 9, PWN, Warsaw, pp. 277–287.
644. Skordev, D.G. (1992). Computability in combinatory spaces. An algebraic generalization of abstract first order computability, *Mathematics and its Applications* (East European Series), 55 Kluwer Academic Publishers Group, Dordrecht.
645. Sloman, A. (2002). *The Irrelevance of Turing machines to AI*, http://www.cs.bham.ac.uk/ axs/.
646. Smith, C.H. (1994). *A Recursive Introduction to the Theory of Computation*, Springer-Verlag, New York.
647. Smith, C.H. and Gasarch, W.I. (1995). Recursion-theoretic models of learning: Some results and intuitions, *Ann. Math. Artificial Intelligence*, v. 15, no. 2, pp. 151–166.
648. Smith, W. (1993). Church's Thesis Meets the N-body Problem, Manuscript http://www.neci.nec.com/homepages/wds)
649. Smith, W. (1999). Church's Thesis Meets Quantum Mechanics, Manuscript http://www.neci.nec.com/homepages/wds)
650. Smullian, R.M. (1962). *Theory of Formal Systems*, Princeton University Press.
651. Solomonoff, R.J. (1960). *A Preliminary Report on a General Theory of Inductive Inference*, Technical Report ZTB-138, Zator Company, Cambridge, Mass.
652. Solomonoff, R.J (1964). A formal theory of inductive inference, *Information and Control*, v. 7, no. 1, pp. 1–22; no. 2, pp. 224–254.
653. Spector C. (1959). Inductively Defined Sets of Natural Numbers, in *Infinitistic Methods*, New York, Pergamon Press, pp. 97–102.
654. Spielmann, M., Tyszkiewicz, J. and Van den Bussche, J. (2002). Distributed Computation of Web Queries Using Automata, in *Proceedings of the Twenty-first ACM SIGACT-SIGMOD-SIGART Symposium on Principles of Database Systems*, Madison, Wisconsin, ACM, pp. 97–108.
655. Spielmann, M., Tyszkiewicz, J. and Van den Bussche, J. (2002). Distributed computation of web queries using automata, in *Proceedings of 21th ACM Symposium on Principles of Database Systems (PODS 2002)*, ACM Press.
656. Stahl, G. (1981). Character and acceptability of Church's thesis, *Rep. Math. Logic*, no. 11, pp. 63–67.
657. Steinhart, E. (2002). Logically possible machines, *Minds and Machines*, v. 12, pp. 259–280.
658. Stannett, M. (1990). X-machines and the halting problem: Building a super-Turing machine, *Formal Aspects of Computing*, v. 2, pp. 331–341.
659. Stewart, I. (1991). The dynamic of impossible devices, *Nonlinear Science Today*, v. 1, no. 4, pp. 8–9.
660. Steele, G.L. (1990). *Common Lisp the Language*, Thinking Machines, Inc., Digital Press.
661. Stockmeyer, L. (1987). Classifying the computational complexity of problems, *Journal of Symbolic Logic*, v. 52, no. 1, pp. 1–43.

662. Suppes, P. and Han B. (2000). Brain-wave representation of words by superposition of a few sine waves, *Proceedings of the National Academy of Sciences*, v. 97, pp. 8738–8743.

663. Suppes, P., Han, B., Epelboim, J., and Lu Z.-L. (1999). Invariance between subjects of brain wave representations of language, *Proceedings of the National Academy of Sciences*, v. 96, pp. 12953–12958.

664. Suppes, P., Han, B., Epelboim, J., and Lu, Z.-L. (1999a). Invariance of brain-wave representations of simple visual images and their names, *Proceedings of the National Academy of Sciences*, v. 96, pp. 14658–14663.

665. Talbot, S. (2001). Beyond the Algorithmic Mind, in *CT 2001, Lecture Notes in AI 2117*, Springer-Verlag, Berlin, Heidelberg, pp. 190–202.

666. Teuscher, C. and Sipper, M. (2002). Hypercomputation: Hype or computation? *Communications ACM*, v. 45, no. 8, pp. 23–24.

667. Thomas, W.J. (1973). Doubts about some standard arguments for Church's thesis, *Logic, language, and Probability* (Selected Papers, 4th Internat. Congress on Logic, Methodology, and Philosophy of Science, Bucharest, 1971), Synthese Library, Vol. 51, Reidel, Dordrecht, pp. 55–62.

668. Thomson, J.F. (1954–55). Tasks and super-tasks, *Analysis*, v. 15, pp. 1–13.

669. Thornburg, M. and Casselman, S. (1994). Transformable computers, *International Parallel Processing Symposium* (IPPS94), pp. 674–679.

670. Tiuryn, J. (1979). A survey of the logic of effective definitions, *Lecture Notes in Comp. Science*, v. 125, pp. 198–245.

671. Toffoli, T. and Margolus, N. (1987). *Cellular Automata Machines*, The MIT Press, Cambridge, Massachusetts.

672. Tourlakis, I. (2000). Time-Space Lower Bounds for SAT on Uniform and Non-Uniform Machines, in *Proceedings of 15th Annual IEEE Conference on Computational Complexity* (CoCo'00), Florence, Italy, pp. 22–28.

673. Trahtenbrot, B.A. (1956). Signalizing functions and table operators, *Research Notices of the Pensa Pedagogical Institute*, v. 4, pp. 75–87 (in Russian).

674. Trahtenbrot, B.A. (1974). *Algorithms and Computing Automata*, Moscow, Sovetskoye Radio (in Russian).

675. Trahtenbrot, B.A. and Barzdin, J.M. (1970). *Finite Automata: Behavior and Synthesis*, Moscow, Nauka (in Russian).

676. Traub, J.F. and Wozniakowski, H. (1991). Information-based complexity: New questions for mathematicians, *Mathematical Intelligencer*, v. 13, pp. 34–43.

677. Trnková, V. (1974). On minimal realizations of behavior maps in categorical automata theory, *Comment. Math. Univ. Carolinae*, v. 15, pp. 555–566.

678. Tsichritzis, D. (2001). Forget the past to win the future, *Communications of the ACM*, v. 44, no. 3, pp. 100–101.

679. Tucker, J.V. and Zukker, J.I. (1988). *Program Correctness over Abstract Data Types*, CWI Monographs, v. 6, North Holland.

680. Tucker, J.V. and Zukker, J.I. (1992). Deterministic and nondeterministic computation and Horn programs on abstract data types, *Journal Logic Prog.*, v. 13, pp. 23–55.

681. Tucker, J.V. and Zukker, J.I. (2002). Abstract computability and algebraic specification, *ACM Transactions on Computational Logic*, v. 3, no. 2, pp. 279–333.

682. Turing, A. (1936). On computable numbers with an application to the Entscheidungsproblem, *Proc. Lond. Math. Soc.*, Ser.2, v. 42, pp. 230–265.

683. Turing, A. (1937). Computability and λ-definability, *Journal of Symbolic Logic*, v. 2, pp. 153–163.

684. Turing, A. (1939). Systems of logic based on ordinals, *Proc. Lond. Math. Soc.*, Ser. 2, v. 45, pp. 161–228.

685. Turing, A.M. (1951). Can digital computers think?, in *Machine Intelligence*, v. 15, Oxford University Press, Oxford, 1999.

686. Turing, A.M. (1951a). Intelligent machinery, a heretical theory, in Machine Intelligence, v. 15, Oxford University Press, Oxford, 1999.

687. Turney, P. (1990). Problems with complexity in Gold's paradigm of induction – Part I, Dynamic complexity, *International Journal of General Systems*, v. 17, pp. 329–342.

688. Turney, P. (1990). Problems with complexity in Gold's paradigm of induction – Part II: Static complexity, *International Journal of General Systems*, v. 17, pp. 343–358.

689. Uspensky, V. and Semenov, A. (1993). *Algorithms: Main Ideas and Applications*, Kluwer.

690. Valk, R. (1978). Self-Modifying Nets – A Natural Extension of Petri Nets, in *Automata, Languages, and Programming*, Lecture Notes in Computer Science, 62, Springer-Verlag, New York, Berlin.

691. Van der Waerden, B.L. (1971). *Algebra*, Springer-Verlag, New York, Berlin.

692. Van Lambagen, (1989). Algorithmic information theory, *Journal for Symbolic Logic*, v. 54, pp. 1389–1400.

693. Van Leeuwen, J. and Wiedermann, J. (1985). Array Processing Machines, in *Fundamentals of Computation Theory*, Lecture Notes in Computer Science, 199, Springer-Verlag, New York, Berlin, pp. 99–113.

694. Van Leeuwen, J. and Wiedermann J. (2000). A computational model of interaction, Techn. Rep. Dept. of Computer Science, Utrecht University, Utrecht.

695. Van Leeuwen, J. and Wiedermann, J. (2000a). Breaking the Turing Barrier: The case of the Internet, Techn. Report, Inst. of Computer Science, Academy of Sciences of the Czech. Rep., Prague.

696. Van Leeuwen, J. and Wiedermann, J. (2000b). On the power of interactive computing, *Proceedings of the IFIP Theoretical Computer Science 2000*, pp. 619–623.

697. Van Leeuwen, J. and Wiedermann, J. (2001). The Turing Machine Paradigm in Contemporary Computing, in *Mathematics Unlimited: 2001 and Beyond*, Springer, New York.

698. Vardi, M.Y. and Wolper P. (1986). An automata-theoretic approach to automatic program verification, in *Proceedings of the 1st Annual Symposium on Logic in Computer Science*, pp. 322–331.

699. Vardi, M.Y. and Wolper, P. (1994). Reasoning about infinite computations, *Information and Computation*, v. 115, no. 1, pp. 1–37.

700. Virtual Computer Corporation (2000). *Fixed Hardware and Reconfigurable Hardware*, http://www.vcc.com/intro2.html.

701. Viscek, T. (2002). Complexity: The bigger picture, *Nature*, v. 418, pp. 131–132.

702. Vitanyi, P. (2001). Quantum Kolmogorov complexity using classical descriptions, *IEEE Trans. Inform. Theory*, v. 47, no. 6, pp. 2464–2479.

703. Vizard, M. (2001). Alan Cooper of Cooper Interactive Design sees planning as key to downstream dividends, *InfoWorld.com* (06/14/01).

704. Vyugin, V.V. (1981). Algorithmic entropy (complexity). of finite objects and its application to a definition of randomness and quantity of information, *Semiotics and Informatics*, v. 16, pp. 14–43.

705. Wakefield, J. (2001). Complexity's business model, *Scientific American*, vol. 284, no. 1, p. 31.

706. Waldrop, M.M. (2002). Grid computing, *Technology Review* (05/02), vol. 105, no. 4, p. 31.

707. Wang, H. (1995). On computabilism and physicalism: Some sub-problems, in *Nature's Imagination: The Frontiers of Scientific Vision*, Oxford University Press, Oxford, pp. 161–189.

708. Watanabe, O. (ed.). (1992). *Kolmogorov Complexity and Computational Complexity*, Springer-Verlag, New York, Berlin.
709. Waxman, M.J. (1996). *On Problem Complexity*, (unpublished work).
710. Welch, P.D. (2000). Eventually infinite time Turing machines degrees: Infinite time decidable reals, *Journal of Symbolic Logic*, v. 65, no. 3, pp. 1193–1203.
711. Wegner, P. (1995). Interactive foundations of object-based programming, *IEEE Computer*, v. 28, no. 10, pp. 70–72.
712. Wegner, P. (1995a). Interaction as a basis for empirical computer science, *Comput. Surv.*, v. 27, pp. 45–48.
713. Wegner, P. (1997). Why interaction is more powerful than algorithms, *Communications of the ACM*, v. 40, pp. 80–91.
714. Wegner, P. (1998). Interactive foundations of computing. Theoretical aspects of coordination languages, *Theoret. Comput. Sci.*, v. 192, no. 2, pp. 315–351.
715. Wegner, P. and Goldin, D. (1999). Coinductive models of finite computing agents, in *CMCS'99 Coalgebraic Methods in Computer Science* (Amsterdam, 1999), 21 pp. *Electron. Notes Theor. Comput. Sci.*, 19, Elsevier, Amsterdam.
716. Wegner, P. and Goldin, D. (1999a). Interaction as a framework for modeling, in *Conceptual Modeling – Current Issues and Future Directions*, Lecture Notes in Computer Science, Vol. 1565, Springer-Verlag, Berlin, pp. 243–257.
717. Wegner, P. and Goldin, D. (2003). Computation beyond Turing machines, *Communications of the ACM*, v. 46, no. 4, pp. 100–102.
718. Weihrauch, K. (2000). *Introduction to Computable Analysis*, Springer, Berlin.
719. Werbos, P. (1974). *Beyond regression: New tools for prediction and analysis in the behavioural sciences*, Ph.D. thesis, Harvard University, Cambridge, MA.
720. Wiedermann, J. (2000). *Fuzzy Computations Are More Powerful than Crisp Ones*, Techn. Report no. 28, Inst. of Computer Science, Academy of Sciences of the Czech. Rep., Prague.
721. Wiedermann, J. (2000). Fuzzy Turing machines revised, *Computing and Informatics*, v. 21, pp. 251–263.
722. Wigner, F. (1960). The unreasonable effectiveness of mathematics, *Communications in Pure and Applied Mathematics*, v. 13, pp. 1–14.
723. Wilkinson, B. and Allen, M. (1999). *Parallel Programming – Techniques and Applications Using Networked Workstations and Parallel Computers*, Prentice Hall, New Jersey.
724. Williams, C.P. and Hogg, T. (1992). *Using Deep Structure to Locate Hard Problems*, Proceedings 10th National Conference on Artificial Intelligence, pp. 472–477.
725. Williams, C.P. and Hogg, T. (1993). *Extending Deep Structure*, Proceedings 11th National Conference on Artificial Intelligence, pp. 152–157.
726. Williams, C.P. and Hogg, T. (1994). *Expected Gains from Parallelizing Constraint Solving for Hard Problems*, Proceedings 12th National Conference on Artificial Intelligence, pp. 1310–1315.
727. Winograd, T. (1997). The Design of Interaction, in *Beyond Calculation: The Next Fifty Years of Computing*, Copernicus, pp. 149–161.
728. Winskel, G. (1986). Category Theory and Models for Parallel Computation, in *Lecture Notes in Computer Science*, Vol. 240; *Category Theory and Computer Programming*, pp. 266–281.
729. Wolfram, S. (1983). Cellular automata, *Los Alamos Science*, v. 9, pp. 2–21.
730. Wolfram, S. (1984). Computation theory of cellular automata, *Communications in Mathematical Physics*, v. 96, pp. 15–57.
731. Wolfram, S. (1984). Universality and complexity in cellular automata, *Physica* 10D, no. 1.

732. Wolfram, S. (2002). *A New Kind of Science*, Champaign, Ill., Wolfram Media.
733. Wolverton, R.W. (1974). The cost of developing large-scale software, *IEEE Transactions on Computer*, Volume C-23, no. 6, pp. 615–636.
734. Woodhouse, D., Johnstone, G., and McDougall, A. (1984). *Computer Science*, 2nd Edition, The Jacaranda Press.
735. Xia, Z. (1992). The existence of noncollision singularities in the n-body problem, *Ann. Math.*, v. 135, no. 3, pp. 411–468.
736. Yamada, H. (1960). *Counting by a Class of Growing Automata*, Doctoral Thesis, University of Pennsylvania, Philadelphia, Pa.
737. Yamada, H. (1962). Real-time computation and recursive functions not real-time computable, *IRE Trans.*, EC-II, pp. 753–760.
738. Yanovskaya, S.A. (1959). Mathematical Logic and Foundations of Mathematics, in *Mathematics in the USSR for Forty Years*, v. 1, Fismatgiz, Moscow (in Russian).
739. Yamada, H. and Amoroso, S. (1969). Tesselation automata, *Information and Control*, v. 14, pp. 299–317.
740. Yamada, H. and Amoroso, S. (1971). Structural and behavioral equivalence of tesselation automata, *Information and Control*, v. 14, pp. 1–31.
741. Yao, A.C. (1979). Some complexity questions related to distributed computing, *Proc. 11th ACM STOC*, pp. 209–213.
742. Yao, A.C. (2003). Classical physics and the Church-Turing Thesis, *Journal of the ACM*, v. 50, no. 1, pp. 100–105.
743. Yates, F.E. (1978). Complexity and the limits to knowledge, *American Journal of Physiology*, v. 235, R201–R204.
744. Zadeh, L.A. (1969). Fuzzy algorithms, *Information and Control*, v. 19, pp. 94–102.
745. Zadeh, L.A. (1973). *The Concept of a Linguistic Variable and its Application to Approximate Reasoning*, Memorandum ERL-M 411, Berkeley.
746. Zervos, C. (1977). *Colored Petri Nets: Their Properties and Applications*, Technical Report 107, System Engineering Laboratory, University of Michigan, Ann Arbor, Michigan.
747. Zurek, W.H. (1991). Algorithmic Information Content, Church-Turing Thesis, Physical Entropy, and Maxwell's Demon, in *Complexity, Entropy and the Physics of Information* (Zurek, W.H., ed.), Addison-Wesley, pp. 73–89.
748. Zurek, W.H. (1991a). *Complexity, Entropy and the Physics of Information*, Addison-Wesley.
749. Zuse, H. (1998). *History of Software Measurement*, Berlin.
750. Zvonkin, A.K. and Levin, L.A. (1970). The complexity of finite objects and the development of the concepts of information and randomness by means of the theory of algorithms, *Russian Mathematics Surveys*, v. 256, pp. 83–124.

Index